**Monitoring Ecological Impacts**
Concepts and practice in flowing waters

T0282707

*Monitoring Ecological Impacts* provides the tools needed by professional ecologists, other scientists, engineers, planners and managers to design assessment programs that can reliably monitor, detect and allow management of human impacts on the natural environment. The procedures described are well-grounded in inferential logic, and statistical models needed to analyse complex data are given. Step-by-step guidelines and flow diagrams provide the reader with clear and usable protocols, which can be applied in any region of the world and to a wide range of human impacts. In addition, real examples are used to show how the theory can be put into practice. Although the context of this book is flowing water environments, especially rivers and streams, the advice for designing assessment programs can be applied to any ecosystem.

**Barbara J. Downes** is a Senior Lecturer in Ecology in the School of Anthropology and Environmental Studies at the University of Melbourne, Australia. She is an aquatic ecologist, with research experience in upland streams and coastal marine environments.

**Leon A. Barmuta** is a Lecturer in Zoology at the University of Tasmania, Australia. He is a freshwater ecologist with extensive experience in basic and applied ecology in Australia and the United States of America.

**Peter G. Fairweather** is Professor of Marine Biology at the Flinders University of South Australia. He has worked in marine, estuarine and freshwater ecosystems in Australia and USA, and has edited the *Australian Journal of Ecology*.

**Daniel P. Faith** is a Principal Research Scientist at the Australian Museum, with research interests in systematics, biodiversity conservation and biological monitoring. He is an Associate Editor of *Systematic Biology*.

**Michael J. Keough** is a Reader in Zoology at the University of Melbourne. His research interests include the ecology of natural and human-induced disturbances in coastal habitats. He is co-author of *Experimental Design and Data Analysis for Biologists*, Cambridge University Press, 2002.

**P. S. Lake** is Professor in Ecology at Monash University, Australia. He is currently Chief Ecologist in the Cooperative Research Centre for Freshwater Ecology and is on the editorial board for *Freshwater Biology*.

**Bruce D. Mapstone** is a Senior Principal Research Fellow and Program Leader of the Sustainable Industries Program with the Cooperative Research Centre for the Great Barrier Reef World Heritage Area. He is based at James Cook University, Australia, and has research interests in tropical fisheries and their management and assessment of human impacts on natural systems.

**Gerry P. Quinn** is a Senior Lecturer in Biological Sciences at Monash University. He is an aquatic ecologist with interests in coastal marine habitats, and rivers and their floodplain wetlands. He is co-author of *Experimental Design and Data Analysis for Biologists*, Cambridge University Press, 2002.

# Monitoring Ecological Impacts

## Concepts and practice in flowing waters

BARBARA J. DOWNES
University of Melbourne

LEON A. BARMUTA
University of Tasmania

PETER G. FAIRWEATHER
Deakin University

DANIEL P. FAITH
The Australian Museum

MICHAEL J. KEOUGH
University of Melbourne

P.S. LAKE
Monash University

BRUCE D. MAPSTONE
CRC Reef Research Centre and
James Cook University

GERRY P. QUINN
Monash University

CAMBRIDGE
UNIVERSITY PRESS

CAMBRIDGE UNIVERSITY PRESS
Cambridge, New York, Melbourne, Madrid, Cape Town, Singapore, São Paulo

Cambridge University Press
The Edinburgh Building, Cambridge CB2 8RU, UK

Published in the United States of America by Cambridge University Press, New York

www.cambridge.org
Information on this title: www.cambridge.org/9780521771573

First published 2002
Reprinted 2003
This digitally printed version 2008

A catalogue record for this publication is available from the British Library

Library of Congress Cataloguing in Publication data

Assessing ecological impacts: applications to flowing waters / Barbara
J. Downes ... [et al.].
    p.   cm.
Includes bibliographical references (p.   ).
ISBN 0 521 77157 9
1. Water quality biological assessment.   2. Stream ecology.   I. Downes,
Barbara J., 1958–
QH96.8.B5 A76 2002
577.6′4′0287–dc21   2001043778

ISBN 978-0-521-77157-3 hardback
ISBN 978-0-521-06529-0 paperback

To Monty Python and the pursuit of Grails, holy or otherwise

and

To the noble but dying art of fellmongery

# Contents

# Preface and Acknowledgements

This book would not have been possible to write without a grant from xi The Land and Water Resources Research and Development Corporation (Canberra, Australia), awarded through its National River Health Program. We are grateful to both Dr Peter Davies (Director, NRHP) and Dr Nick Schofield (Program Manager – Water Resources) for their support for a fairly unusual project.

Various of our colleagues were asked to read and critique sections of the book. Thank you all for your candid and constructive criticism and for giving your opinions so freely. Your input made a difference:

Dr Kevin Cash (Environment Canada, Canada)

Dr Bruce Chessman (NSW Department of Land and Water Conservation, Australia)

Prof. David Dudgeon (University of Hong Kong, Hong Kong)

Dr Rob Goudey (Victorian Environmental Protection Authority, Australia)

Dr Stuart Halse (WA Department of Conservation and Land Management, Australia)

Dr Chris Humphrey (NT Environmental Research Institute of the Supervising Scientist, Australia)

Dr Richard Marchant (Museum of Victoria, Australia)

Mr Leon Metzeling (Victorian Environmental Protection Authority, Australia)

Dr Steve Ormerod (Cardiff University, Wales)

Dr Vince Resh (University of California at Berkeley, USA)

Dr Mike Winterbourn (University of Canterbury, New Zealand)

We thank also an anonymous referee who commented on the entire manuscript and made many practical and valuable suggestions.

Steve Ormerod and David Bradley of the Catchment Research Group at Cardiff University made available data from a long-running experiment in ecological restoration, which has been used as an example throughout this book. We thank them very much for making their data available to us, and for participating in e-mail discussions about their experiment.

PSL and GQ thank the CRC for Freshwater Ecology for financial support during some periods of writing. BJD thanks the Department of Biological Sciences at Monash University for providing a desk and writing space during a six-month sabbatical. Additionally, various people helped in the production of the manuscript: Chandra Jayasuriya and Andrew Hardie (School of Anthropology, Geography and Environmental Studies, University of Melbourne) drew or redrew the figures. Jim Thomson carried out substantial literature searches, and Leanne Matheson and Bill Dixon located some critical references. Alice King did much of the editing and formatting of the final manuscript, and Jodie Street assisted in compiling the final reference list. We are grateful to Alan Crowden (Cambridge University Press) for his support of the original book proposal and for considerable patience.

We thank numerous colleagues, partners, family members and friends – particularly Menna Jones and Gillian Napier – who have collectively provided generous support as well as, in some cases, being sounding-boards for some of the more controversial material.

Finally, a word about authorship. The structure, philosophy, content and opinions in this book arose largely from intensive debate among the authors during a series of three-day workshops. All chapters were shaped by our discussions and were co-written to some degree. As such, this is a truly collaborative, eight-authored book. The order of authorship does *not* indicate individual levels of contribution.

**Twenty important issues in good monitoring design**

Effective monitoring programs comprise multiple elements that collectively provide sound use of logic and result in rational decision-making without a profligate use of resources.

| Issue | Primary chapter(s) |
|---|---|
| Current levels of human use and abuse of water resources mean we need to implement good monitoring design as an essential – not a luxury – requirement for their further use and management. | 1 |
| Good monitoring design requires us to understand how ecosystems work. In flowing-water ecosystems, structure and function are strongly dependent on the operation of longitudinal and predominantly unidirectional linkages (upstream–downstream), and on lateral linkages (channel–floodplain); these linkages affect how we can apply monitoring designs. | 2 and 8 |
| *Perturbation* of a system consists of two sequential events: the *disturbance* to the system and the *response* of the system to that disturbance. Effective monitoring requires understanding the nature, and temporal and spatial scales, of both the disturbance and the response. | 3 |
| Monitoring may be done for different purposes, and these serve different management needs as well as posing different questions to be answered. | 3 |
| The logical principles of designing a monitoring program to detect the effects of human activities apply irrespective of whether a frequentist or a Bayesian approach to statistics is adopted. Currently, hypothesis-testing via frequentist statistics offers the best-developed and most widely used tools for making decisions about impacts but we expect more development of Bayesian approaches. | 4 |

| Issue | Primary chapter(s) |
|---|---|
| The key strategy for inference of impacts is to find some evidence for impact that cannot easily be explained away by various other processes, such as natural variation in the system. Support for an impact hypothesis is only found if the probability of that outcome is small, under normal circumstances, in the *absence* of impact. This pursuit of improbability provides the rationale for specific aspects of monitoring design. | 4 |
| These design aspects include sampling control and impact locations, both before and after putative impact (so-called Before–After–Control–Impact (BACI) designs) together with proper replication of each of these four elements, where possible. Replicated BACI-type designs allow us to separate, with relatively high confidence, human-caused effects from natural processes. | 5 |
| It is important to recognize, from the outset, that deficient monitoring designs usually cannot be rescued regardless of the quantity and sophistication of statistical analysis applied to the data. | 5 |
| We can illustrate the importance of good monitoring design by looking at how river biologists have addressed questions about human impacts in the past. Historically, there has been a number of issues that have prevented river biologists from implementing designs with the strongest possible inferential base. Some of these issues have been within the control of biologists, and some are external constraints imposed either by the geographical peculiarities of the river under study, or by socioeconomic factors. | 6 |
| There are different types of BACI designs, which result in distinctly different analytical models that address different questions. These conceptual models are all justifiable approaches to the detection of impacts under particular circumstances, but it is *essential* that anyone implementing one of these designs be aware of the differences between them, and of the important characteristics of each of them. | 7 |

| Issue | Primary chapter(s) |
|---|---|
| Different BACI designs lie along a gradient of inferential certainty from relatively strong to relatively weak, rather than providing either perfect or zero inference about human impacts. | 7 and 9 |
| Applying BACI requires tailoring designs to the specifics of the system (its size and uniqueness) and impact to hand (point or non-point impact), and includes developing criteria that help ensure the comparability of control locations to each other and to the impact location(s). There is a dilemma in that, the more narrowly we define characteristics of control locations, the more similar they are likely to be to each other but the fewer will be the number of places likely to meet our criteria. | 8 |
| We can implement a structured 'levels-of-evidence' approach to improve the inferential strength of monitoring designs. This approach uses causal criteria (effectively, a set of circumstantial arguments), which have been developed particularly well in the field of epidemiology. Making a levels-of-evidence case is especially important when elements of BACI designs are missing. | 9 |
| Variables chosen for monitoring should be efficacious: relevant to the questions asked; strongly associated with the putative impact; ecologically and/or socially significant; efficient to measure. | 10 |
| The magnitude and form of unacceptable environmental changes ('effect sizes') should be negotiated and defined ahead of beginning a monitoring program; it is impossible to prescribe universal effect sizes for biological variables. | 11 |
| Negotiations should involve all stakeholders, and defining important changes should include societal wishes as well as scientific input regarding the implications and risks associated with those changes. | 11 |

| Issue | Primary chapter(s) |
|---|---|
| Decision-making should consider the risks of making two sorts of errors: detection of an environmental change that is not actually real (statistically speaking, a Type I error); and failure to detect an important change (statistically speaking, a Type II error). The risks of making either sort of decision error should be balanced in the decision-making process in inverse relation to the respective costs of committing that error. | 12 |
| Effective monitoring programs should be optimized, in which the number of samples required is compared to the resources available, and trade-offs made that reduce monetary costs without compromising the inferential strength of the program. | 13 |
| The emerging discipline of ecological restoration requires effective monitoring to ensure that its goals are being reached. As well as general design considerations promoted throughout this book (especially the ability to determine if any change has occurred by comparing with unrestored controls), monitoring for restoration requires an additional treatment set. In this we compare putatively restored areas to the ideal that we wish the ecosystem to be restored to, what we refer to as a 'reference' condition. Only this more elaborate monitoring with three sorts of states (restored, control and reference) can allow us to know when our restorative goal is attained. | 14 |
| Monitoring programs must be linked to management decision-making, such that particular triggers (e.g. an effect being detected) will result in some action being taken. No one source of information, including this book, has all the answers, and a degree of flexibility of approach is required as the best way of designing monitoring programs; such best practice will continue to evolve and progress. | 15 |

# Part I

# Introduction to the nature of monitoring problems and to rivers

# 1

# Why we need well-designed monitoring programs

How serious are the current and future problems created by human activities on flowing waters? Impacts on ecological processes from altered flows, salinization, organic pollution, exotic species and the like can be very difficult to assess. These pressures being placed on global water sources can be seen convincingly in some statistics on the proportions of stream and river flow used by human beings.

First, fresh water comprises about only 2.5% of the Earth's total volume of water. After subtracting the volumes of fresh water locked up in ice caps and glaciers, only 0.77% ( $\sim$ 10 665 000 km$^3$) is left as free fresh water (i.e. in aquifers, soil pores, wetlands, streams etc.; Postel *et al.* 1996). Flowing waters are only a very small fraction of the world's fresh water, with global annual runoff being about 40 700 km$^3$ (Postel *et al.* 1996) or $\sim$ 0.003% of the Earth's total volume of water. However, like atmospheric water, the surface movement of fresh water into and out of rivers and streams is high and the residence time is low (*c.* 8–14 days).

Human beings use or affect a high proportion of this water. Human-made impoundments now have a major influence on the flow and ecology of many rivers, storing about 14% (5500 km$^3$) of total annual runoff. Human use of fresh waters is rapidly rising as populations increase and standards of living rise (e.g. daily consumption of rural peasants may be 50 L per person compared with 400 L per head for an affluent householder; Newson 1994). Worldwide total use of fresh water, including irrigation and industrial demands, is about 1800 L per person per day (Pimentel *et al.* 1997). Levels of consumption are rapidly rising because both industry and irrigation require large amounts of fresh water (e.g. 2100 L of water to produce 1 kg of steel (Newson 1994); 2000 L to produce 1 kg of soybeans, and 100 000 L to produce 1 kg of beef

(Pimentel *et al.* 1997)). These needs, even if constrained by sensible planning aimed at achieving some level of ecological sustainability, are already placing, and will increasingly place, immense pressures on the fresh waters of the world, especially flowing waters (Pimentel *et al.* 1997; Postel 1997; Postel *et al.* 1996). By the end of the year 2000, human water use in Europe was expected to have risen to 673 km$^3$ (Shiklomanov 1989) or 63% of riverine base flow. Human beings now directly consume ~ 18% of available runoff. Withdrawals from water bodies combined with water for in-stream flow purposes (e.g. pollution dilution and treatment) account for a further 6780 km$^3$ or 36% of available runoff (Postel *et al.* 1996). Thus 54% of available runoff or 30% of the total accessible renewable fresh water supply of the world is now used by humans (Postel *et al.* 1996). The problem of increasing human demands for water is particularly stark in the world's driest regions. With current rates of population growth, arid and semi-arid regions of Africa and Asia are likely to host > 75% of the world's population by 2025, but per capita water supplies are inadequate now to secure food self-sufficiency (Falkenmark 1997).

To meet these increasing human needs, more dams and more diversion schemes will be built. Postel *et al.* (1996) estimated that for the next 30 years about 350 new dams ( > 15 m high) per year will be built. In addition, there will have to be an emphasis on the conservative use of water and on the recycling of polluted water to meet new demands. Many rivers, as a result of dam construction and water abstraction, are now only a trickle of their original flow. Stark examples of this demise of once-impressive rivers include the Colorado River in USA and the Murray River in Australia. Currently, 10 684 GL per year or 78% of mean annual flow is withdrawn from the Murray River, with more than 95% of the diverted water being used in irrigation (Australian Department of Industry, Science and Tourism 1996). Clearly, the capacity of rivers and their biota to maintain any substantial degree of ecological integrity and to perform ecosystem services, such as pollution dilution and water quality protection (Postel & Carpenter 1997), are going to be under immense pressure from large diversions and regulation.

Along with the pressures from direct consumption and use of water by humans, flowing waters face myriad other stressors, many of which will increase in spatial scale and strength with increasing human populations. Increasing development and activity on stream catchments will increase such degrading forces as salinization, sedimentation, eutrophication and pollution from point and non-point sources. Habitat loss and fragmentation due to activities such as drainage, channelization, river regulation and the removal of in-stream and riparian vegetation and woody debris will lead to loss of biodiversity. In streams native biota may have to face pressure from the proliferation of exotic biota, be they pathogens (new viruses, bacteria and fungi) or competitors and predators. Finally, at a large scale, both temporally and spatially, there are the deleterious changes exerted on streams and their catchments by climatic warming that include trends such as increasing water temperatures, increases and decreases in stream discharge and an increase in the intensity and frequency of floods and droughts (Houghton *et al.* 1996; Karl *et al.* 1997). The number of streams already in a damaged state is large. For example, of the 5 200 000 km of streams in the contiguous states of USA, only 2% were considered to be of high quality (Benke 1990). Of the 139 large rivers (i.e. virgin mean annual discharge $> 350 \, \mathrm{m^3 s^{-1}}$) in North America north of Mexico, Europe and Russia, 61% were affected by flow regulation and fragmentation (Dynesius & Nilsson 1994).

Faced with these rapidly expanding human-generated pressures, not only will it become increasingly difficult to protect and maintain the diverse biota of flowing waters, it will become much more difficult to protect and conserve the water resources of rivers for human use. Ongoing, rigorous environmental monitoring that allows us to detect, assess and manage human impacts properly is vitally important.

## 1.2 THE NEED FOR THIS BOOK

This book is about rigorous impact assessment. It is about the logic and ways we may make strong inferences that lie behind the design of effective monitoring programs – those that allow us to detect and assess, with some confidence, whether specified human activities are causing unacceptable changes to the environment (Box 1.1). We consider impact assessment primarily from an ecological perspective – that is, we are interested ultimately in effects upon biota, but we do consider some physicochemical variables as well. The book does not deal with methods for carrying out compliance monitoring (i.e. comparing end-of-pipe pollutant levels with established water quality standards) nor with monitoring the general state or 'health' of environments, although some of the principles we discuss can cross over into these sorts of monitoring programs as well. These other two types of monitoring are carried out for different purposes and are described in detail in other publications (see chapter 3).

Good impact assessment studies are complicated to design properly – that is, in such a way as to minimize decision errors. Monitoring

---

**Box 1.1** What are 'unacceptable environmental changes'?

Impacts can be manifest as changes in a variable of interest in either direction, either increase or decrease in value. It is then a social decision whether that detected change is deemed to be desired and/or acceptable, or their converse.

Not all impacts need be deemed unacceptable (e.g. increased nutrients may increase biomass of macrophytes that in turn provide more shelter for a wider range of biota). What is acceptable or unacceptable is a value judgement that ought to be decided by the wider public, not just scientists. Within this, scientists have a duty to identify the changes going on and to judge how unusual they are (subjects that are covered by this book). Obviously society may choose to put more resources into detecting changes that are deemed unacceptable, an issue we return to in chapter 11.

For these reasons, when we use the word 'impact' in this book, we don't limit our meaning to a negative connotation. Nor do we presume that an impact is present, even when we omit the adjective 'putative'. All sorts of impacts should be detected to understand the full range of consequences of human actions.

---

needs to detect significant human impacts when they are present, but it also needs to guard against imputing impacts when they, in fact, do not exist. Such essential studies may range from dealing with a single impact at a specific point to large-scale, catchment-wide studies that address multiple human impacts. In all such studies, monitoring of environmental variables (like water or sediment quality), biota or ecological processes is carried out. Monitoring to achieve the necessary sensitivity may be both time-consuming and expensive. Thus, there must always be a trade-off between the resources available to be invested in monitoring and the reliability required of any data upon which management decisions are based. It should also be borne in mind that ecological monitoring at both the spatial scale and sensitivity required for effective management may require the establishment and maintenance of large collaborative and interdisciplinary teams. This would be a new step for many management agencies and academic institutions (Carpenter 1998).

The above paragraphs describe the sorts of human impacts we see on streams and rivers and why detecting and measuring such impacts are difficult exercises. The variety and incidence of human impacts are likely to increase, and it is becoming increasingly imperative that monitoring studies be well designed. Poorly designed monitoring programs are costly, both in unacceptable impacts that go initially undetected as well as in the great costs of eventual restoration and rehabilitation of sites, which can be prohibitively high. Money may actually be wasted in particularly poor monitoring. Moreover, as we shall discuss, well-designed monitoring programs incorporate the relative financial costs of different designs. Trade-offs need to be understood by all involved parties – representatives of industry, government departments and conservation groups, and interested members of the public – but especially by those doing the monitoring. Although most monitoring programs will require compromises in terms of numbers of samples or levels of replication (usually driven by financial considerations), some compromises are far more sensible, relative to the information that can be gained, than are others. Some compromises are completely unacceptable in that the resulting design cannot produce data that can result in any definitive or objective decisions about impacts. Clearly, it is important that those involved in conducting monitoring programs have a sound grasp of the conceptual bases of monitoring design and that considerations of design precede the sorts of decisions that would affect data collection.

Obtaining an adequate grasp of all of the issues relating to good design is difficult. The literature is spread over a great variety of journals, and some of the discussion has become so technical as to be beyond the immediate grasp of many professionals. Opinions have also become entrenched within some sub-disciplines and effective interchange of new ideas seldom occurs. Some important topics have had little discussion or debate. Thus, the objectives of this book are threefold:

1.  To provide a logical framework for making decisions about the existence, sizes and effects of human impacts on flowing waters
2.  To provide designs that are well-grounded theoretically but also offer practical solutions to real problems
3.  To initiate discussion and offer some solutions to problems that have traditionally been neglected.

## 1.3 THE SCOPE, APPROACH AND INTENDED AUDIENCES OF THIS BOOK

Philosophically we believe that it is our duty to monitor impacts that humans have on the environment, but then the onus is also upon us to do this in an efficient manner. Mindless monitoring, which uses practices just because they 'are the way it's always been done', is an ineffective use of resources unless designs are evaluated and carefully considered for the return of information versus the costs of their doing. Indeed some monitoring is merely a waste of scarce resources rather than anything worth doing for itself.

This book is written by a larger number of authors than many earlier works in this area. We consciously did this to pool our resources and experience with a range of monitoring schemes in, of course, fresh waters but also other sorts of habitats. We also were mindful that many of the topics we sought to tackle (e.g. effect sizes to detect) did not come from a large body of work with any clear consensus. Hence we had to push such ideas along further than before. Each author took responsibility for beginning one or more chapters but each has been scrutinized and altered by all eight of us, during a series of workshops and rewriting sessions, by the time you see this in print.

This book is aimed at several audiences:

- Ecologists, especially those working in freshwater ecosystems but also those working in other habitats who are interested in the principles of good monitoring design
- Other scientists, social scientists or academics in the humanities with an interest in human impacts on the environment (e.g. environmental chemists, hydrologists and geomorphologists, engineers, economists, human geographers)
- Professionals in the water industry and government, including those in managerial decision-making positions
- Postgraduate or upper undergraduate students.

As such, we will usually assume readers have at least some familiarity with basic lotic ecology (and its associated jargon) as well as basic statistics. Given the often highly technical nature of the material, this book is not intended to be particularly accessible to lay persons with an interest in human impacts. This is not because we think that goal is not worthwhile – to the contrary. However, our aim is to produce a practical manual that can be used by those with the responsibility to design and manage monitoring programs. It is not possible to meet this goal (with its attendant requirements of precision of language and scholarly citation) without becoming highly technical and fairly detailed. We believe a book that is fully accessible to lay persons with an interest and desire to get involved in monitoring and managing human impacts is very much needed, but it needs to be a separate publication.

## 1.4  THE STRUCTURE OF THE BOOK AND THE PURPOSE OF EACH OF THE CHAPTERS

Designing an effective monitoring program requires that we make decisions about many things: the characteristics of the particular human activity at hand; its likely effects; what goals we are trying to achieve; how we will decide an impact has occurred; and so forth. These decisions need to be addressed in a sensible order, however, if we are to design an effective monitoring program that answers the questions we want it to answer. Some aspects of designing a monitoring program *cannot* be addressed before others have been dealt with satisfactorily. An ineffective, or inefficient, monitoring program is a likely outcome when design decisions are made in the wrong order. For example, in our experience it is common for people to want to discuss what variables to monitor well ahead of deciding what questions the monitoring program is actually supposed to answer. That sort of a sequence can lead to inappropriate choices of variables and result in a monitoring program that doesn't answer any of the questions management considered important.

So, the crux of designing an effective, efficient monitoring program is not only to understand *what* decisions about design must be made, but to address them *in the right order*. Figure 1.1 sets out the order we suggest for these design decisions, together with the relevant chapters where the relevant issues are discussed.

The chapters of this book are divided into three parts. The first part, 'Introduction to the nature of monitoring problems and to rivers' (chapters 1–3), sets out the nature of the problems we face and the need for a better understanding of principles of good monitoring design (chapter 1), as well as the background and philosophy for the whole tome. Chapter 2 describes some of the main features of flowing water ecosystems as a refresher, a source of further reading and to highlight how characteristics of rivers and streams affect our monitoring designs. Chapter 3 describes the different classes of perturbations wrought by humans on the environment, and the different aims and approaches of monitoring programs designed to meet different sorts of management needs and questions. In particular, we describe the type of monitoring – impact assessment – that is the topic of this book. In our second part, 'Principles of inference and design' (chapters 4–7), we first set out the necessity for a formal, logical framework for making statistical decisions, explaining the basics of hypothesis-testing, including some novel additions to the framework for assessing human impact studies (chapter 4). Chapter 5 describes what is logically required for ideal monitoring

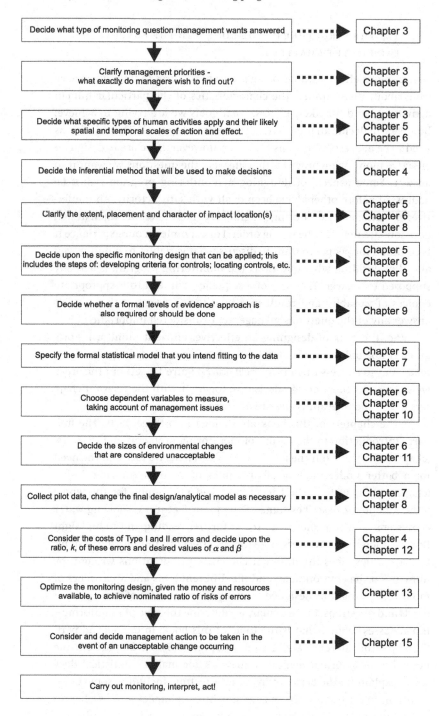

| | |
|---|---|
| Decide what type of monitoring question management wants answered | Chapter 3 |
| Clarify management priorities - what exactly do managers wish to find out? | Chapter 3<br>Chapter 6 |
| Decide what specific types of human activities apply and their likely spatial and temporal scales of action and effect. | Chapter 3<br>Chapter 5<br>Chapter 6 |
| Decide the inferential method that will be used to make decisions | Chapter 4 |
| Clarify the extent, placement and character of impact location(s) | Chapter 5<br>Chapter 6<br>Chapter 8 |
| Decide upon the specific monitoring design that can be applied; this includes the steps of: developing criteria for controls; locating controls, etc. | Chapter 5<br>Chapter 6<br>Chapter 8 |
| Decide whether a formal 'levels of evidence' approach is also required or should be done | Chapter 9 |
| Specify the formal statistical model that you intend fitting to the data | Chapter 5<br>Chapter 7 |
| Choose dependent variables to measure, taking account of management issues | Chapter 6<br>Chapter 9<br>Chapter 10 |
| Decide the sizes of environmental changes that are considered unacceptable | Chapter 6<br>Chapter 11 |
| Collect pilot data, change the final design/analytical model as necessary | Chapter 7<br>Chapter 8 |
| Consider the costs of Type I and II errors and decide upon the ratio, $k$, of these errors and desired values of $\alpha$ and $\beta$ | Chapter 4<br>Chapter 12 |
| Optimize the monitoring design, given the money and resources available, to achieve nominated ratio of risks of errors | Chapter 13 |
| Consider and decide management action to be taken in the event of an unacceptable change occurring | Chapter 15 |
| Carry out monitoring, interpret, act! | |

designs, and is perhaps the most important chapter in the whole book. It sets out a theoretical ideal (in language as precise as we can make it) without considering any of the compromises we are always forced to make for real monitoring programs. Chapter 5 should be read with this aim in mind, because the rest of the book is about how we deal with the compromises we are forced to make that usually prevent us from ever reaching this ideal. Chapter 6 then delineates the reasons why, in the past, we have rarely achieved ideal monitoring designs for rivers and streams and the consequences of the compromises we have made. Chapter 6 may be fairly safely skipped by those wishing to move immediately to analytical and practical solutions, but it does illustrate further why good designs are essential, and how and why rivers and streams pose particular problems. Chapter 7 then tackles directly the specific choices for designing monitoring programs and the analytical models available to analyse the data, together with the decisions required, and consequences of, applying those models. In our third part, 'Applying principles of inference and design' (chapters 8–15), we consider all of the practical constraints that limit monitoring programs, with examples from rivers and streams. In chapter 8, we discuss how to apply the designs discussed in chapter 7 for both small- and large-scale impacts. Chapter 9 describes how, using a levels of evidence approach, we can improve the inferential strength of our monitoring programs, particularly for situations where whole elements of good design (such as controls or data from before human activities started) are missing. Chapter 10 examines briefly the criteria that drive choices of variables used to detect effects of, and responses to, impacts and provides an overview of the sorts of ecological variables that can be used. Chapter 11 focuses on the little-examined but critically important issue of deciding what constitutes an 'important' change – statistically speaking, an important effect size – and makes some suggestions for ways in which effect sizes

---

**Fig. 1.1** (*opposite*) A flow diagram that indicates a logical and efficient order in which to make decisions about the design and implementation of monitoring programs to detect human impacts. In each case, dotted lines point to the main chapters where issues regarding that decision are discussed. Where multiple chapters are given, they collectively cover theoretical issues regarding that decision, past experiences, and suggested, practical ways of making that decision. Note that the flow diagram applies also to monitoring programs designed to examine restoration efforts, with issues specific to those programs addressed in chapter 14.

could be rationally decided. Chapters 12 and 13 consider another neglected issue: the ways in which designs can be sensibly modified to optimize information content relative to financial cost. This is done by firstly examining how decisions require us to trade-off risks (chapter 12), and then by giving a guide to optimization procedures relative to financial costs (chapter 13). While our major concern is with detecting and understanding deleterious impacts, restoration of streams and their catchments is rapidly gaining momentum. Thus, monitoring exercises, similar to those used to assess impacts, can and should be used to follow and evaluate the success, or otherwise, of restoration efforts. The discipline of restoration ecology is a young one and does not have a background of accumulated knowledge or tested principles. Thus, its development will depend upon the accumulation of well-documented cases, cases that we wish to see done with properly executed monitoring and assessment (Hobbs & Norton 1996; Palmer *et al.* 1997). The principles we discuss throughout this book apply to restoration projects as well, something we emphasize in chapter 14. Finally, chapter 15 considers what happens after a monitoring design is in place; it describes the criteria by which monitoring programs might be evaluated and considers further research that could complement monitoring programs.

We recommend that each of the chapters is read in sequence, but those who wish to can skip some of them. For example, experienced freshwater ecologists could omit chapter 2 but, on the other hand, should be interested in how we apply the monitoring models to streams (chapter 8). In contrast, their students or ecologists working in other ecosystems may profit more from our summary of how rivers work (chapter 2). Managers may not be interested in rivers as ecosystems but should benefit from the gentle introductions to complex inferential issues provided in chapters 3–6. Statisticians may also be interested in how their inferential models are applied to this real-world situation, but probably will not need to read chapter 4.

As an aid to readers, we end each chapter with a selection of key points as fundamental considerations for good monitoring design. The most significant of these **Important issues**, as we term them, are gathered together into Box 15.1, and are repeated in the end papers of the book. Here is the first important issue:

1.5 IMPORTANT ISSUE

• Current levels of human use and abuse of water resources mean
  we need to implement good monitoring design as an essential –
  not luxury – requirement for their further use and management.

## 2

---

# The ecological nature of flowing waters

14    In this chapter we provide a very brief introduction to some major concepts in stream or lotic ecology. This chapter is not meant to be comprehensive, but serves to explain some ideas and terms referred to later in this book. Those readers unfamiliar with the basic nature and coverage of stream ecology are recommended to consult the general textbooks of Allan (1995), Giller & Malmqvist (1998) and Boulton & Brock (1999). Readers wanting to gain an understanding of stream hydrology are recommended to consult Gordon *et al.* (1992).

## 2.1 RIVERS AND THEIR CATCHMENTS

Flowing waters or lotic systems comprise a large array of intergrading types of water channels, be they springs, rills, runnels, becks, burns, brooks, creeks, streams, drains, tributaries, distributaries or floodplain rivers. Even though streams and rivers only make up a small fraction of the surface area in most landscapes, flowing waters are a vital environmental component, even in deserts. Running waters have shaped and continue to shape many landscapes, as they transport water, sediments, chemicals, detritus and biota from headwaters to floodplains and estuaries, and finally to the seas. Rivers supply water to both terrestrial organisms (from trees to humans) and the fully aquatic biota, as well as the biota that inhabits terrestrial systems that are intermittently flooded. Streams in the natural state serve as corridors for the movement of the aquatic biota, and their riparian fringes serve as valuable habitat for, and as potential corridors for movements of, the terrestrial biota.

Streams are surrounded by their drainage or catchment area, which covers the area from which the stream derives its water. Adjacent catchments are separated from each other by the drainage divide, which usually follows the highest points between the catchments (Fig. 2.1a).

*(a)*

*(b)*

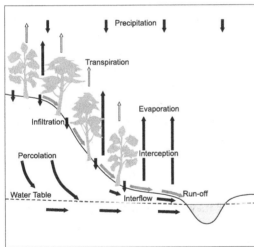

**Fig. 2.1** (a) Plan of an upland stream catchment showing the drainage divide (- - -) and the dendritic form of the drainage pattern; (b) pathways of movement of water in an upland stream catchment.

The amount of water entering the catchment as precipitation that ends up in the stream depends greatly on characteristics of the catchment, such as catchment geomorphology, geology, soil type and development and vegetation types and extent of cover. A large part of the precipitation entering a catchment is returned to the atmosphere by transpiration and interception by plants, and by evaporation (Fig. 2.1b). Water falling on the ground may infiltrate the soil or run overland. Overland runoff may be considerable in catchments with impervious surfaces, such as natural rock, or roads and carparks. In catchments with permeable surfaces, water infiltrating the soil percolates down to the water table. Streams arise where the land surface intersects the water table and groundwater from the water table usually comprises a major part of the stream discharge. Perennial streams are maintained by the groundwater flow during times of little or no rainfall.

A major issue to consider in monitoring and measuring human impacts is that streams and rivers are inextricably linked with their catchments. Hydrologists and geomorphologists have realized the close links between flowing waters and their catchments for some time, while ecologists have come to appreciate the strong links only relatively recently (Hynes 1975). Similarly, up until quite recently, planning and management of rivers were largely concentrated on the channels themselves, and it is only now that we see the advent of concepts such as integrated catchment management. However, there are still many aspects of channel – catchment linkages, such as the ecological functioning of riparian zones, that we understand only poorly.

Streams receive water and materials (both dissolved and particulate) and are provided with physical substrata (from silts to boulders) from their catchments. Gases, such as oxygen, enter the streams directly from the atmosphere. Many chemicals are found in very small amounts in rainfall and then enter streams in larger amounts, usually via catchment processes, such as solution and erosion. Major cations of biological importance are sodium, potassium, calcium and magnesium, and important anions include chloride, sulphate and bicarbonate. Other elements in very small amounts may be critical for the metabolism of particular biota. For example, silica derived from the erosion of rocks is essential for the growth of diatoms. The various forms of nitrogen and phosphorus are vital nutrients that move from the catchment into flowing waters. Because of the unidirectional flow, nutrients are displaced downstream while moving between various biotic and abiotic compartments. The nutrient pathways of streams are open rather than cyclic (as they are viewed in standing waters) and the nutrients are seen as 'spiralling' downstream (Elwood et al. 1983). Thus, the movement of nutrients is a function of the rate of downstream water movement, both surface and subsurface (hyporheic) water, and the rates of uptake, retention and release of nutrients by the biotic and abiotic components of the stream environment.

Ecological processes within catchments exert a strong control on the inputs of organic and inorganic chemicals, both particulate and dissolved, into the downslope stream (e.g. Fisher *et al.* 1998; Likens *et al.* 1977). The riparian zone with its vegetation may exert a strong role in regulating inputs of materials from the catchment into streams (Gregory *et al.* 1991; Naiman & Décamps 1997). The disruption of inputs from the catchment is a reliable signal of disturbance. Natural disturbing forces on catchments include fire, cyclones, defoliation by insects etc., while human-generated disturbances consist of forces such as acid rain, timber harvesting, livestock grazing and land clearance. Logging catchments with associated roading may increase the inputs of nutrients, such as nitrogen and phosphorus, and of sediments into streams (Waters 1995), and alter the inputs of organic matter (Webster *et al.* 1992). Salinization of rivers in Australia, especially Western Australia, is a stark example of the effects of catchment disturbance (Williams 1987). Dryland salinity arises because land clearance greatly reduces native vegetation cover, reducing transpiration and allowing saline groundwater to rise to the surface and salinize surface waters, including headwater streams.

Streams interact with and shape their catchments by moving sediments through three major and interrelated processes: erosion, transportation and deposition (Leopold *et al.* 1964). Geomorphologically, streams and rivers can be divided into three intergrading longitudinal zones. The upland headwater streams are dominated by erosion (production zone of Schumm (1977)), followed by the zone where sediment transportation is dominant (transfer zone of Schumm) and finally on the floodplain sediment deposition dominates (storage zone of Schumm). In the erosional or production zone, the water is generally clear, well oxygenated and turbulent. Habitats encountered here consist of riffles (sections of relatively shallow, rapid and turbulent flow over coarse substrate), pools (sections of relatively deep, slow flow often with eddies), along with cascades, rapids, chutes, runs or glides, and debris dams. The major habitat division is between riffles and pools. In the transport or transfer zone, the riffle–pool sequence is maintained at a larger scale, and lateral and mid-channel bars are evident. The channel may be braided and large log-jams may have an important influence on channel morphology. In the depositional or storage zone, the river is large, deep and turbid with a distinct floodplain. Due to aggradation by sediment deposition, the river now meanders across the floodplain. The meandering course steadily changes with time to produce cutoffs that become oxbow lakes or billabongs and these may give rise to a complex array of floodplain wetlands. The channel bed is made up of fairly homogeneous fine sediments and there may be a range of different types of bars. Large log-jams and persistent snags may produce valuable habitat for the biota.

## 2.2 THE BIOTA OF RIVERS AND STREAMS

The biota of flowing waters is both taxonomically diverse and species rich. This is somewhat surprising given the small surface area of the land occupied by flowing waters, but not surprising given the high spatial and temporal heterogeneity both within, and between, streams (Giller & Malmqvist 1998). Important groups within the biota include bacteria, fungi, micro- and macroalgae, bryophytes (liverworts, mosses), macrophytes, microinvertebrates (e.g. Protozoa, Rotifera, Nematoda), macroinvertebrates (Mollusca, Crustacea, Insecta, Acarina, oligochaete annelids) and vertebrates, notably fish. Taxonomic knowledge is very unevenly spread with only scanty knowledge being available on microscopic components of the biota, such as bacteria, fungi and the microinvertebrates.

Relatively few organisms live in the water column; most of the biota is benthic – dwelling on the bottom. On solid surfaces of the bottom, bacteria and algae may be abundant and they produce biofilms over the bottom substrata. Detritus particles, from fine to coarse, on the stream bed are colonized by fungi, predominantly hyphomycetes, and by bacteria. Both the biofilms, with their algae and bacteria, and the detritus particles are utilized by consumers. Most of the lotic macrobenthos, on and in the bottom sediments, are insects and most of these have an aquatic larval or nymphal stage and a terrestrial adult stage. Exceptions are adults of aquatic Hemiptera and some Coleoptera (e.g. Elmidae, Gyrinidae). Within the bottom substrata, such as sand and gravel, there is the hyporheic zone. This habitat is linked with the surface water by upwelling and downwelling zones, and is also linked with the groundwater of the water table. In this zone, there is usually an abundant microbial and microfaunal biota, collectively called the hyporheos. The hyporheos depends upon detritus as its collective food source.

Nekton is the collective term for that part of the biota that spend a major part of their lives in the water column. This is a demanding mode of life, especially in turbulent upland streams, and requires streamlining of the body and well-developed musculature. Most river-dwelling fish are nektonic but there are very few nektonic invertebrates. There is a specialized component of the biota called the pleuston associated with the air–water film of the water surface. These comprise microscopic organisms, such as bacteria, algae and protozoans, and macroinvertebrates – invariably scavenging hemipterans.

In the past 25 years, there have been considerable advances in understanding the metabolism and trophic structure of streams, with catchment processes being shown to exert a major influence on the structure and processes of stream ecosystems. Some streams have sparse riparian vegetation and/or plenty of sunlight and nutrients, which give rise to food webs in which in-stream algal growth is an important energy source. However, many perennial upland or headwater streams are heavily shaded by the surrounding catchment and riparian vegetation, and often by the steep terrain itself. Such streams are strongly heterotrophic, with the metabolic functioning of the biota dependent on the allochthonous inputs from the catchment of both dissolved and particulate detritus (Cummins 1973, 1974; Hynes 1975). The dissolved component (e.g. humic acids) may be at considerable levels (Mulholland 1997), but overall the particulate input is metabolically more important. The particulate input may range from small leaf fragments, to leaves, sticks and logs. Particles, such as leaves, are broken down fairly rapidly and metabolized, whereas large pieces, such as branches and logs, break down slowly and may form debris dams and snags along the river, and thus generate important habitat.

With the discovery of the importance of allochthonous detritus as the major source of metabolic energy for the biota of upland shaded streams, came the grouping together of macroinvertebrates into functional feeding groups (FFGs; Cummins 1973, 1974). The groups are defined by the size and type of food particle consumed, and by mode of feeding. The major FFGs consist of shredders that feed on large detritus particles ( > 1 mm), collectors feeding on small detritus particles ( < 1 mm), scrapers that scrape microbial–algal layers off solid surfaces, piercers that attack living macrophytes, and predators. While the delineation of FFGs has had considerable heuristic value in the development of stream ecology, it has also generated difficulties by masking interesting patterns of trophic variability in many biotas and across different stream ecosystems. The value of assigning FFG roles to consumers to elucidate the trophic functioning of flowing waters has been questioned (e.g. Lake 1995; Mihuc 1997). Riverine fish have been classified into trophic guilds (e.g. Horwitz 1978; Poff & Allan 1995) ranging from omnivores to specialized piscivores. In contrast with invertebrate FFG, fish guilds currently appear to be rather flexibly defined (Austen et al. 1994).

## 2.3 CONCEPTS OF RIVER STRUCTURE AND FUNCTIONING

As mentioned previously, rivers have been divided into different zones using geomorphological criteria (e.g. Leopold *et al.* 1964; Schumm 1977). Linked with the physical zonation of rivers, many biologically based schemes of longitudinal zonation have been proposed (Cummins 1972; Hawkes 1975; Hynes 1970). For example, in northern Europe, Huet (1949) recognized four major fish-defined zones along the upland–floodplain transition. Illies (1962), mindful of invertebrate distribution, delineated three zones in rivers: the crenon or source-spring zone; the rhithron, the upland stream zone; and the potamon, the lowland river zone. However, the evidence for such distinct faunal zonation is poor and the concept of distinct zonation has been viewed as being 'a naïve stage' in the development of stream ecology (Townsend 1996).

In 1980 with the publication of a paper by Vannote and colleagues, the idea of a continuum rather than strict zonation was proposed – the River Continuum Concept or RCC (Vannote *et al.* 1980). Basically, this idea proposes that geomorphological–hydrological attributes of flowing waters form a fundamental template that determines key attributes of community structure and ecosystem function. Key attributes that change longitudinally along rivers include the types and levels of organic matter inputs and transport (levels and proportions of allochthonous versus autochthonous organic matter), the structure of the invertebrate communities, the representation of functional feeding groups and patterns of resource partitioning. Despite many criticisms being levelled at the RCC (e.g. Statzner & Higler 1985; Winterbourn *et al.* 1981), the concept still remains influential in stream ecology (Cummins *et al.* 1995).

The RCC was proposed to explain patterns of river attributes in a longitudinal direction – it did not address lateral movements of water in rivers, especially those of floodplain rivers. When in flood, rivers may move out of their channels onto the surrounding land, the riparian zone or the floodplain. This inundation may be of considerable duration and is now regarded as a key event structuring and maintaining the biota of both the river and its floodplain, which may contain a great variety of wetlands. Such flooding may be very important for the maintenance of river fisheries (Welcomme 1979). The inundation is called the flood pulse and the idea linking the river dynamically with its floodplain is known as the Flood Pulse Concept (Junk *et al.* 1989). The floods may be regular and seasonal, for example the annual inundation of floodplains in the Wet Tropics (Dudgeon 1999; Lewis *et al.* 1990; Payne 1986), or irregular and somewhat unpredictable as in the case of rivers in arid regions (Puckridge *et al.* 1998; Walker *et al.* 1995). Floodplain rivers are 'flood-dependent ecological systems' (Lewis *et al.* 1990) and human activities that alter the flood regime, such as damming and water diversion, impair the ecological functioning and integrity of the river, the floodplain and their collective biota (Poff *et al.* 1997; Rosenberg *et al.* 1997). The combination of the River Continuum Concept, which emphasizes catchment–stream channel links and upstream–downstream links, with the Flood Pulse Concept, which emphasizes the two-way linkages between the river channel and the floodplain, gives rise to the important concept that rivers are dynamic systems strongly dependent for their functioning on the maintenance of hydrological connectivity with upstream–downstream and channel–floodplain linkages (Townsend 1996).

Furthermore, an attempt has been made to weld together these concepts into one grand concept, that of the Fluvial Hydrosystem (Petts & Amoros 1996). Rivers are viewed as 'three-dimensional systems ... dependent on longitudinal, lateral and vertical transfers of energy, material and biota'. Longitudinal fluxes, lateral movements and vertical exchanges with groundwater are major dynamic pathways for river ecosystem functioning. The concept differs from previous ones in that an explicit attempt is made to set levels of spatial scale, ranging from the drainage basin to the 'mesohabitat' (Petts & Amoros 1996).

The advent of the Flood Pulse Concept, apart from its explanatory value for floodplain ecology, has also served to highlight the fact that at present there is an incomplete understanding of what floods and their inverse, droughts, do ecologically to fluvial biota and their functioning (Giller 1996; Lake 2000). Floods as rapid pulse disturbances in constrained streams can alter habitat availability and deplete both food resources and biota. Surprisingly, after most floods the recovery of the biota can be fairly rapid. In floodplain rivers, floods may constitute a major means of wetland replenishment (Sparks *et al.* 1990). Similarly, in intermittent streams floods may constitute a major means of supplying the water for habitat inundation and replenishment (Fisher & Grimm 1988; Gasith & Resh 1999). Far less is known about the slowly developing disturbance of droughts (Lake 2000).

Rivers differ greatly in their annual flow regimes (Haines *et al.* 1988) and in their levels of flow variability (Poff & Ward 1990a; Fig. 2.2). While rivers differ greatly in their flow regimes, there is a poor understanding of the ecological effects of flow variability *per se*. However, it has become increasingly obvious that the reduction in flow variability, due to river regulation, has been accompanied by the loss of important biota

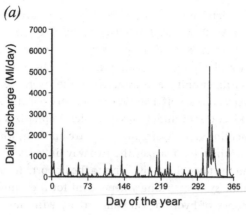

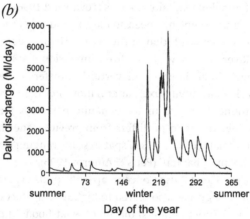

**Fig. 2.2** Two extremes of river flow regimes. (a) An annual flow regime for a groundwater-fed stream where monthly runoff is fairly uniform year round (Group 1: Uniform of Haines *et al.* 1988), in this case the Hanapepe River, Hawai'i (Source: US Geological Survey) and (b) an annual flow regime of an intermittent stream in a Mediterranean-type climate region where monthly runoff is clearly concentrated into the winter period (Group 13: Extreme Winter of Haines *et al.* 1988), here the Buckland River, Victoria, Australia (Source: Department of Natural Resources & Environment, Melbourne, Australia). In each case, data from the year 1955 have been plotted. Discharge data have not been corrected for differing catchment areas (46.6 and 303 km$^2$, respectively).

and by major changes in the structure of the biotic communities (Poff *et al.* 1997). Additionally, it should be realized that most of the rivers of interest to humans, be they consumers or ecologists, are perennial. However, in many parts of the world, especially Africa and Australia, intermittent rivers are common (Gasith & Resh 1999; Lake 1995; McMahon *et al.* 1992; Puckridge *et al.* 1998). These rivers may flow regularly but only in some seasons (Fig. 2.2b), or may flow entirely unpredictably. Again, we understand little of the effects of flow duration or timing on these systems.

## 2.4 ISSUES OF SCALE AND PATCHINESS IN FLOWING WATERS

Consideration of scales – both spatial and temporal – is a major issue for monitoring studies. Scale can be examined from many viewpoints, from the viewpoint of particular populations, organisms, groups of species, ecological processes or particular habitats, as well as different human impacts. However, a concern for scale has been only a recent advent in stream ecology and more attention has been paid to spatial scaling than temporal. Indeed, investigators seem to prefer doing large spatial-scale studies in a short time frame rather than small spatial-scale studies over a long time.

For any ecological investigation, the detection of spatial or temporal patterns is a function of the study's extent and grain. Extent or range is the area over which the investigation occurs, whereas grain or resolution is the size of the individual samples (Mac Nally & Quinn 1998; Wiens 1989). Thus, in a population survey of a mayfly nymph, the extent may be a 1 km section of river 10 m wide ($10\,000\,\mathrm{m}^2$), while the grain may be that of a $0.1\,\mathrm{m}^2$ quadrat sampler. The spatial scope of a survey is the ratio of the extent to grain, $1 \times 10^5$ in this case, and the sampling fraction is the ratio of the number of samples taken to the scope or potential number (Schneider 1994). The number of unsampled units relative to the sampled units is the magnification factor or the level of scaling-up required to make an estimate for the population in the selected extent.

Like other systems, the grain size used in sampling stream benthos can have a strong bearing on the levels of accuracy and precision of the monitoring exercise (e.g. Andrew & Mapstone 1987). Stream beds, because of the small-scale spatial heterogeneity of flow conditions and of substrates, can have a high level of small-scale patchiness in any variable (e.g. Palmer *et al.* 1997). Many benthic samplers (e.g. standard Sürber and Hess samplers) enclose quite arbitrarily sized segments of the stream bottom (e.g. $0.1\,\mathrm{m}^2$) that may comprise a variety of flow conditions and of substrates. As many biota respond in density to particular flow and substrate conditions (e.g. simuliids, chironomids), and because large-sized samples may encompass a range (usually unquantified) of

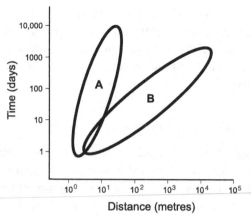

**Fig. 2.3** Imaginary scope diagrams for (A) adult crayfish (e.g. *Euastacus*) and (B) fish (e.g. carp) in lowland rivers.

conditions, such large samples may have high variances and low precision and therefore provide poor predictability. It is clear that grain size and grain (habitat) specificity should receive specific attention in designing a monitoring exercise.

Natural phenomena, be they organisms or debris dams or floods, have their distinctive scope both temporally and spatially. In this situation, the scope of a natural phenomenon is defined as the ratio of the upper to the lower limit of the space in which the phenomenon may occur, or the extent to the grain of the phenomenon (Schneider 1994). Scopes of natural phenomena may also be defined in temporal terms. The two scale axes of space and time may be combined in scope diagrams (Schneider 1994). For example, for adult fish the scope diagram (Fig. 2.3) may depict the range of movements from short-term localized foraging movements to long-term migratory movements. It should be pointed out that at this stage of our understanding of most riverine biota, knowledge of their scopes is rudimentary at best. For many stream invertebrates, notably insects, determination of appropriate spatial scales for any length of time may be particularly difficult, because the life history consists of both an aquatic and a terrestrial stage each with different scopes (Downes & Keough 1998).

In environmental investigations, there should be an awareness of the need to consider explicitly the scopes of the various components of the study (Mac Nally & Quinn 1998). For example, if an environmental impact on a river is to be assessed by monitoring, there must be some preliminary quantification of the scope of the impending disturbance matched with the scopes of organisms to be monitored. In turn, these two types of scope need to be compared with the feasible or affordable scopes of the survey. The latter will be substantially set by the constraints of economic and political concerns on logistics, scientific resources and desired levels of prediction. 'Scope diagrams can be used to make explicit the competing factors that determine the scope of applied studies' (Schneider 1994).

For studies of important ecological phenomena one must be aware of Wiens's (1989) important point about predictability and space–time scaling. With increasing spatial scale, the time-scales of many phenomena also increase. Thus, as the range of spatial scales of observation increases so does the time required to understand that system. Situations in which the dynamics of a large-scale system are studied without an increase in the temporal scope are likely to result in poor predictions. This is a common fault of most environmental surveys and environmental impact assessments, and arises because of a great mismatch between the spatio-temporal scopes of the objects or phenomena under study and those used to collect the data.

Similarly, studies at a small spatial scale over a long period of time may also be flawed, in this case by the problem of scaling-up. Scaling-up is an immense problem in both basic and applied ecology (Turner *et al.* 1989). As Schneider (1994) points out, scaling-up is a very common yet unrecognized problem as ecologists 'so routinely gather data and fit them to models with unstated scopes'. This is a pressing problem because predicting or determining the environmental effects of human activities invariably requires estimates to be made at much larger scales than the scale of sampling or measurement. There are a number of scaling-up strategies, ranging from simple multiplication to the use of hierarchy theory. There is no consensus on ecologically applicable strategies and understandably this has been identified as a crucial research area. In stream ecology, in spite of the high and variable levels of heterogeneity of streams (Giller *et al.* 1994), scaling-up has been and is very common. It usually takes the form whereby from a limited scale of measurements, projections are made for whole stream sections, if not entire stream systems. To remedy this difficulty, if possible, multi-scale sampling of a pattern or process may provide indications of how scaling-up may be applied (Cooper *et al.* 1998).

Scaling-down is also a common occurrence in ecology. In stream ecology a form of scaling-down comes in the application of hydrological equations for large-scale aggregates to small-scale patches. The mismatch can be instructive; for an example see the study of Downes *et al.* (1997), in which it was shown that commonly used hydraulic equations for predicting stream substrate movement with high water events that operate at the reach scale provided very poor predictions of actual substrate movements at small scales. In flowing waters, as in all other types of ecosystems, large-scale aggregates are made up of different types of patches. The heterogeneity of patches at one spatial scale can make it very difficult to scale up or scale down with any degree of certainty. In stream ecology, as in many other areas of ecology, there is a great need to develop feasible methods to detect patterns of spatial heterogeneity (Cooper *et al.* 1997, 1998) and to elucidate whether there are discernible trends in spatial aggregation. The use of the patch dynamics concept in stream ecology may be premature given our very poor empirical knowledge of patterns of spatial heterogeneity and of how patches are created and maintained in streams.

However, in spite of the dangers of scaling-up or scaling-down, in stream ecology a number of hierarchical spatio-temporal scale schemes have been devised. Some of these have been proposed to conceptualize the operations of environmental variables and ecological processes across scales (e.g. Minshall 1988; Petts & Amoros 1996). Such schemes have served to illustrate the key point that processes at the small spatial scale may operate much more rapidly than processes at the larger spatial scale. Other schemes appear to have been largely devised to provide a hierarchy of spatial units for management (e.g. Frissell *et al.* 1986; Lotspeich 1980). From the practical point of view such hierarchical schemes are useful, but from an ecological process point of view, such schemes may prove to be a poor depiction of the actual situation.

## 2.5 IMPORTANT ISSUES

- Flowing waters are intrinsically linked with their catchments, both structurally and functionally.
- The functioning of flowing water ecosystems is strongly dependent on the operation of longitudinal and predominantly unidirectional linkages (upstream–downstream), and on lateral linkages (channel–floodplain).
- Flowing waters harbour a rich, diverse and unique biota specialized to dwell in this very dynamic environment.
- Flowing waters comprise a very distinctive type of ecosystem with their unidirectionality, their integration with the catchment, their highly dynamic nature and their unique biota.
- The unidirectionality and the high level of spatial and temporal

heterogeneity of streams and rivers make the assessment of ecological impacts a challenging task.

- An informed awareness of issues of spatial and temporal scale is essential for effective sampling and monitoring of flowing waters.

# 3

## Assessment of perturbation

28   Rivers are usually perturbed by natural disturbances. Human activities, either in the catchments of streams or in their channels or in both, can generate disturbances that change streams and their biota. To understand the impacts to streams of disturbance, regardless of origin, it is necessary to monitor. It is also crucial that there is a clear understanding both of the type of disturbance and of the purpose(s) of monitoring.

### 3.1 TYPES OF DISTURBANCE

The major reason for monitoring flowing-water ecosystems is to detect the effects of perturbation. A **perturbation** to a population, community or ecosystem occurs when there is a distinct and abnormal change to properties of the system due to disturbance (Bender *et al.* 1984). A perturbation consists of two events: the **disturbance**, which is the application of the disturbing force (e.g. flow reduction) or agent (e.g. pollution) to the biota of the system; and the **response** of the affected biota to the disturbance (Glasby & Underwood 1996). It is important to separate the application of a disturbance from the consequential biotic responses (Glasby & Underwood 1996; Lake 1990, 2000). This allows comparisons to be made of the differential responses of individuals, populations and communities of organisms to similar disturbances and to different types of disturbance. It is also important to note that there may be a considerable time lag between the disturbance and the consequential response by the biota.

    Disturbances may damage rivers over a range of temporal and spatial scales and they can be characterized by their size, intensity and frequency (Petraitis *et al.* 1989), as well as by their predictability, duration, mode of application and the extent of physicochemical alteration

*(a)*          *(b)*

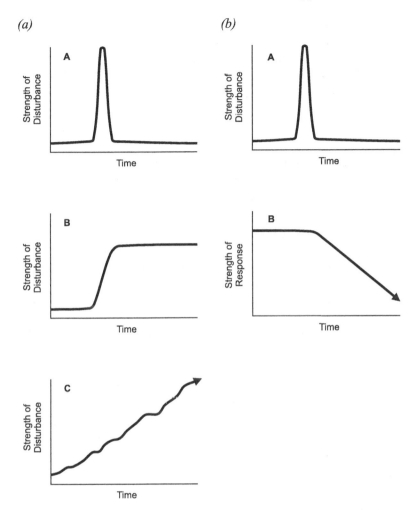

**Fig. 3.1** (a) The three types of disturbance: (A) pulse, (B) press and (C) ramp. (b) A perturbation comprising the application of a disturbance, in this instance a pulse disturbance, that gives rise to a ramp response. In each figure, the arrow indicates that still further change of some kind is possible.

to the affected ecosystem. In terms of duration and levels of intensity, disturbance appears to have three forms: pulse, press and ramp. **Pulses** are short-term events that have a sharp peak in intensity, **presses** are long-term events that are constant in strength (Bender *et al.* 1984), and **ramps** are long-term events that with time change in intensity (Lake 2000; Fig. 3.1a). The response of the biota may vary from being a pulse to an extended ramp (Glasby & Underwood 1996; Lake 2000; Fig. 3.1b).

Thus, in terms of monitoring there is a partition between monitoring to detect long-term trends and changes that are not linked with an obvious and discrete disturbance event, and monitoring to assess the actual or potential impact of spatially and/or temporally distinct disturbances.

Disturbances affecting flowing waters can either be due to natural or to human activities. The latter can be either human-caused amplifications of natural processes, for example a sharp increase in sedimentation, or novel forms of disturbance, such as the release of human-created chemicals (e.g. chlorinated hydrocarbons). It is also important to distinguish between **directly** and **indirectly** applied disturbances. In rivers, disturbances may either be applied directly to the river channel or indirectly via the catchment and the riparian zone. In flowing waters, floods, droughts and human activities, such as damming or channel dredging, may be direct, while disturbances such as high nutrient or sediment inputs due to catchment land-use changes may be deemed to be indirect. The division is somewhat arbitrary but has important management implications in determining the ultimate source(s) of disturbance to river systems, especially if management authorities are not only seeking to identify sources of disturbance, but also aim to curtail the effects of disturbance. The latter particularly applies to river restoration efforts that in particular instances may be largely concerned with restoring the catchment rather than the river channel itself.

In aquatic ecology, disturbance has most commonly been conceived as being due to physicochemical factors. It has also been recognized, especially in marine systems, that the activities of animals, such as foraging within soft bottom substrata, constitute a source of disturbance. The introduction of new organisms and the unpredictable irruption of pathogens are other sources of disturbance. Such invasions or irruptions can have a devastating effect on native biota (Lodge *et al.* 1998). One example is the devastating effects on local native fish populations of the introduction of Nile perch into Lake Victoria (Achieng 1990). Another example is the unplanned arrival of the zebra mussel, *Dreissena polymorpha*, into the American Great Lakes and Hudson River and its subsequent effects on phytoplankton, zooplankton and benthos (Strayer *et al.* 1999). The effects of introduced biota may not be as severe as the above examples; indeed the effects may be indirectly mediated. Introduced trout in New Zealand streams suppress the foraging activities of grazing insects and this indirectly augments algal growth (Flecker & Townsend 1994). Examples of the damaging effects of pathogen outbreaks come from the devastating effects that a microsporidian parasite, *Cougourdella* sp., had on populations of the caddisfly *Glossosoma nigrior* in

Michigan (Kohler & Wiley 1992), and the depletion of native crayfish populations in Europe by the introduced crayfish plague, the fungus *Aphanomyces astaci* (Alderman 1993; Cerenius *et al.* 1988).

Natural pulse disturbances in flowing waters include events such as floods, substratum and habitat disruption, landslides, cyclones, deoxygenation events, earthquakes and volcanic eruptions. Examples of large-scale natural pulse disturbances on streams include the effects of the Mount St Helens volcanic eruption (Hawkins & Sedell 1990), and the effects of the 1988 Yellowstone fires (Minshall *et al.* 1989). Human-generated pulse disturbances include such events as spills of rapidly degradable or dilutable chemicals, rapid thermal pollution, rapid changes in flow with river regulation or diversion, short-term substrate movements including removal of substrates from their channel altogether, and the input in spring of acid water from polluted snowpacks. It should also be noted that the failure of a predictable event, such as a seasonal flood, is also a disturbance. The failure of a seasonal flood to occur in a river can have damaging consequences for biota, not only of the channel, but also of the riparian zones and of flood-dependent wetlands (Poff *et al.* 1997).

The inputs of pollutants, nutrients and sediments into rivers as a result of human activities may be either from **point** or **non-point** sources (British Ecological Society 1990; Carpenter *et al.* 1998a,b). Point inputs include discharges from sewage treatment works, industrial plants, mining operations and fish farms, and the study of their effects has traditionally constituted the major concern of pollution ecology (e.g. Hynes 1960; Warren 1971). In terms of pollution abatement, point sources have understandably received a lot of attention, especially those point discharges that are continuous. More difficult to control are those point sources that discharge intermittently, such as the pulsed discharges of urban runoff into rivers through culverts after rain. Non-point inputs from terrestrial sources are transported overland, underground or via the atmosphere to rivers. Non-point inputs may be continuous, but in most cases they are intermittent, being closely linked with precipitation and runoff events. Non-point pollution, such as acidification and nutrient inputs, impairs the ecological health of many rivers and may affect considerable lengths of rivers. In the USA, non-point inputs of nitrogen and phosphorus are the dominant source of such nutrients for most reaches of the rivers (Newman 1995) and non-point inputs constitute the major source of pollution of water bodies (Carpenter *et al.* 1998b). Unfortunately non-point pollution is both difficult to assess and to regulate. Remediation is also difficult because it usually requires measures to be implemented over a large spatial scale and recovery from such damaging processes as eutrophication may be slow (Carpenter *et al.* 1998b).

Natural press disturbances in streams include situations such as stream-damming (by animals, such as beavers, or by bank slumping and landslides), changes in sediment and nutrient inputs due to natural catchment disturbance (e.g. fire, destruction of vegetation by pathogens) and changes in light and temperature due to changes in riparian vegetation, for example after forest defoliation by cyclones (Vogt *et al.* 1996). Human-induced press disturbances are many and include channelization, dredging, removal of woody debris, removal or reduction of riparian vegetation, persistent point and non-point pollution, persistent fishing pressure, increased sedimentation, detritus and nutrient inputs due to catchment alterations, and the planned introduction of biota ranging from fish to riparian plants.

Ramp disturbances, in which the intensity of a disturbance is steadily rising (or declining), invariably occur over large spatial scales because they act over long time frames. Natural examples include such phenomena as the strength of droughts as they build in duration. As they occur for the most part at large spatial–temporal scales, ramp disturbances may include small-scale pulse and press disturbances. For example, the steady deterioration of a stream by sedimentation due to poor catchment management and land-use changes is a ramp disturbance. Nested within this long-term disturbance there may be pulses, such as floods. The intensity of these floods on the stream biota is then exacerbated by the shifting, imported sediment from the catchment (O'Connor & Lake 1994).

Many human activities result in ramp disturbances, which, if unchecked, involve a steady increase in the intensity or size of disturbance. Examples include phenomena such as:

- The increased diversion of water from rivers
- The increasing imposition of barriers along rivers
- The increased frequency of fires in catchments
- The processes of land clearing, tillage and settlement that give rise to effects such as increased sedimentation, salinization, nutrients, toxic chemicals, and invasion of exotic biota
- The manifold processes of urbanization that result in, for example, increased flow variability, increased stream channelization and decreasing stream vegetation cover.

On a larger scale, there are the potential impacts of global climate change, which, for flowing waters, may result directly in increasing temperatures and marked changes in regional flows (e.g. Firth & Fisher 1992). Associated with the projected climate change are potential indirect effects such as increased sedimentation, nutrient inputs, salinization and the spread of exotic biota.

The biotic responses to disturbance can be divided into **resistance**, the capacity to withstand a disturbance, and **resilience**, the capacity to recover from disturbance (Kelly & Harwell 1990; Lake & Barmuta 1986). These two properties can be specified for the various aggregates of ecological organization, from populations to ecosystem processes.

It is difficult to compare resistance between different species, populations or communities because the strengths of the disturbances themselves are rarely quantified. For most forms of natural and human disturbance in flowing waters, it appears that the resistance of the biota is moderate to low. Resistance is influenced by the life histories, life forms and behaviour of the biota, their physiological capabilities, their level of past exposure to the disturbing force and by the availability of refugia (Lake 2000). Different biota, even closely related species, may respond quite differently to the same disturbance. For example, in a Sonoran desert stream, invertebrates were greatly depleted by severe flooding while fish populations were unaffected (Meffe & Minckley 1987). Resistance varies greatly with the form of disturbance; for example, heavy-metal contamination may eliminate crustaceans and molluscs from contaminated sections of river but not insects (Norris *et al.* 1982). Resistance to the same form of disturbance also varies with stream type – the invertebrate fauna of an intermittent prairie stream was more resistant to floods than the fauna of a nearby perennial stream (Miller & Golladay 1996).

In general, the capacity to recover from disturbances is relatively high (i.e. the resilience of flowing-water biota and communities is high; Giller 1996; Yount & Niemi 1990). From a survey of 129 cases of disturbance to lotic systems, Yount & Niemi (1990) concluded that the rapid recovery rates were mainly due to four attributes:

> (1) Life history characteristics that allowed rapid recolonization and repopulation of the affected areas, (2) the availability and accessibility of unaffected upstream and downstream areas and internal refugia to serve as sources of organisms for repopulation, (3) the high flushing rates of lotic systems that allowed them to quickly dilute or replace polluted waters, and (4) the fact that lotic systems are naturally subjected to a variety of disturbances and the biota have evolved life history characteristics that favour flexibility or adaptability.

They further noted that slow recovery rates were found in disturbances that had resulted from major alterations in physical conditions. Examples of the latter include channelization (e.g. Arner *et al.* 1976; Hortle & Lake 1982, 1983) and sedimentation of the channel (Metzeling *et al.* 1995; O'Connor & Lake 1994; Waters 1995). Analysing similar instances of natural and human-generated disturbance, Niemi *et al.* (1990) found that in most cases the recovery time was less than three years. A similar trend in recovery time was found by Milner (1994) with an important difference being that recovery after watershed disturbances, in the form of logging and mining, usually took from five to more than 20 years. Such long recovery times usually occur when the disturbance leads to persistent changes in habitat structure, when pollutants persist in the affected system, or when the affected system is cut off from sources of colonizing biota. A major avenue of recolonization of disturbed sections is by stream drift (Williams 1981). Persistent changes in physical structure include the construction and maintenance of barriers, such as dams (Petts 1984; Poff *et al.* 1997). Such barriers may eliminate normal stream drift and thus hamper recolonization of disturbed areas downstream. Examples of the persistence of contaminants preventing recovery include the slow recovery ( > 10 years) of invertebrates in a Montana stream after cessation of the input of pollutants (Chadwick *et al.* 1986), and the case of the Molonglo River near Canberra, Australia, where in spite of cessation of mining and expensive rehabilitation works, the invertebrate fauna was still impoverished some 20 years later (Norris 1986).

While disturbances in terms of their major forces may be classified into pulses, presses or ramps, it should be realized that it is uncommon for a single agent or force to be the source of a disturbance. For example, under natural conditions, a disturbance such as drought with low water levels and greatly reduced flows is accompanied by disturbances such as deteriorating water quality (e.g. deoxygenation), elevated stream temperatures, decreased food supplies and decreased habitat availability (Lake 2000). Similarly, many human-generated disturbances are also multifactorial. The numerous effects of river regulation by dam construction and operation are stark examples of this (Petts 1984; Poff *et al.* 1997).

Global change is predicted to create large-scale disturbances to the biota of the Earth (Sala *et al.* 2000; Vitousek *et al.* 1997b). Major forces for global change in biodiversity and ecosystem structure include land-use change, climate change, changes in biogeochemical processes, and biotic exchange.

Land-use change in the case of freshwater systems covers changes both in catchment land use and in use of river channels. Both changes to catchments and to the rivers themselves disrupt biogeochemical processes. For example, in creating impoundments, dams greatly alter flow regimes and quantities (Poff *et al.* 1997; Rosenberg *et al.* 2000) and catchment land use changes may generate non-point pollution by nutrients (Carpenter *et al.* 1998b). Global climate change is due to the effects of both the ozone hole and the Greenhouse phenomenon. The ozone hole is giving rise to increased levels of UV radiation in high latitudes that could deleteriously affect algae and benthos in shallow streams (Kiffney *et al.* 1997), especially those at high altitude. Greenhouse effects include increased temperatures (more in temperate than tropical latitudes), altered flow volumes and water availability (Arnell *et al.* 1996) and increased frequency of extreme hydrological events (both floods and droughts; Arnell *et al.* 1996; Fowler & Hennessy 1995). Biotic exchange refers to the deliberate or accidental introduction of biota to river systems. With increasing international trade and with the increasing speed of trade transactions, biotic exchange is an increasing but underrated threat to freshwater biodiversity. The detection of such large-scale effects over long periods of time is a great challenge for ecologists, for funding agencies and for river and catchment managers (Stow *et al.* 1998).

The biodiversity of freshwater ecosystems, due to the impacts of the forces of global change, appears to be decreasing at greater rates than those recorded for terrestrial ecosystems (Ricciardi & Rasmussen 1999). For streams and rivers, depletions in biodiversity by 2100 are projected to be largely driven by land-use changes, climate change and biotic exchange (Sala *et al.* 2000).

### 3.2 THE PURPOSES OF MONITORING

In assessing the effects of human-induced perturbation on the ecological state of rivers, there are basically four major aims serving different circumstances. These aims address different ecological questions and serve to meet different management needs. However in a substantial and thorough investigation of any river, all four aims may be addressed at various times and localities during the investigation.

In listing these four aims it should be borne in mind that addressing the third aim, assessment of human-induced disturbances on flowing-water systems, is the major task of this book.

### 3.2.1 To assess the ecological state of ecosystems

Investigations with this general aim may range from studies at the population level through to those concerned with ecosystem properties. At the population level, the questions may involve projects such as the assessment of the population viability of a species of high conservation value, the stock density of a species of recreational or commercial importance, the status of a pest species, or the distribution of a species and whether its range is expanding or contracting. Similar types of projects may be undertaken for groups of co-occurring species or communities. All projects of this type may gain considerably in value and accuracy if reliable data are available from past assessments. In many cases, past data are available but unfortunately in most cases the quality of such data is relatively low and not amenable to even reasonable levels of statistical analysis. The reasons for carrying out such assessment tasks may range from the conservation of a species, community or particular type of habitat to tasks such as environmental audits and 'state of environment' reports. Carrying out such assessments may also serve the important aims of identifying threatening processes, and of detecting and providing warning of impending disturbances. Such survey exercises are decidedly not suitable for assessing the impact of future disturbances, even though such exercises have in the past been, and continue to be, regarded as 'environment impact statements'.

In flowing-water systems an important aim of management authorities can be to have an assessment of the ecological integrity or ecological health (see Box 3.1) of the rivers under their responsibility (Karr 1991). This aim may be partly met by assessment of water quality data (e.g. detection of acidification and eutrophication), especially if the data have been collected frequently for a long time. However, it is widely accepted that to gain an adequate appraisal of ecological health, it is imperative to monitor the biota.

In terms of ecological health assessment, biological monitoring has concentrated on biota such as invertebrates and fish along with, on some occasions, biota such as microbes, algae and riparian vegetation. Very rarely have ecosystem processes, such as decomposition, been monitored. Both habitat attributes and the biota may be monitored. In the case of biota, a great emphasis has traditionally been placed on indicator species and indices based on knowledge on the distribution of the biota and their physiological tolerances. The well-known Saprobien-system (vide Sladecek 1973) is an example of such an approach. With time, and often for particular regions, a plethora of biotic indices has

**Box 3.1** Ecological integrity and ecosystem health

Ecological integrity as defined by the major proponent of the concept 'refers to the capacity to support and maintain a balanced, integrated, adaptive biologic system having the full range of elements and processes expected in the natural habitat of a region' (Karr 1995). For a river in a region this would mean that ecological processes such as production, decomposition, nutrient dynamics and movement of biota would be identical with those of natural rivers in the same region. The river would be exposed to natural disturbances and have capacity for effective recovery or resilience. Systems of high ecological integrity are those formed by natural evolutionary and biogeographical forces and maintained as ecologically sustainable systems by natural ecological forces largely devoid of human intervention. However this concept has a limited application in these times when most rivers are affected to some extent by damaging human forces, be it toxic pollution, catchment land-use change or global climate change.

   Thus, as ecological integrity does not readily encompass human uses of, and pressures on ecosystems, some ecologists (e.g. Karr 1995; Meyer 1997; Rapport 1989) have proposed the notion of ecological health to incorporate human activities into ecosystem properties and dynamics. Ecological health, as opposed to integrity, is not a scientific concept as it also incorporates human uses and values. Thus, as good health medically refers to the fully functional, unstressed condition, the good ecological health of a river refers to a fully functional, unstressed river ecosystem. The river may also be providing important goods (e.g. biota) and ecosystem services for humans but it does not show, as judged from monitoring key indicators, significant signs of stress. A healthy river has the capacity to recover from stress induced by disturbance. Stress may be revealed from the monitoring of indicators, and in healthy rivers recovery subsequently occurs while in unhealthy rivers resilience may be low and indeed the river may be approaching or at the point of no return (Loehle 1991).

**Box 3.2** The distinction between 'rapid' assessment techniques and the techniques described in this book

There are several schemes for rapid biological assessment that would seem, initially, to provide conceptually easier and cheaper alternatives to the methods we advocate in this book. However, these rapid methods address different issues from those addressed in this book and their application to monitoring and assessment must be appreciated as implying a sacrifice in inferential power in order to gain rapid applicability.

The philosophy behind rapid bioassessment techniques is to measure the deviation of the composition of the biota of a location (e.g. a reach of a river) from some notional ideal – often called the reference condition (see Box 5.1). This information is then used to assess the status of the location (section 3.2.1) rather than assess a specific disturbance (section 3.2.3) or restoration activity (section 3.2.4). That philosophy of assessment does have some fundamental similarities to the philosophical approach highlighted in this book. Some observational evidence for disturbance is obtained, and this evidence then is judged 'impressive' only if it is improbable that it arose as a chance observation under normal conditions. The methods advocated in this book pin down those important notions of 'improbability', 'chance' and 'normal' through aspects of design that critically include replication (to get a handle on improbability) and controls (to get a handle on what is 'normal'). For rapid bioassessment techniques, the reference condition is to provide this information.

Advocates have proposed that such rapid bioassessment methods could be useful for:

- 'Status reporting' such as those exercises required by legislation or treaty obligations (e.g. State of the Environment Reporting)
- A source of information for planning, where broad-scale, coarse-resolution information is required to identify rivers or streams that are either in potentially very good or very poor condition
- Prosecution where a severe disturbance has likely resulted in an impact that is sufficiently large to be detected by a rapid-assessment method.

Basically, such status reports from rapid bioassessment techniques may alert us to whether *something* has gone wrong, but may not be able to distinguish or diagnose exactly *what* has gone wrong. Moreover, because these methods are rapid they are analogous to rapid techniques of measuring the status or health of humans: measuring temperature and blood pressure of a patient could alert us to a serious illness, but by themselves would not diagnose the precise nature of the illness; nor would these methods detect more subtle conditions by themselves.

The place of rapid assessment methods in the broader context of environmental monitoring and assessment is, we feel, still open for debate. In general these rapid bioassessment methods suffer from the following drawbacks when used for the types of monitoring that we concentrate on in this book (sections 3.2.3 and 3.2.4):

- The reference condition may not be well defined
- Temporal replication may be limited, with the potential to confound natural changes through time with perceived impacts
- The flora or fauna is treated as a multivariate syndrome which may be difficult to interpret ecologically and complex to explain to a lay audience
- Subtle impacts may not be detected because of the 'rapid' nature of the method employed (e.g. coarse taxonomic resolution, limited spatial or temporal scales of sampling).

Of course the methods advocated in this book do not totally escape these kinds of problems. We note that the problem of definition may apply to putative controls, temporal replication may be limited, multivariate interpretation may be a problem, and coarse taxonomic resolution or limited sampling may reduce power.

We would like to emphasize that we do not view rapid assessment methods as a cheap alternative to the methods advocated in this book. Differences in inferential power must be considered as well. At least one jurisdiction where rapid assessment methods have been included (ANZECC & ARMCANZ 2001) adopts a similar stance where the uses of rapid assessment methods are clearly prescribed and they are not viewed as interchangeable with methods based on formal controls in space and time.

been developed. These indices invariably attempt to reduce or summarize a collection of data into a single index or set of indices. Some indices (e.g. Index of Biological Integrity (IBI); Karr 1999) are based on the biota, whereas others are based on combination of habitat and biotic attributes (e.g. Index of Stream Condition (ISC); Ladson *et al.* 1999).

In recent times, the use of multivariate statistical descriptions of river biota has increased greatly (e.g. Barbour *et al.* 1995; Norris 1995; Reynoldson *et al.* 1997; Zampella & Bunnell 1998). Such approaches are based on the idea that at sites affected by human disturbance, the biota is different from that found at unimpacted sites that are broadly similar to the impacted sites. This approach has given rise to surveillance programs such as RIVPACS (Wright 1995) and AusRivAS (Simpson *et al.* 1997) to assess ecological health in rivers.

We do not view programs devised to assess ecological health, such as RIVPACS or the IBI, as an alternative to impact assessment (detailed in section 3.2.3 below), but such schemes may certainly complement some impact assessments (Box 3.2).

### 3.2.2  To assess whether regulated performance criteria have been exceeded

In many instances regulated standards are set for the levels of contaminants (abiotic and biotic) in flowing-water ecosystems. The assessment of adherence to the standards is carried out by regulatory or compliance monitoring. There are many standards set, such as concentrations of bacteria in water and levels of persistent toxic chemicals in water, sediments and biota. Chemicals of concern may include heavy metals, radioactive isotopes and organochlorine compounds (e.g. pesticides, dioxin, poly-chlorinated biphenyls (PCBs)). Thus monitoring may involve measuring variables, such as density of biota in the water column, or measuring concentrations of chemicals in biota.

Compliance monitoring may be focused on a particular activity being carried out at a particular site, such as a point discharge, or may be more widespread, such as the monitoring of pesticide levels in biota over a large region. Whatever the spatial focus and extent of monitoring, the basic question remains: Is the chosen variable being measured within the limits of the regulatory criteria? If values exceed limits, then this may trigger further investigation, especially in regional studies, to identify the source(s) of the higher than acceptable levels.

### 3.2.3 To detect and assess the impacts of human-generated disturbance(s)

The rigorous assessment of the potential impacts of a new, or pre-existing source of disturbance to flowing waters is the crux of this book. When a new development is planned to be sited in a river (e.g. a dam) or on the catchment of a river, there is always the risk that disturbances may be generated by the new development and damage the river. Thus the basic question for monitoring becomes: *Will the potential disturbance(s) generated by the new development significantly damage the biota and ecological processes of the receiving river?* The new developments may range from being particular point sources, such as a factory with a single discharge pipe for its treated waste, to a large-scale occurrence such as urbanization. The new development may range from only producing a single type of disturbance, such as hot water from a thermal power station, to projects such as a mining operation that may produce various types of chemical pollution from different parts of the operation.

The aim of the monitoring project is to assess the scale and magnitude of the disturbance and the responses of the variables (biotic and abiotic) selected for monitoring to such a disturbance. If detectable damage does occur then this could trigger management action to ameliorate the disturbance or even serve as the basis for compensation.

Successful monitoring projects require several elements: the availability of controls, the collection of data before any potential impacts have occurred from both the control and impact locations, and proper replication. The before-impact data collection must be carried out for sufficient time and at an appropriate spatial scale; we discuss each of these elements explicitly in chapter 5. The success of such projects in allowing adequate protection of the targeted river depends on maintenance of the monitoring effort after the implementation of the putative disturbance. Ideally environmental impact assessments should follow this protocol, but in many cases the gathering of the before-impact data is perfunctory and the gathering of after-impact data is either a once-off exercise or is done very infrequently.

Unfortunately in many cases the river in question has already been damaged by disturbance and/or is continuing to be damaged. The value of the data, as evidence, may be limited by the lack, or the poor quality, of before-impact data or control localities and hence have only weak inferential strength (issues that we consider specifically in chapter 9).

### 3.2.4 To assess the responses to restoration efforts

Increasingly efforts are being made to attempt to restore streams and their catchments. In many such projects evaluation of success is assessed by monitoring (e.g. Kondolf & Micheli 1995; Lockwood & Pimm 1999; Westman 1991). Successful monitoring in restoration projects needs the requisites outlined in section 3.2.3 of controls, before data and suitable replication. Further, in restoration there is the setting of goals as achievable endpoints. Thus, monitoring must be designed to evaluate whether the project is approaching the set targets. Monitoring in restoration efforts is addressed again in chapter 14, especially regarding the particular problems of design and interpretation.

3.3 IMPORTANT ISSUES

- *Perturbation* of a freshwater system consists of two sequential events: the *disturbance* to the system and the *response* of the system to the disturbance.
- Disturbances consist of three types: pulse, press and ramp.
- Human activities may disturb streams by acting on catchments, or in the channels, or on both areas simultaneously.
- Human-generated disturbances may vary from the application of physicochemical forces, such as building dams and changing river flows, to the introduction of exotic biota.
- The responses of the biota to disturbances vary in relation to the strength of *resistance* – the capacity to withstand the disturbance – and the level of *resilience* – the capacity of the biota to recover.
- Monitoring abiotic and biotic components of streams serves four main aims:

1. To assess the ecological state of ecosystems
2. To assess whether regulated performance criteria have been exceeded
3. To detect and assess the impacts of human-generated disturbance(s)
4. To assess the responses to restoration efforts.

# Part II

## Principles of inference and design

# 4

# Inferential issues for monitoring

This book will recommend a variety of statistical designs and analyses for detecting human impacts in rivers and streams. The aim of this chapter is to provide the basic principles of statistical inference for understanding these designs. We assume some prior statistical background, and suggest you read the first few chapters of a good introductory applied statistics text, such as Sokal and Rohlf (1995). We will take a classical, frequentist approach in this chapter that covers confidence intervals and $P$ (probability) values from statistical tests of hypotheses. We also consider the alternative Bayesian approach for interpreting probabilities and making decisions. The logic of designing a rigorous monitoring program to detect the effects of specific human activities in fresh water is, however, relatively independent of the debate between frequentists and Bayesians and our preference is for the classical methods. We emphasize the importance of developing statistical models that match the design chosen for monitoring flowing waters. The statistical model acts as a guide for analysis. With it, we can construct an analysis of variance (ANOVA) table and try to determine how much of the variation in a response variable (such as nutrient concentration, species diversity etc.) is accounted for by the influence of predictor variables (such as current velocity, human activity, time etc.). These statistical models also allow us to test specific hypotheses of interest; for example, is there a difference in mean species richness between locations upstream and downstream of a sewage discharge? The analyses we will use are very dependent on underlying assumptions and we stress the importance of checking those assumptions before drawing conclusions from the analyses. We also distinguish between univariate and multivariate analyses, pointing out that both may be relevant for any monitoring program but will require different tools and have different limitations.

**Box 4.1** Mining in Kakadu National Park, Northern Territory, Australia

The Alligator Rivers Region (ARR) lies in tropical northern Australia, $\sim 200$ km west of the city of Darwin. It is an area that incorporates Kakadu National Park, a World Heritage ecosystem, has a rich cultural heritage, and contains mineral reserves of uranium, palladium, gold and platinum. Organized mining and milling operations in the area began in 1979, with the opening of the Ranger Uranium Mine near Jabiru. However, community concern about the environmental impacts of mining meant that the Federal Government passed legislation specific to the region in 1978 establishing the Alligator Rivers Region Research Institute (ARRRI), to be managed by the Office of the Supervising Scientist for the ARR. The role of the ARRRI, under the direction of the Supervising Scientist, was, and is, to carry out monitoring programs that test the adequacy of regulatory controls placed upon mining operations to ensure that there are no 'observable effects' upon the natural environment. The region lies about 13° S of the Equator and has a tropical climate with seasonal, monsoonal rains. Excess water can accumulate in mine sites and lead to either deliberate or inadvertent releases of water containing harmful chemicals, such as metals and hydrocarbons, into downstream locations. It is this potential damage from wastewaters that has formed the main focus of regulatory and monitoring programs, which have focused primarily on two streams. Much of **Magela Creek** and its floodplain lies downstream of the Ranger Uranium Mine, and the **South Alligator River** flows past mining activities at Coronation Hill. In both these cases, locations upstream of mining activities have been compared to putatively affected locations downstream, using a BACIP design (see chapter 7). The ARRRI adopted BACIP designs in the first instance, in part because the mines produce specific changes not otherwise observed naturally and in part because of concerns about whether neighbouring streams could act as controls. See Humphrey *et al.* (1995) and Faith *et al.* (1995) and references therein for more information.

Consider the monitoring program to assess the effects of the proposed gold–platinum–palladium mine at Coronation Hill within the Kakadu National Park in northern Australia on aquatic biota in the South Alligator River (see Box 4.1). The program used eight locations on the river, two upstream of Coronation Hill and six downstream. Within each location, a Sürber sampler was used to collect four 25 cm$^2$ sampling units and the abundance of all species of macroinvertebrates was recorded. Note that it is difficult to separate the effects of an impact (the mine) from longitudinal changes in the biota of a river using a simple upstream–downstream design such as this and we will propose better designs later (chapter 7). This design is, however, suitable for our purposes of illustrating basic statistical principles and ideas.

## 4.1 SAMPLING

Suppose we wish to know the average species richness of macroinvertebrates from 25 cm$^2$ sampling units from one location on the South Alligator River. We could measure the species richness from all possible 25 cm$^2$ sampling units at the site but there are so many possible sampling units that it is impractical, and unnecessary, to measure them all. Sometimes, sampling units may be natural habitat units, such as stones in a riffle of an upland stream or pieces of woody debris in a lowland river. In sampling terminology, the species richness of each sampling unit is an observation and the species richness on all the possible sampling units is known as our statistical population and comprises all the observations about which we are interested in making inferences. This can also be termed the 'sampling universe' or the 'universe of inference or application' (Walters & Green 1997). Note that we must define precisely the spatial and temporal boundaries of our statistical population at the start of any study. For example, our population might be the possible sampling units at a location that extends 50 m along a river on 25 December 1998 and our formal statistical inferences are restricted to that population, although we may speculate across spatial and temporal scales.

Because the statistical population is too large to record all observations, we must estimate the average species richness per sampling unit in the population from the average species richness per sampling unit from a sample. A sample is a subset of sampling units and observations from the population. There are a number of methods for choosing which observations will be included in a sample, although random sampling, where each sampling unit has an equal chance of being selected, is commonly used and has many statistical advantages. Primarily, random sampling allows us to make reliable, generalized inferences about the population from our sample. If a population is spatially heterogeneous, then stratified random sampling is often used where the population is separated into spatially more-homogeneous strata, and one or more random samples are taken from within each stratum. Sampling can also be adaptive, whereby the method for selecting sampling units is modified depending on early results (Thompson 1992).

True random sampling is often difficult in practice; one method is to allocate all possible sampling units a number and sampling units are then chosen using a set of randomly generated numbers. In practice, we often sample haphazardly, whereby each sampling unit in a sample is chosen in a simple manner that attempts to avoid systematic bias. For example, to choose which possible $25\,cm^2$ sampling units would be incorporated in a sample, an ecologist might stand in a section of river with their eyes closed, throw an object and place the sampling device closest to where the object lands. This process would be repeated for the number of sampling units required for the sample. Note also that the term 'sample' can be used in two different contexts in aquatic ecology. Strictly, a sample is a collection of observations from the population of interest, such as a collection of sampling units. However, the material from a single sampling unit (such as a Sürber sampler) is also sometimes called a 'sample'. This latter use is misleading and we will try to restrict our use of the word sample in this book to the first meaning.

## 4.2 UNCERTAINTY AND PROBABILITY

Later chapters in this book will describe monitoring designs for detecting impacts of human activities under conditions of uncertainty, and examine how we make decisions about those impacts. This usually involves attaching probability statements to particular outcomes, such as the probability that the observed species richness of macroinvertebrates in a river or stream is a product of natural variation. Because our decision-making process relies heavily on probability statements associated with particular outcomes, it is very important that we understand the interpretation of probability in a statistical sense, and the role of such probabilities in inference.

Probability can be viewed simply as a quantification of uncertainty. Say we take two samples, each consisting of the same number of sampling units, from a section of river and calculate the average number of species of macroinvertebrates per sampling unit. These two averages will almost certainly be different, despite the samples coming from the same section of river. There are two general causes of this uncertainty in measurements we might take (Hilborn & Mangel 1997); that is, two reasons why two measures of species richness might be different. **Process** uncertainty results from the actual number of species being different when the second sample is taken compared with the first; such temporal changes in biotic variables, even over very short time-scales, are common in ecological systems. **Observation** uncertainty results from sampling error; the average value in a sample is simply an imperfect estimate of the average value in the population (all the possible sampling units) and, because of natural variability between sampling units, different samples will nearly always produce different averages. Observation uncertainty can also result from **measurement error**, where the machine or human measuring device is imperfect, such as humans making mistakes in identifications or a pH meter malfunctioning.

The implication of sampling variability is that we can never be certain about the value of a variable in a population unless we census the whole population by recording all the possible sampling units. This means that any conclusions we draw from our analyses of monitoring designs, and any decisions resulting from those conclusions, have to be probabilistic. It is important that we understand the interpretation of probabilities associated with our statistical conclusions.

How do statistical probabilities play a role in deciding whether a human activity has caused an impact? Uncertainty leads naturally to the use of probabilities, suggesting that we might assess an impact as probable or not. But probabilities will play a quite different role. Rather than seeking a high probability as support for an hypothesis of impact, it is a low probability, an improbability, that is the key to inferring an impact. To understand this, it is useful to acknowledge that uncertainty in monitoring extends beyond that related to sampling variability to include inferential uncertainty. Strictly speaking, the detection of an impact of a specific human activity does not mean that we observe it to be true, but only strongly infer its presence. This recalls the idea in philosophy of science (Popper 1963) that we cannot prove an hypothesis. Any supposed evidence for impact can always be explained away to some degree by conceivable no-impact scenarios. How does inference of impact then proceed?

Let us first consider one tempting strategy. It is well known that Popper and other philosophers offer 'falsification' as a tactic, and this has been linked to monitoring (e.g. Underwood 1990, and discussed below). Certainly, we might be able to falsify an hypothesis of impact –

for example, if we know that the impact implies that something else must be true (say, that a pipe must be open) and that condition is blatantly false (the pipe is not open). But that neat, negative inference may not be typical in monitoring. Most of our real-world efforts in monitoring must involve learning as much as we can under conditions where we cannot falsify the impact hypothesis and cannot verify it either. Fortunately, outright falsification is not the only inferential pathway.

Clearly, failure to falsify does not, on its own, provide strong inference. Just because our evidence (e.g. the observed data from our monitoring program) is not a counter-instance, it cannot be interpreted automatically as good support for our hypothesis (Popper 1983). This is the motivation for a desire for an improbability. According to Popper, we should not be very impressed with some purported evidence $e$ (e.g. the observations from our monitoring program) for hypothesis $h$ (e.g. the hypothesis that there is an impact), if in fact we don't really need $h$ at all to have had a good chance of observing $e$. There is strong support (Popper's 'corroboration') for $h$ only if it is improbable to obtain $e$ in the absence of $h$, given only our background knowledge, $b$, about other ways in which evidence like $e$ can arise. This will be the key strategy for inference of impacts – finding some evidence for impact that cannot easily be explained away by various other processes, such as natural variation in the system. Summarizing all those other factors as background knowledge means that we want evidence, $e$, to be improbable given only that knowledge, $b$ – or a low $P(e,b)$, the probability of $e$, given $b$.

We will link $P(e,b)$ with the examination of tail probabilities in statistics (e.g. in testing null hypotheses). For Popper, background knowledge was quite general, consisting of any knowledge accepted provisionally while testing the hypothesis. But the special case of substituting a null or other statistical model (e.g. the null hypothesis) as a specific instance of background knowledge, $b$, and as the basis for calculating $P(e,b)$, is well established (for discussion see Faith 1999). This framework for arguing for support for an hypothesis is further developed through the linking of corroboration or 'severe tests' to basic statistical tests (Mayo 1996). Low $P(e,b)$ is given by a small tail probability in the conventional test of a null hypothesis. Mayo (1996, p. 193) says, 'by rejecting the null hypothesis $H_0$ only when the significance level is low, we automatically ensure that any such rejection constitutes a case where the non-chance hypothesis $H$ passes a severe test.' If support for the hypothesis $h$ is gained by rejecting a null hypothesis, we might ask why corroboration/severity is needed – this would appear to be just standard Popperian falsification of a (null) hypothesis (as argued by Underwood 1990). However, we are not falsifying the null but arguing that it is merely improbable as a way to account for this evidence. We are not dealing strictly with falsification, but rather with an hypothesis that has resisted falsification and may or may not have gained some support (corroboration) as a consequence.

We have now made a philosophical link between the inevitable use of probabilities of particular outcomes and the evaluation of hypotheses of impact. The outcome of interest is one that is interpreted as possible evidence for impact. Support for that impact hypothesis is only found if the probability of that outcome is small under normal circumstances, in the absence of impact. In the next chapter, this pursuit of improbability provides the rationale for specific aspects of monitoring design including controls.

Interpretation of probabilities also goes to the heart of one of the long-standing debates in statistical science. The classical interpretation of probability is the relative **frequency** of an event that we would expect in the long run, or in a long sequence of trials. For example, let us hypothesize that the species richness of macroinvertebrates in a section of river affected by gold mining is 100 species per sampling unit, because that is the approximate number of species found in similar types of river affected by this type of mining, as reported in the literature. We then take a number of sampling units (a sample) from this section of river and find that the average species richness per sampling unit is 50. A sensible question might then be: What is the probability of a sample from this section of river having an average of 50 species per sampling unit if the real average number of species per sampling unit is actually 100? If that probability is low, then we might conclude that the true average species richness is not 100. This probability (i.e. the probability of getting a sample with an average of 50 species when the real species richness is 100 species per sampling unit) has a frequentist interpretation. It is a measure of the frequency of occurrences of samples with an average of 50 species if we repeatedly sample from a river with 100 species per sampling unit. In practice, of course, we don't actually do repeated sampling so we have to determine the probability of getting a sample with an average of 50 species from a population with 100 species per sampling unit in other ways. Note that this frequentist probability is the probability of the data (or sample) given a particular hypothesis (i.e. the probability of a sample with an average of 50 if the population has an average of 100). This is often written $P(\text{data}|\text{hypothesis})$ and can clearly be interpreted as a frequency because we can, in theory, take lots of samples and calculate the number that have an average of 50 species. If the population really does have 100 species, then this frequency is the probability of getting a sample with an average of 50 species from a population with an average of 100 species. This long-run frequency interpretation is the classical statistical interpretation of probabilities and is how we usually interpret confidence intervals and $P$-values from statistical tests.

An alternative to the frequentist approach to statistical analysis is **Bayesian** methodology. Bayesian analyses differ from classical analyses in a number of ways and we can only briefly summarize them here – see Ellison (1996), Hilborn and Mangel (1997) and Lee (1997) for more details. The first difference is allowing a more subjective, non-frequentist interpretation of probability. For Bayesians, probability represents some quantification of our opinion about whether something is true or an event will happen. Our opinion may be derived from previous observations, theoretical considerations, knowledge of the particular event under consideration etc. For example, we might want to know the probability that the species richness in a section of river is 100 species per sampling unit. This probability has no frequentist interpretation because the species richness does not depend on any long-run frequency of sampling outcomes. Our opinion of the probability that the species richness of the river is 100 is more subjective, based on previous information, our knowledge of this stream etc. Using probabilities in this way has been criticized as being too subjective, although we use such non-frequentist probabilities commonly in real life, such as the probability of it raining tomorrow or the probability of a particular horse winning a race. A second difference is that Bayesians want to measure the probability of a particular hypothesis, given the data ($P$(hypothesis|data)). For example, what is the probability that the average species richness per sampling unit is 100, given that we have a sample with an average of 50 species per sampling unit. This is the converse of the frequentist probability, which would be the probability of a sample with an average of 50 species per sampling unit if we assume that the real species richness is 100 species per sampling unit. A third difference between the frequentist and the Bayesian approaches is that Bayesians would usually try to incorporate prior information or opinions about how likely an outcome is into their analysis, whereas classically we base our statistical conclusions only on the sample data (this is not strictly correct because we use tail probabilities; more about this in section 4.7). So a Bayesian analyst who had strong prior belief that the species richness in a river affected by mining would be much less than 100 could incorporate this belief (prior probability) into their analysis and this would modify the final (posterior) probability. In simple terms, this means a frequentist and a Bayesian might draw different conclusions, even with identical sample data, because the Bayesian might have strong prior opinions about whether the species richness is likely to be 100 or not. This comparison is a little contrived, however, because the questions being asked are different. The frequentist is asking what is the long-run probability of getting a particular value in a sample if the population value is fixed and specified. In contrast, the Bayesian is asking what is the probability of any population value, given that we have a sample value, and also given that we have incorporated our prior opinions into the calculation of the probability.

In monitoring impacts, both approaches can be brought to bear on the same hypothesis – that there was an anthropogenic impact at this particular site. The frequentist, of course, is not arguing that there will 'in the long run' be an impact, but is using those long-run probabilities to examine whether the evidence for impact can be judged quite improbable in the absence of impact. The Bayesian too is interested in impact at just this site, but will introduce prior probabilities from previous experience (say experience with potential mining impacts of this sort). The two approaches to evaluating the same hypothesis can give quite different answers. In fact, the mode of frequentist inference to be explored in this book has been criticized by Bayesians, and it is useful to translate one of their proposed counter-examples (Howson 1997) into the monitoring context (Box 4.2).

Suppose a large change in abundance of a pollution-sensitive species has been observed at the site where there is mining activity and (based on an appropriate ideal experimental design) the null hypothesis of no change is rejected at the 0.05 level. Impact would be inferred. But now suppose also that the probability of an impact at such a mine site is 0.001 (as estimated say from many previous mines of this type). Then a standard application of Bayes' theorem (Box 4.2) indicates that the chance of there actually being an impact when we observe this change in our sensitive species is less than one time in 50! The observed change was somewhat improbable based on background knowledge, but, far from indicating that there is an impact, the results of the ideal monitoring scheme mean that the chance of there actually being an impact is very small. Clearly the mining company whose activities might be stopped as a consequence of inferred impact would see a big difference in the two results. Why stop mining when there is an actual impact less than one time in 50 when an impact is supposedly 'detected'?

---

**Box 4.2** Howson's 'dilemma' and the inference of impacts

Suppose the null hypothesis of no change is rejected at the 0.05 level and impact is inferred at a mine site. Suppose also that the probability of an impact at such a mine site is 0.001 (as estimated say from many previous mines of this type). We can apply Bayes' theorem (see also equation 4.6), stated below in the symbols of section 4.2, to this problem.

$$P(h,eb) = \frac{P(e,hb)P(h,b)}{P(e,b)}$$

In comparing this formula to equation 4.6, $e$ is equivalent to the 'data' and each term now includes background knowledge, $b$, as given. For example, $P(h,eb)$ is the probability of the hypothesis given both $e$ and $b$.

The probability of the hypothesis given the data and $P(e,hb)$ is approximately equal to 1, $P(h,b)$ is equal to 0.001, and $P(e,b)$ is equal to 0.05. Thus, $P(h,eb)$ is approximately equal to $0.001/0.05 = 1/50$.

This result indicates that the chance of there actually being an impact when we find such evidence is less than one time in 50!

---

One response to this conundrum is to reconcile the two results as not all that incompatible. If impacts occur only 1 in 1000 times, then even a finding that our evidence (the change in that sensitive species) could occur as often as 49 out of 50 times without impact could call for follow-up investigation (but perhaps not a halt to mining). The observed evidence, after all, means that what once had a probability of only 1 in 1000, now has a probability of 1 in 50. The Bayesian priors therefore may have a place in a frequentist, adaptive monitoring program.

### 4.3 VARIABLES

In our mining example, based on Faith *et al.* (1995; Box 4.1), species richness is called a variable (the characteristic of the population in which we are interested) and is considered a random variable because its value is not known until we sample. In monitoring programs in fresh-water systems, random variables can be physical (current flow, depth etc.), chemical (pH, phosphorus concentration etc.), biological (species richness, abundance, etc.) or a more complex index (such as Shannon–

Weiner diversity). Many variables recorded in monitoring programs are continuous random variables and can take any value above a detection limit imposed by the measuring device. We can also measure discrete random variables, which can take only certain, usually integer, values, such as counts of organisms, presence/absence of a compound, alive or dead after a toxicological experiment. Abundances, although count variables and therefore strictly discrete, behave more like continuous variables because they can take such a wide range of integer values from zero upwards when the sampling unit is larger than the size of the objects being counted. Note that counts of organisms and most measurement variables in freshwater monitoring cannot include negative values (i.e. we cannot have a negative number of organisms nor a negative pH or phosphorus concentration).

Another useful dichotomy is to distinguish response ('dependent') variables from predictor ('independent') variables. The former are outcome variables from monitoring programs or from experiments and are nearly always random variables. Predictor variables can be random or might be fixed in advance by the investigator. They are used to explain variability in one or more response variables, usually in a statistical modelling context (see section 4.5). In the mining example where the sampling units were 25 cm$^2$ and taken using a Sürber sampler, species richness might be one response variable, and random predictor variables might be depth, current velocity, sediment grain size etc. If we designed our sampling program so that we took sampling units from predefined depths or distances from the bank, then depth and distance would be fixed predictor variables. Fixed predictor variables are commonly used in field and laboratory experiments, where they are termed factors (or effects).

Any random variable has a probability distribution that is the distribution of relative frequencies or probabilities of all possible values of the variable (Fig. 4.1). These probability distributions are important when using samples to estimate population parameters. In most monitoring situations, there are a number of response variables that might be of interest (e.g. species richness, abundance of a particular species, some physical or chemical measurement such as pH or dissolved oxygen; see chapter 10). Ideally, we would estimate each variable independently of each other (i.e. using different samples). In practice, however, that is unrealistic. For example, if we knew there were $p$ species in a section of river, we would never contemplate taking $p$ separate samples (each with $n$ sampling units) to independently estimate the abundance of each species. Rather, we would estimate the abundance of each

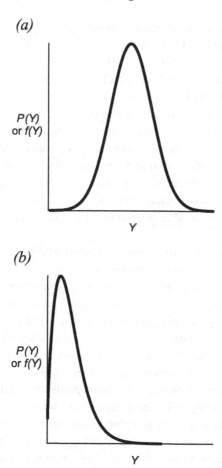

*(a)*

P(Y)
or f(Y)

Y

*(b)*

P(Y)
or f(Y)

Y

**Fig. 4.1** Two common probability distributions for continuous variables: (a) a normal distribution that is symmetrical; (b) a log-normal distribution that is positively skewed. P(Y) is the probability of Y and f(Y) is the frequency of Y.

species from a single sample (with $n$ sampling units). There will, however, be a degree of biological and statistical dependence between our estimates of abundance of each species and our interpretation must be conditional on this lack of independence. Some types of statistical analyses make use of this dependence (correlations) between variables recorded from the same sampling units (multivariate analyses, such as principal components analysis, discriminant function analysis etc.; see for example James & McCulloch (1990), Manly (1994) and section 4.10).

## 4.4 ESTIMATION

When we take a sample from a population of possible sampling units, we do so to estimate some characteristic of our variable of interest for that population; characteristics of statistical populations are termed parameters. What characteristic of the variable should we be interested in? For example, do we wish to estimate the average species richness per sampling unit in a river or the variability in species richness between sampling units? For most monitoring programs, the two population characteristics of most interest are some measure of the middle of the distribution (central tendency) of a variable, such as the mean or average, and some measure of its spread or variability between sampling units, such as the variance or standard deviation (Box 4.3). As we will see in later chapters, both can be used to measure the effect of human impacts on biota; for example, the difference between the means of species richness in polluted and unpolluted sites, or the percentage of variance in algal biomass explained by the nutrient concentration. Characteristics of samples are termed statistics and the appropriate statistic provides an estimate of the population parameter. For example, what sample statistic provides the best estimator of the population mean or average? The answer to this question depends on how we define a good estimate. The best estimator of a parameter should be the most precise; that is, repeated sampling (note the frequentist connotations) should produce values of the statistic closer to each other compared to any other statistic we might use. The precision of a statistic is measured by its standard error (Box 4.3). There are other requirements for estimators of population parameters (see for example Hays 1994) but these do not concern us for this discussion. There are a number of possible statistics we could use to estimate the population mean. It turns out that the best estimator of the population mean is the sample mean (or average). A single value of a sample statistic which estimates a population parameter is a point estimate, so the sample mean is a point estimate of a population mean.

Point estimates are limited in value unless we have some idea of how good they are, especially how close they are likely to be to the population parameter. What we need is an interval estimate of a parameter, which is a range of values that will include the parameter with some level of confidence. We can use the sample variance and standard error to determine the confidence we have in our estimate of the mean (i.e. how confident we are that our sample mean is close to the true, but unknown, population mean; Box 4.4). Confidence intervals for

**Box 4.3** Common statistics and parameters

Consider a random sample of $n$ observations $(y_1, y_2, \ldots, y_i, \ldots, y_n)$ from a clearly defined population. We can calculate the following statistics from this sample as estimates of particular parameters of the population:

| Characteristic | Sample statistic | Equation | Population parameter | Notes |
|---|---|---|---|---|
| Mean | $\bar{y}$ | $\dfrac{\sum\limits_{i=1}^{n} y_i}{n}$ | $\mu$ | Average value. |
| Median | Sample median | $\dfrac{y_{n+1}}{2}$ if $n$ odd<br><br>$\dfrac{y_{n/2} + y_{(n/2)+1}}{2}$ if $n$ even | Population median | Middle value of rank-ordered observations.<br><br>Equals mean for normal distribution but sample median less precise estimator of $\mu$ compared with sample mean.<br><br>Better represents centre of skewed distribution. |

| Variance | $s^2$ | $\dfrac{\sum\limits_{i=1}^{n}(y_i - \bar{y})^2}{n - 1}$ | $\sigma^2$ | Numerator termed sum-of-squares (SS). Denominator $n - 1$ rather than $n$ so $s^2$ unbiased estimator of $\sigma^2$. |
|---|---|---|---|---|
| Standard deviation | $s$ | $\sqrt{\dfrac{\sum\limits_{i=1}^{n}(y_i - \bar{y})^2}{n - 1}}$ | $\sigma$ | Square root of variance so units are same as variable. |
| Coefficient of variation | Sample CV | $\dfrac{s}{\bar{y}}$ | Population CV | Expresses standard deviation independent of measurement units. |

**Box 4.4** Illustration and interpretation of confidence intervals for a population mean

If we sample repeatedly from a population, we can determine numerous sample means, one for each sample we take. If those sample means are very different, then it suggests that any one of them may not be close to the true population mean; if they are very similar, it suggests that they are all close to the true population mean. If we have many sample means from a single population, a measure of the variability of those sample means is their standard deviation. The standard deviation of a statistic, like the sample mean, is called the standard error; so the standard deviation of the sample means is called the standard error of the mean. If the standard error of the mean is small, it indicates that any sample mean is likely to be close to the true population mean; if the standard error of the mean is large, it indicates that any single sample mean may not be close to the true population mean.

In reality, we do not take lots of samples, but a single sample of size $n$. It turns out that we can estimate the standard error of the mean from a single sample:

$$s_{\bar{y}} = \frac{s}{\sqrt{n}} \tag{4.4.1}$$

where $s$ is the standard deviation of our response variable in our single sample and $n$ is the size of the sample. So the standard error of the mean is a measure of how confident we are that our single sample mean is close to the true population mean. Can we improve on this measure of confidence?

An additional step is to convert the standard error, which relates to the sample mean, to a confidence interval, which relates to the population mean. Although we won't go into the relatively simple arithmetic behind the development of a confidence interval for $\mu$ (see Sokal & Rohlf 1995), it is important to understand its interpretation. The confidence interval for the true, but unknown, population mean is based on the mean and standard error from our single sample and the $t$ distribution. For example, the 95% confidence interval for $\mu$ is:

$$\bar{y} - t_{0.05(n-1)} \frac{s}{\sqrt{n}} \leq \mu \leq \bar{y} + t_{0.05(n-1)} \frac{s}{\sqrt{n}} \tag{4.4.2}$$

where $\bar{y}$ is the sample mean, $s/\sqrt{n}$ is the sample standard error of the mean and $t_{0.05(n-1)}$ is the value from the $t$ distribution (with $n-1$ degrees of freedom) between which 95% of all $t$ values occur. Note that all the components of this equation are available from our single sample and from the $t$ distribution. How do we interpret this interval? It is telling us that there is a 95% probability that any single confidence interval (from any single sample) will include the population mean. It is not a 95% probability that the specific confidence interval determined from our sample will contain the population mean; the population mean is a fixed value so a specific confidence interval either contains it or it doesn't. The 95% probability here must be interpreted in the long-run frequency context. However, we shouldn't worry too much about statistical pedantics; this interval gives us an idea of the range within which the true population mean lies. We can determine intervals for other levels of confidence by simply using different $t$ values; a 99% confidence interval would be wider than a 95% interval.

parameters are often misinterpreted. A confidence interval is a range of values that we are confident (conventionally 95% confident) includes the unknown population parameter. The probability associated with confidence intervals has an interpretation via long-run frequencies. For 95% confidence intervals, repeated samples from the population would result in 95% of the intervals containing the unknown parameter and 5% not.

We have focused on standard errors and confidence intervals for means but we can determine standard errors for other statistics and confidence intervals for the relevant parameters using similar logic, as long as the sampling distribution of the statistic is approximately normal and we know the formula for its standard error. In most cases, the confidence interval is based on the $t$ distribution. If these two conditions don't hold, resampling methods (e.g. bootstrap) permit us to generate a sampling distribution for a statistic specific to the sample data at hand and thus determine a standard error for that statistic (see Crowley 1992; Dixon 1993; Manly 1997).

### 4.5 STATISTICAL MODELS

While we nearly always use a sample mean or variance to estimate a population mean or variance, we usually do so as part of a more sophisticated process of fitting a statistical model to our monitoring data. This statistical model, the structure of which should be determined at the design stage, is an attempt to represent biological relationships between a response variable (Y) and one or more predictor or explanatory variables (Xs) in a mathematical form. We will focus on linear models, although non-linear models are also possible (Box 4.5, Fig. 4.2).

### 4.5.1 Regression models

As a simple example, we might wish to model species richness (our response variable Y) from sampling units against two continuous predictor variables, depth ($X_1$) and current velocity ($X_2$), for a single location along the river. Note that the response and predictor variables should be measured at commensurate scales, in this case the scale of a sampling unit. Our biological model might look like:

species richness = effect of depth + effect of current velocity
+ unexplained error                                                      (4.1)

The left-hand side of the model always represents our response variable. The right-hand side of the model represents our predictor variables and a measure of uncertainty (Hilborn & Mangel 1997) or error – this error is measured by how much our observed values of species richness differ from those predicted by our model.

Our formal statistical model representing this biological model is:

$$y_i = \beta_0 \beta_1 x_{i1} + \beta_2 x_{i2} + \varepsilon_i \qquad (4.2)$$

where
   $y_i$ is the value for the response variable (species richness) on the $i$th sampling unit
   $x_{i1}$ is the value for the first predictor variable (depth) on the $i$th sampling unit
   $x_{i2}$ is the value for the second predictor variable (current velocity) on the $i$th sampling unit
   $\beta_0$ represents the value of species richness when both depth and current velocity are zero. This is commonly called the Y-intercept
   $\beta_1$ measures the strength of the relationship between species richness and depth independent of current velocity. Formally, $\beta_1$ is the slope of a linear regression line between Y and $X_1$, the change in Y for a unit change in $X_1$, holding $X_2$ constant and is termed a partial regression slope
   $\beta_2$ measures the strength of the relationship between species richness

and current velocity independent of depth. Formally, $\beta_2$ is the slope of a linear regression line between $Y$ and $X_2$, the change in $Y$ for a unit change in $X_2$, holding $X_1$ constant and is also a partial regression slope. $\varepsilon_i$ is an error term representing the difference between the observed values of the response variable and the values predicted by the true regression model (i.e. it measures the variation in the response variable not explained by our chosen predictor variables).

This is a population model because it represents the relationship between the response and the predictor variables in the population of possible sampling units. As such, $\beta_0$, $\beta_1$ and $\beta_2$ are parameters we must estimate from our sample data. Note that the predictor variables do not have to be independent of each other, indeed current velocity may well be greater in shallow water than deep water, although high correlations between predictors (collinearity) cause difficulties in estimation and hypothesis-testing.

When all predictor variables are continuous, as in this example, the model is usually referred to as a regression model. The effect of each predictor variable on the response variable is measured by the slope of a linear regression line. When all predictor variables are categorical, the models are sometimes referred to as analysis of variance (ANOVA) models and are fundamental to the monitoring designs we will recommend in this book. In practice, the distinction between regression and ANOVA models is artificial; linear models can include both continuous and categorical predictors, and the procedure for fitting the linear statistical model is the same. However, the terminology distinguishing regression and ANOVA models is entrenched in the literature and usually unavoidable. The monitoring programs we will recommend in this book are based primarily around categorical predictor variables such as polluted vs. unpolluted or before activity vs. after activity.

### 4.5.2 Analysis of variance (ANOVA) models

The statistical literature on the theory and application of ANOVA models is extensive and Sokal & Rohlf (1995) and Underwood (1997) provide thorough introductions from a biological perspective. One problem with reading this literature is that a range of terminology and symbols is used and this can be confusing for the novice. We have tried to summarize the main points in general statistical terms in Box 4.6 and Tables 4.1–4.3, and we will illustrate these different models based on the mining example in Kakadu National Park from Faith *et al.* (1995; Box 4.1),

**Box 4.5** Linear versus non-linear models

The term 'linear' has two meanings when applied to statistical models, particularly regression models. If the shape of the relationship between a predictor variable and a response variable approximates a straight line (with either a positive or negative slope), then this relationship is often termed linear. This use of the term linear applies specifically to regression analyses in which the predictor variables are usually continuous. More generally, however, a linear model is one in which the parameters appear simply because none is an exponent, multiplied or divided by another parameter (Neter *et al.* 1996). Linear models do not have to represent linear (straight-line) relationships, although they usually do. For example:

$$y_i = \beta_0 + \beta_1 x_{i1} + \varepsilon_i \qquad (4.5.1)$$

is a linear model that represents a straight-line relationship (Fig. 4.2a) where $\beta_0$ is the intercept and $\beta_1$ the slope of the straight-line relationship. The following is also a linear model:

$$y_i = \beta_0 + \beta_1 x_{i1} + \beta_2 x_{i1} + \varepsilon_i \qquad (4.5.2)$$

but the relationship is now a curvilinear one (Fig. 4.2b). These models, where a single predictor variable is also raised to the second or higher power, are sometimes called polynomial models.

A non-linear model is one in which the parameters themselves are no longer simple and represents a relationship that cannot be depicted by a straight line. For example:

$$y_i = \beta_0 e^{\beta_1 x_i} \varepsilon_i \qquad (4.5.3)$$

is an exponential growth model where one parameter ($\beta_1$) appears as an exponent. The relationship is not a straight line (Fig. 4.2c) but this model can be made linear by a transformation of Y to logs:

$$\log(y_i) = \log(\beta_0) + \beta_1 x_i + \log(\varepsilon_i) \qquad (4.5.4)$$

Note that Y, the intercept ($\beta_0$) and the error terms ($\varepsilon_i$) are now in log units but the relationship is now a straight line and the model is linear.

*(a)*

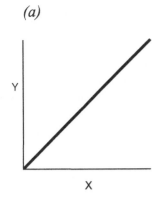

*(b)*

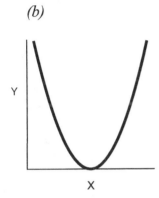

*(c)*

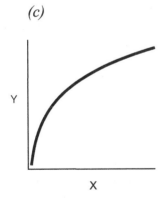

**Fig. 4.2** (a) Straight-line relationship from linear model; (b) curvilinear relationship from linear polynomial model; (c) curvilinear relationship from non-linear exponential growth model (e.g. X is time, Y is size).

modified slightly to expand the focus of this book beyond macroinvertebrates.

Consider a design where we wished to examine the effects of a mining operation on algal biomass on submerged pieces of wood (coarse woody debris or 'snags') in the South Alligator River. If we choose a number of (control) locations upstream of the mine and a number of (impacted) locations downstream of the mine and recorded the average algal biomass on snags within each location, our biological model might be:

$$\text{algal biomass} = \text{upstream vs. downstream} + \text{unexplained error}$$
(4.3)

Here, our basic sampling unit is a location, although we may have a number of subsampling units (individual snags) within each location. Note also that such a design does not necessarily allow us to infer effects of the mine, just whether upstream locations are different from downstream locations. This issue of inferring impacts from upstream versus downstream comparisons will be considered in detail in chapter 5. When our predictor variables are categorical, we generally measure the effects of the predictor variable as deviations of the mean of each category (i.e. level) from the overall mean for the response variable. Our statistical model is represented by equation 4.6.1 in Box 4.6. This model is sometimes called a single-factor (one-way) ANOVA model. There are only two categories of the predictor variable (factor) in this example, upstream and downstream, although there are commonly more than two categories for factors, such as multiple sampling times. Note that the terminology used for these models in the literature, especially different subscripting etc., can be confusing (Box 4.6). For the remainder of this chapter, we will just discuss the statistical model but remember that these statistical models are simply representations of biological models that indicate the predictor variables that we think influence the response variable.

In many cases, we record our response variable on subsampling units within each main sampling unit. For example, we might record algal biomass on a number of snags within each location. Our model is now represented by equation 4.6.2 in Box 4.6. Notice that we have two spatial scales in this design (between locations and between snags within locations) and this model is termed a two-factor nested model. The $\beta_{i(j)}$ term is called a nested factor, where locations are nested within upstream vs. downstream groups. Nested spatial and temporal scales of subsampling are common in the designs we will describe in chapters 5

**Box 4.6** Analysis of variance models and notation

Consider a linear model relating a response variable $Y$ (e.g. average algal biomass per snag) to a single categorical predictor variable, A (e.g. upstream vs. downstream of mine), with $a$ categories ($i = 1$ to $a$). There are $n$ replicate observations (e.g. locations along river) of $Y$ within each category, any observation denoted as $k$.

$$y_{ik} = \mu + A_i + \varepsilon_{ik} \qquad\qquad (4.6.1)$$

where

$y_{ik}$     denotes a single observation (number $k$ in that category) belonging to category $i$ of factor A (e.g. the average algal biomass for the $j$th location upstream or downstream)

$\mu$     denotes the overall population mean of all possible observations of interest (e.g. the overall mean of algal biomass)

$A_i$     is the effect of being in category $i$ of factor A, expressed as a deviation from $\mu$ (e.g. the effect of the $i$th level (or category) of the upstream–downstream factor (i.e. being upstream or downstream) on algal biomass)

$\varepsilon_{ik}$     is an error term that represents the influence of other variables not recorded and therefore unexplained causes of variation that result in the values of $Y$ not being identical in each category of A (e.g. the variation in average algal biomass per snag between upstream locations and between downstream locations). These error terms are estimated from the differences between the observed values of algal biomass and those predicted by the model.

Fitting this model to the data using (ordinary) least squares only requires that the observations within each category be sampled randomly from the population of all possible observations. Obtaining reliable confidence intervals and tests of hypotheses about model parameters also requires that the residuals from the model are normally distributed, have equal variance within each category and are independent of each other, both within a category and between categories. If the residuals meet these assumptions, then so must the observations, $y_{ik}$.

    We can also have models that include additional, nested factors that usually represent levels of subsampling. Imagine a model relating $Y$ (e.g. algal biomass per snag) to factor A (e.g.

upstream vs. downstream of mine), with $a$ categories ($i = 1$ to $a$), factor B (e.g. locations upstream and downstream) with $b$ categories ($j = 1$ to $b$) nested within each category of factor A. The actual categories of B are different within each category of A (e.g. locations upstream are different from locations downstream). There are $n$ replicate observations (subsampling units; e.g. snags within locations) within each combination (cell) of A and B, and any observation is denoted as $k$:

$$y_{ijk} = \mu + A_i + B_{i(j)} + \varepsilon_{ijk} \qquad (4.6.2)$$

where

$y_{ijk}$    denotes a single observation (number $k$ in that cell) belonging to category $i$ of factor A and category $j$ of factor B nested within factor A (e.g. the algal biomass for the $k$th snag in the $j$th location upstream or downstream)

$\mu$    denotes the overall population mean of all possible observations of interest (e.g. the overall mean of algal biomass)

$A_i$    is the main effect of being in category $i$ of factor A, averaged across the categories of B within category $i$ and expressed as a deviation from $\mu$. This main effect is measured as the difference between the mean of the $i$th category of A and the overall mean, averaging the levels of B within each level of A. In our example, this is the effect of the $i$th level (or category) of the upstream–downstream factor (i.e. being upstream vs. downstream) on algal biomass

$B_{i(j)}$    is the nested effect of being in category $j$ of factor B, nested within category $i$ of factor A, expressed as a deviation from $\mu_i$ (the mean of category $i$ of factor A). This effect is measured as the difference in means between levels of B within each level of A. In our example, this is the effect of the $j$th location within the $i$th category of upstream vs. downstream on algal biomass

$\varepsilon_{ijk}$    is an error term that represents the influence of other variables not recorded and therefore unexplained causes of variation that result in the values of $Y$ not being identical in each category of A (e.g. the variation in algal biomass between upstream snags in each location). These error terms are estimated from the differences between the observed values of algal biomass and those predicted by the model.

Now let's include an additional predictor variable B (time: before vs. after mining) that is crossed with factor A (upstream vs. downstream of mine). This means that all categories of A occur for each category of B and vice versa; all combinations of A and B are used. There are $a$ categories of factor A ($i = 1$ to $a$) and $b$ categories of factor B ($j = 1$ to $b$) and $n$ replicate observations within each combination (cell) of A and B, any observation denoted as $k$:

$$y_{ijk} = \mu + A_i + B_j + AB_{ij} + \varepsilon_{ijk} \qquad (4.6.3)$$

where

$y_{ijk}$   denotes a single observation (number $k$ in that cell) belonging to category $i$ of factor A and category $j$ of factor B (e.g. the algal biomass for the $k$th location in the $i$th upstream vs. downstream category and the $j$th before–after category)

$\mu$   denotes the overall population mean of all possible observations of interest (e.g. the overall mean of algal biomass per snag)

$A_i$   is the main effect of being in category $i$ of factor A, averaged across the categories of B and expressed as a deviation from $\mu$. This is termed a main effect because it is measured as the difference between the mean of the $i$th level of A and the overall mean, pooling across levels of B. In our example, it is the main effect of the $i$th level (or category) of the upstream–downstream factor on algal biomass, pooling times

$B_j$   is the main effect of being in category $j$ of factor B, averaged across the categories of A and expressed as a deviation from $\mu$. This is also termed a main effect because it is measured as the difference between the mean of the $j$th level of B and the overall mean, pooling across levels of A. In our example, it is the effect of the $j$th level (or category) of the time factor on algal biomass, pooling upstream and downstream

$AB_{ij}$   is the combined (synergistic) effects of factors A and B acting together (interaction between A and B). The presence of an interaction effect indicates the effects of A are not independent of the level of B and vice versa. In our example, it is the interactive effect of the $i$th level (or category) of upstream vs. downstream and the $j$th level of time on algal biomass. This interaction measures how much the differences between the means of each upstream vs. downstream category vary between times and vice versa, how much the differences between the mean of each time vary between each upstream vs. downstream category. This is the fundamental measure of impact in this model: is the difference between upstream and downstream locations the same after mining as before?

$\varepsilon_{ijk}$    is an error term that represents the influence of other variables not recorded and therefore unexplained causes of variation that result in the values of $Y$ not being identical in each cell (combination of A and B; e.g. the variation in algal biomass between snags in each combination of before–after and upstream–downstream). These error terms are estimated from the differences between the observed values of algal biomass and those predicted by the model.

In multifactor models, such as the crossed model above, the main effects of A and B are averaged across the levels of the other factor, and it is often more convenient to represent the effects of A in the model above as $A_{i..}$, with the notation $i..$ indicating category $i$ of factor A summed (or averaged) across the levels of B, with the individual observations also summed. With this notation, the linear model for the two factor crossed design becomes:

$$y_{ijk} = \mu + A_{i..} + B_{.j.} + AB_{ij.} + \varepsilon_{ijk} \qquad (4.6.4)$$

The analysis of variance partitions the total observed variation in our response variable ($Y$) into its components due to the different terms in our model. We initially partition the variation as sum-of-squares, but it is statistically more convenient to use mean squares (i.e. variances) because these take into account the number of items (measured as degrees of freedom) that comprise each source of variation. The mean squares associated with each term in the model are sample variances and they estimate components of the variance in the populations (one population of $Y$ in each category or cell in our design) from which we have sampled. The components of the population variation estimated by these sample variances are termed expected mean squares or mean square estimates and are provided for common linear models in Tables 4.1, 4.2 and 4.3. By examining ANOVA tables that include expected mean squares, we can see that when we estimate the variance in Y associated with a particular term in our linear model, our estimates are composed of different variance components. For example, the expected value of the mean square associated with factor A in a nested model (Table 4.2) consists of three components. One represents the variation between replicate observations within each category of B nested within each category of A, another represents the variation between the means for each category of B within each category of A and a third represents the variation between the means for each category of A. Clearly, only the last component is a measure of the variation due to factor A by itself.

The expected mean squares depend on whether each factor in our design, and therefore whether each term in our model, is fixed or random. This can influence our hypothesis tests for specific terms. In particular, the inclusion of one or more random factors in our design can alter the nature of the hypothesis tests for terms involving fixed factors and interactions in our model (Table 4.3). The 'variance component' associated with a fixed factor (e.g. factor A) is measured as the squared effects of individual categories of factor A and is usually designated as:

$$\delta_a^2 = \sum_{i=1}^{a} \frac{(\mu_i - \mu)^2}{a - 1} \tag{4.6.5}$$

This is a variance between the population means for the specific categories of factor A used in our design and our conclusions are restricted to those levels. If factor A is random, it has a true variance component measured as a variance between the means of all the possible categories of factor A we could have used in our design, $\sigma_A^2$. Although both $\delta_A^2$ and $\sigma_A^2$ represent variances, the former is about fixed category effects and the latter about all possible category effects of which we only have a random sample. Sometimes $\sigma_A^2$ is used for both A fixed or A random and it is important to remember that when A is fixed, this quantity is actually measuring the fixed treatment effects in 4.5.5, not a random variance component.

**Box 4.7** Ordinary least squares (OLS) and maximum likelihood (ML) methods for estimation of parameters

Consider a random sample of $n$ observations $(y_1, y_2, \ldots, y_i, \ldots, y_n)$ from a clearly defined population. The value of a sample statistic that provides the (ordinary) least squares (OLS) estimate of a population parameter is the one that minimizes:

$$\sum_{i=1}^{n} [y_i - \theta]^2 \qquad (4.7.1)$$

where $\theta$ is the estimated value of the parameter (i.e. the sample statistic). For example, the OLS estimate of the population mean is the value of $\bar{y}$ that minimizes:

$$\sum_{i=1}^{n} [y_i - \bar{y}]^2 \qquad (4.7.2)$$

It turns out that this value is the sample mean from Box 4.3. OLS estimation can be applied to a variety of estimation problems, including estimating parameters for linear models, although derivation of confidence intervals and tests of hypotheses require specific assumptions (e.g. normality) to be met.

The value of a sample statistic that provides the maximum likelihood (ML) estimate of a population parameter is the one that maximizes the likelihood of observing our sample data. ML estimation assumes an underlying distribution (e.g. normal) for the data and the ML estimate maximizes the likelihood function:

$$L(y_i; \theta) = \prod_{i=1}^{n} f(y_i; \theta) \qquad (4.7.3)$$

where $L(y_i; \theta)$ is the likelihood of observing our sample data $(y_i)$ for possible estimated values $(\theta)$ of the parameter, $f(y_i; \theta)$ is the probability distribution of $y_i$ for possible values $(\theta)$ of the parameter and $\prod$ = indicates multiplication. So we can try different values of $\theta$ and see which one maximizes the product of the densities from the probability distributions of $y_i$ for each value of $\theta$. For example, to estimate the population mean via ML, we might calculate the probability density of $y_i$ for different values of $\theta$ (our estimate of $\mu$) in a normal probability distribution (for a given $\sigma$). The product of these densities is the likelihood function and the ML estimate of $\mu$ is the value of $\theta$ that maximizes this likelihood function.

In practice, the calculations are easier if we use the log-likelihood function:

$$LL(y_i;\theta) = \sum_{i=1}^{n} \ln[f(y_i;\theta)] \qquad (4.7.4)$$

For many linear models, ML provides the same exact estimators as OLS when the specific assumptions of OLS hold. ML estimation is more reliable in situations when those assumptions are not met (e.g. for GLMs or non-linear models) or for more complex estimation problems. In these circumstances, ML estimators are usually found by iterative procedures that basically try different values of the parameter(s).

and 7. The single-factor model can be derived from the nested model if we simply average the data for snags within each location in the second design and use those average values as the observations for the single-factor model.

One way to be more confident that any upstream–downstream difference we observe is actually due to the mining effect would be to include measurements before and after mining has occurred. This is the logic behind the range of Before–After–Control–Impact (BACI) designs that will be discussed extensively in chapters 5 and 7. We still have the mine on a river in Kakadu National Park, but now we record algal biomass on snags before and after mining, as well as having putative impact locations downstream of mining and control locations upstream. We now have two predictor variables: upstream vs. downstream (two categories: upstream and downstream of the mine) and time (two categories: before mining commences and after mining commences). Within each of the four combinations of upstream vs. downstream and time, we will choose $n$ locations where $n > 1$. The average algal biomass per snag per location is our response variable. Our linear model is now equation 4.6.3 in Box 4.6. This model is termed a two-factor crossed (or factorial) model. Interactions are particularly important for models with categorical predictor variables. In most of the monitoring designs we will recommend in chapters 5 and 7, as in this simple example, the effect of a particular human disturbance is measured by the interaction between 'control' (e.g. upstream) and 'impact' (e.g. downstream) locations and before and after the impact.

If the number of replicate observations in each cell (factor combination) of these multifactor designs is equal (balanced designs), then the effects of the factors on the response variable are independent of each other. With unbalanced designs (unequal numbers of observations per cell), then the effects of the factors are not independent and the analyses are more complex.

### 4.5.3 Fitting models

There are two main statistical methods for estimating the parameters of a linear model from our sample data. The simplest approach that is described in most linear models textbooks is termed (ordinary) least squares (OLS) estimation. An OLS estimator is one that minimizes the sum of squared deviations (sum-of-squares) between the observed sample values of the response variable and the values predicted by the model (Box 4.7). This sum-of-squares (SS) is termed the residual or error SS and the smaller the residual SS, the better the model fits (i.e. explains Y). Using OLS to fit a model to sample data provides a number of useful outputs:

- Point estimates of the parameters of the model, so the model can then be used for predictive purposes
- Interval estimates of the parameters, so we can determine our confidence in the parameter estimates and also test specific hypotheses about those parameters
- Measures (sum-of-squares and variances) of the variation in the response variable, how much is explained by the model and how much is unexplained (uncertainty or error).

OLS point estimators are simple to calculate, can be used for a range of linear models, and are easily interpreted because the sums-of-squares can be converted to variances that have known probability distributions. The only assumption for OLS estimation is that the sample was taken randomly from the population. However, if we wish an interval estimate of a parameter in our model, or we wish to test an hypothesis about that parameter (see section 4.7), then we have to assume that our response variable has a normal (symmetric) distribution for OLS estimation. We will consider this assumption in more detail in section 4.9, although it turns out that for many regression and ANOVA models, OLS interval estimation and hypothesis-testing is still reliable even if the underlying probability distribution of our response variable is not strictly normal.

An alternative method of estimation for linear (and non-linear) models is termed maximum likelihood (ML) where we find an estimate of the population parameter that maximizes the likelihood of observing our sample data (Box 4.7). ML estimation is computationally more difficult than OLS and we must use an iterative algorithm to derive the estimator; these algorithms are becoming more commonly available in statistical analysis software. ML estimation also allows us to specify the nature of the distribution of the error terms of the model, and therefore the response variable. If it is normal, then we will get identical parameter estimates to OLS; if the distribution is not normal, but we can specify the type of distribution (such as a Poisson distribution for count data), then ML estimation comes into its own. We can also model discrete (categorical) response variables that will rarely be distributed normally. So ML allows us the flexibility of specifying distributions of the response variable other than normal when deriving interval estimates of, or testing hypotheses about, parameters in our model. Fitting linear models using ML when the probability distribution of the response variable is specified is termed generalized linear modelling (GLM; see section 4.9), in contrast to general linear modelling (ANOVA and regression modelling using OLS).

### 4.5.4 Comparing models

In many cases in environmental monitoring, we might have a number of competing models, particularly when we are modelling a single system or process. Usually, we compare a series of nested (hierarchical) models (Hilborn & Mangel 1997) where models represent subsets of other models. For example, we might compare three models to explain algal biomass on snags at river locations upstream and downstream of mining, and before and after mining starts:

Model 1:    $y_{ijk} = \mu + A_i + B_j + AB_{ij} + \varepsilon_{ijk}$
Model 2:    $y_{ijk} = \mu + A_i + B_j + \varepsilon_{ijk}$
Model 3:    $y_{ijk} = \mu + A_i + \varepsilon_{ijk}$

Models 2 and 3 are reduced model subsets of Model 1, which is termed the full model because it contains all the predictor variables we have measured in our study. There are two reasons why a comparison of the fit of these models to our sample data is useful. First, parsimony demands we use the simplest model in any situation if it is as good a predictor of our response variable as more complex models. So if the fit of Models 2 or 3 is as good as Model 1, then we should use one of the simpler models for future prediction. Second, a comparison of each of Models 2 and 3 with Model 1 indicates the importance of each predictor variable separately. For example, comparing the fit of Model 1 with Model 2 tells us the importance of the interaction between upstream vs. downstream and before vs. after on algal biomass on snags in rivers. If Model 2 fits as well as Model 1, then this interaction is clearly not an important predictor and we might conclude there is little impact of mining. Comparisons of the fit of a set of nested linear models, whether by OLS or ML, are a fundamental tool for data analysts and form the basis for regression analyses and analyses of variance. Comparing linear (or non-linear) models is more difficult in practice than this simple intro-duction might suggest. In particular, we need some criteria for deciding whether the reduction in fit by removing a term from a model is large enough for us to conclude that the term is important.

Sometimes we are in the position where we are very unsure of the underlying distribution of our response variable and/or we have not been able to sample our population randomly. For example, we may be modelling with data based on replicate locations determined by political or aesthetic importance, rather than chosen randomly (or haphazardly) from a population of sites. Under these circumstances, neither OLS nor ML estimation may be reliable and we might have to resort to a third method of estimation: computer-intensive, resampling methods such as the jackknife or bootstrap (Crowley 1992; Dixon 1993). These approaches generate a probability distribution of the estimator by resampling, with replacement, from the original data.

## 4.6 ANALYSES OF VARIANCE (ANOVA)

### 4.6.1 Type of factors

We have already distinguished between continuous and categorical predictor variables. The ANOVA designs we described in section 4.5.2 also include nested and crossed factors. To reiterate, nested factors are those whose levels are different, and usually randomly chosen, within each level of the higher factor (e.g. locations nested within areas upstream and downstream of the mine). We can have multiple levels of nesting. For example, Downes *et al.* (1993) examined scales of spatial variability in the distribution of stream invertebrates and sampled stream stones at a number of spatial scales along a river: three randomly chosen sites, two randomly chosen riffles nested within each site, five randomly chosen groups of stones nested within each riffle and three stones randomly chosen from each group. Nested factors generally represent levels of spatial or temporal subsampling in designs to detect human impacts on the environment. Crossed factors are those whose levels are the same within each level of the other factor(s) (e.g. before and after impact for both upstream and downstream of the impact). Crossed designs allow us to measure the interaction between factors, which is not possible for nested designs. Crossed designs can include three or more factors, which means we can measure both two- and three-way interactions between factors. Our recommended monitoring designs will actually combine nested and crossed (factorial) factors (partly nested models) and can also include additional continuous predictor variables (covariates) to try and reduce the unexplained variation in Y. This can result in linear models with many terms.

Predictor variables can also be classified as fixed or random and this distinction has important implications for the way we test hypotheses in linear models and the interpretations we place on our conclusions. We will focus on classical ANOVA models where the predictors are categorical, although the arguments apply equally to regression models and models with combinations of continuous and categorical predictors. A fixed predictor variable, termed a fixed factor, is one where all the possible categories or groups about which we wish to make inferences have been used. If we redid the study, we would use the same groups or levels of that factor. For example, upstream and downstream of a mining operation is a fixed factor because there are only two possible groups (upstream 'control' and downstream 'impacted'). Our conclusions are restricted to those specific groups that we used in our study. A random factor is one where we only used a sample of the possible groups or levels in our study and we wish to use the results from that sample to extrapolate to all the possible groups we could have used. If we redid the study, we would choose another random sample of groups. For example, streams might be a random factor if we chose a number of streams from all the possible streams in a catchment that we could have used. Streams would be fixed if we chose specific streams and did not want them to represent all possible streams. In the nested model above (equation 4.6.2, Box 4.6), locations upstream and downstream is a random factor because we have chosen a number of locations upstream and a number downstream from all the possible upstream and downstream locations. Random factors commonly represent random samples at a particular spatial or temporal scale, and are used to permit extrapolation to sampling units at that scale more generally. Nested factors are nearly always random; crossed factors can be either fixed or random.

### 4.6.2 Partitioning the variance

We have distinguished between regression models, where the predictor variables are continuous, and ANOVA models, where the predictor variables are categorical, although we pointed out that this distinction is artificial as the basic model fitting and testing procedures are identical. Using OLS to estimate model parameters also allows a partitioning of the variation in the response variable into that explained by the different terms in the model and that unexplained (the error or residual variation). To illustrate this process, consider the general, single-factor linear model described in Box 4.6.

We can measure the total variability in $Y$ across all observations with a sum-of-squares (SS). This $SS_{Total}$ can be partitioned into two additive components (Table 4.1):

- $SS_A$ measures the variability between the means of the categories of factor A
- $SS_{Residual}$ measures the variability between the individual observations within A categories, pooled across categories.

The larger the variation in the means of $Y$ between categories relative to the variation between observations within each category, the larger will be $SS_A$ relative to $SS_{Residual}$. Unfortunately, SS are of limited usefulness because they are a measure of variation that is dependent on the number of components contributing to each SS (e.g. the $SS_{Residual}$ is dependent on the number of observations within each category). We can convert each SS to a variance, also called a mean square, by dividing by the degrees of freedom (df; equals the number of observations contributing to the variability minus one). The $df_A$ is the number of categories ($a$) minus one and the $df_{Residual}$ is the number of observations within each category ($n$) minus one, summed for the categories. We usually represent this partitioning of the variance in an ANOVA table (Table 4.1). This ANOVA table shows us that there are two variances, that between category means and that between observations within each category, averaged over the categories. These are sample variances and they estimate components of the variance in the populations (one population of $Y$ in each category) from which we have sampled. The components of the population variation estimated by these sample variances are termed expected mean squares or mean square estimates and are provided in Table 4.1. These mean square estimates are based on a number of assumptions that we will consider below. The most important of these is the homogeneity of variance assumption that the population variance of

Table 4.1. *ANOVA table showing the partitioning of the SS and df, and mean square (MS) estimates, for a balanced (equal number of replicates across groups), single-factor linear model*

| Source of variation | SS | df | MS | MS estimates A fixed | MS estimates A random |
|---|---|---|---|---|---|
| Factor A | $n \sum\limits_{i=1}^{a} (\bar{y}_i - \bar{y})^2$ | $a - 1$ | $\dfrac{n \sum\limits_{i=1}^{a} (\bar{y}_i - \bar{y})^2}{a - 1}$ | $\sigma_\varepsilon^2 + n \dfrac{\sum\limits_{i=1}^{a} (\mu_i - \mu)^2}{a - 1}$ i.e. $\sigma_\varepsilon^2 + n\delta_A^2$ | $\sigma_\varepsilon^2 + n\sigma_A^2$ |
| Residual | $\sum\limits_{i=1k=1}^{a}\sum\limits^{n} (y_{ik} - \bar{y}_i)^2$ | $a(n - 1)$ | $\dfrac{\sum\limits_{i=1k=1}^{a}\sum\limits^{n} (y_{ik} - \bar{y}_i)^2}{a(n - 1)}$ | $\sigma_\varepsilon^2$ | $\sigma_\varepsilon^2$ |
| Total | $\sum\limits_{i=1k=1}^{a}\sum\limits^{n} (y_{ik} - \bar{y})^2$ | $an - 1$ | | | |

*Note:* Factor A has $a$ categories ($i = 1$ to $a$) and there are $n$ replicates ($k = 1$ to $n$) within each category of factor A. $\delta_A^2$ represents the variance between the means of the fixed levels of factor A whereas $\sigma_A^2$ represents the variance between the means of all possible levels of the random factor A that could have been used.

the error terms from the linear model, and therefore the population variance of the $y_{ik}$ is the same for each group:

$$\sigma_{\varepsilon 1}^2 = \sigma_{\varepsilon 2}^2 = \dots = \sigma_{\varepsilon i}^2 = \dots = \sigma_\varepsilon^2 \qquad (4.4)$$

$MS_{Residual}$ estimates this common variance of the error terms. The estimated value of $MS_A$ depends on whether the factors are fixed or random.

Say our factor is upstream vs. downstream and $Y$ is algal biomass per snag per location. There are only two categories we are interested in, upstream and downstream, so this factor is clearly fixed. When factor A is fixed, $MS_A$ estimates the common variance plus an additional component that represents the effect of the categories (such as upstream and downstream) in the populations from which we have sampled. This additional component is the variance between the true means of the specific categories we have used in our monitoring or experimental design. Clearly, if there was an effect of the groups on $Y$, such as an effect of upstream vs. downstream on algal biomass, then we would expect the $MS_A$ to be larger than the $MS_{Residual}$ because the latter estimates an additional component due to the difference between category means. If there was no effect of the groups, then we would expect these two MS to be approximately equal because both estimate the common variance of the error terms.

Now say our factor is stream where we chose a number of streams from all the possible streams within a catchment. We wish to make inferences about the population of possible streams based on our sample of streams, so this factor is clearly random. When factor A is random, $MS_A$ estimates the common error variance plus any additional variance between all the possible categories of factor A. Clearly, if there was an effect of factor A on $Y$, such as variation between all possible streams in algal biomass, then we would expect the $MS_A$ to be larger than the $MS_{Residual}$ because the latter estimates an additional component due to the variance between streams. If there was no effect of factor A, then we would expect these two MS to be approximately equal because both estimate the common variance of the error terms.

Now consider the nested model 4.6.2 in Box 4.6. In nested models, the nested factor B within A is nearly always random, such as randomly chosen locations upstream and randomly chosen locations downstream. Factor A can be fixed or random, but is commonly fixed, like upstream vs. downstream. The $SS_{Total}$ can now be partitioned into $SS_A + SS_{B(A)} + SS_{Residual}$. The MS estimates are given in Table 4.2. Note that $MS_{B(A)}$ estimates the common variance of the error terms plus the

Table 4.2. *ANOVA table showing the partitioning of the df, and the mean square estimates for a balanced, two-factor nested linear model*

| Source of variation | df | MS estimates (A fixed, B random) |
|---|---|---|
| A | $a - 1$ | $\sigma_\varepsilon^2 + n\sigma_{B(A)}^2 + bn\delta_A^2$ |
| B(A) | $a(b - 1)$ | $\sigma_\varepsilon^2 + n\sigma_{B(A)}^2$ |
| Residual | $ab(n - 1)$ | $\sigma_\varepsilon^2$ |
| Total | $abn - 1$ | |

*Note:* Factor A has $a$ categories ($i = 1$ to $a$), factor B has $b$ categories ($j = 1$ to $b$) nested within each category of A and there are $n$ ($k = 1$ to $n$) replicate observations within each AB combination.
$\delta_A^2$ represents the variance between the means of the fixed levels of factor A whereas $\sigma_{B(A)}^2$ represents the variance between the means of all possible levels of the random factor B within each level of A that could have been used.

variance between all possible categories of B within each category of A. $MS_A$ estimates these components plus an additional effect of the specific categories representing factor A.

Finally, consider the two-factor crossed model 4.6.3 in Box 4.6. The $SS_{Total}$ can now be partitioned into $SS_A + SS_B + SS_{AB} + SS_{Residual}$. If both factors are fixed, such as A being upstream vs. downstream and B being before and after mining starts, their interaction is also fixed. Then $MS_A$, $MS_B$ and $MS_{AB}$ estimate the common variance of the error terms plus an effect due to interaction, factor B or factor A, respectively (Table 4.3). This is a simple extension of the single-factor fixed model. However, if we chose a sample of times from a population of possible times, such as randomly chosen times over two years after mining starts, then factor B is random, and the mean square estimates change (Table 4.3). These are termed mixed models (a mixture of fixed and random factors in a model) and the interaction between a fixed and a random factor is considered random. More importantly, however, when factor B is random and factor A is fixed, the estimate for $MS_A$ now includes the variance due to the interaction term. This type of change in mean square estimates when random factors are included in models has a major effect on the way we test hypotheses about parameters in these linear models. There is an alternative derivation (see Ayres & Thomas 1990; Voss 1999) that changes the estimated mean square (EMS) for the random factor but the version we have presented is more common.

The formulae in Tables 4.1, 4.2 and 4.3 are for balanced designs,

Table 4.3. *ANOVA table showing the partitioning of the df and the mean square estimates for a balanced, two-factor crossed linear model*

| Source of variation | df | MS estimates | |
| --- | --- | --- | --- |
| | | A, B fixed | A fixed, B random |
| Factor A | $a - 1$ | $\sigma_\varepsilon^2 + nb\delta_A^2$ | $\sigma_\varepsilon^2 + n\sigma_{AB}^2 + nb\delta_A^2$ |
| Factor B | $b - 1$ | $\sigma_\varepsilon^2 + na\delta_B^2$ | $\sigma_\varepsilon^2 + na\sigma_B^2$ |
| Interaction AB | $(a - 1)(b - 1)$ | $\sigma_\varepsilon^2 + n\delta_{AB}^2$ | $\sigma_\varepsilon^2 + n\sigma_{AB}^2$ |
| Residual | $ab(n - 1)$ | $\sigma_\varepsilon^2$ | $\sigma_\varepsilon^2$ |
| Total | $abn - 1$ | | |

*Note:* Factor A has $a$ categories ($i = 1$ to $a$), factor B has $b$ categories ($j = 1$ to $b$) and there are $n$ ($k = 1$ to $n$) replicate observations within each AB combination. $\delta_A^2$, $\delta_B^2$ and $\delta_{AB}^2$ represent the variance between the means of the fixed levels of factor A, factor B and the AB interaction, respectively, whereas $\sigma_B^2$ and $\sigma_{AB}^2$ represent the variance between the means of all possible levels of the random factor B and the variance among all possible interactions between the fixed factor A and the random factor B.

with equal sample sizes in each group or cell. Unequal sample sizes are common (e.g. Faith *et al.* 1995 had two upstream and six downstream locations) and these formulae can be modified, or SS and hypothesis tests constructed by comparing the fit of full and reduced models (section 4.5.4). Unequal sample sizes can cause some complications (see Underwood 1997) – linear model analyses are more sensitive to violations of assumptions (section 4.9), there is more than one way of calculating SS in factorial designs (Winer *et al.* 1991), and estimation of some parameters, especially variance components, is more difficult.

## 4.7 HYPOTHESIS-TESTING: CLASSICAL APPROACH

Hypothesis-testing has a long history in statistics and has been closely linked to a hypothetico-deductive (falsificationist) approach to scientific method based in part on the work by Karl Popper (Popper 1968; see also James & McCulloch 1985, Scheiner 1993 and Underwood 1990, for ecological perspectives). In our discussion above, we noted that the companion to falsification, the possible degree of corroboration or support for an hypothesis provided by outcomes that do not falsify the hypothesis of interest, is also very relevant to monitoring impacts. We will now explore such hypothesis-testing further in a monitoring context. Continuing the mining example from Faith *et al.* (1995), say we observe (or read) that there is spatial variation in algal biomass on snags between locations along the river. One explanation might be that this pattern is a result of a mining operation, resulting in sites upstream and sites downstream of the mine that differ in algal biomass per snag. Our hypothesis that an impact has occurred and our evidence for the hypothesis will be based on an expectation that, if there is an impact, locations along this river below the mine will have a different algal biomass than locations above the mine. Because it is logically impossible to prove any non-trivial hypothesis (Popper 1968; Underwood 1990), the key evaluation of evidence will address the null hypothesis – as discussed earlier, rejection of the null hypothesis means that the evidence (our data) is very unlikely in the absence of impact. The null hypothesis therefore is an hypothesis of no difference from 'normal' or 'background' expectations or no effect of impact – in this case that there is no difference in algal biomass between locations upstream and downstream of the mine. Note that no difference is a shorthand way of saying that the variation actually observed can be accounted for by our null model, the model that generated the null hypothesis of no effect. The final step is to test this null hypothesis. Ideally, this test would comprise a manipulative field experiment, although for hypotheses involving large-scale anthropogenic activities like mining, such experiments are rarely possible. An alternative, weaker, but more practical test would be a sampling program that simply compared algal biomass at locations upstream of the mine with locations downstream of the mine. Note that such a design cannot easily attribute any differences to the mine activity, it can only test whether upstream locations are different from downstream locations, and so possibly demonstrate that the observed difference is improbable under normal circumstances. Our model for what constitutes evidence, and therefore our null hypothesis, may be more complex than this, often involving temporal processes (such as seasonal and annual patterns, before and after activity starts) and multiple spatial scales (such as multiple unimpacted rivers, locations within rivers).

Nearly all the designs we will describe in later chapters for detecting the effects of human activity in freshwater systems are based around linear models. We will, therefore, introduce hypothesis-testing in the framework of comparing fits of linear models to sample data. Tests of statistical hypotheses are based on four important components:

- Specifying a null hypothesis ($H_0$), which is traditionally an hypothesis of no effect or no relationship (e.g. no relationship between nutrients and fish community structure, no difference in algal biomass upstream and downstream of a mine).

- Comparing a test statistic calculated from our sample data, which provides a measure of the strength of the effect we are testing, to the central theoretical probability distribution for that statistic, which is usually the probability distribution (sometimes called the 'sampling distribution') of the statistic under the null hypothesis (i.e. when the $H_0$ is true).

- Determining the probability of getting our sample test statistic or a value more extreme, and therefore our sample data or data more extreme, if the null hypothesis is true ($P(data|H_0)$). The comparison of our sample statistic to its theoretical probability distribution under the null hypothesis provides this probability. Note that this comparison will often only be valid if the data meet certain assumptions that we will consider in section 4.9.

- Using an a priori decision criterion to decide whether to reject the null hypothesis or not. This criterion is a probability value, often called a significance level; if $P(data|H_0)$ is less than this value, we reject the null hypothesis, otherwise we do not. This significance level is often set, by convention and therefore arbitrarily, at 0.05 or 5%.

The theoretical probability distributions of the common test statistics that we will use in this book are founded on reliable statistical theory. These distributions are mathematically defined and represent sampling distributions: probability distributions of the test statistic under repeated sampling from one or more populations. The distributions vary slightly depending on the sample size, represented statistically by the degrees of freedom. The simplest of these sampling distributions for each statistic is the central distribution and is the probability distribution of a test statistic under the null hypothesis when there is no 'effect'. Central probability distributions for common test statistics are tabled in the back of most statistics books and programmed into the code of most statistical software. Non-central distributions are those for particular effects or differences when the null hypothesis is not true; there are an infinite number of possible non-central distributions for each test statistic, depending on the size of the effect. Algorithms for these non-central distributions are also available in most statistical software.

There are three test statistics with well-defined probability distributions that are used for analysing monitoring designs in this book:

- The $t$ statistic is used to test hypotheses about single parameters (does the slope of the regression model relating species richness to current velocity equal zero?) and about the difference between the equivalent parameter from two populations (is the mean species richness upstream of the mine different from the mean species richness downstream of the mine?).

- The $F$-ratio statistic is the ratio of two sample variances and is used to test hypotheses about the equality of two sources of variation. For example, we can use it to test the null hypothesis that two linear models (full and reduced) explain the variance in our response variable equally well. The $F$-ratio is the basic test statistic we will use for analysing the monitoring designs recommended in later chapters.

- If we are fitting generalized linear models based on ML estimation (see section 4.9), the fit of each model is measured with a log-likelihood statistic. As an analogue to the $F$-ratio, comparing two models under this scenario results in a ratio of two log-likelihoods called the likelihood ratio statistic. If the null hypothesis is true, this statistic can be slightly modified and compared to a theoretical $\chi^2$ distribution to determine probabilities for tests of null hypotheses.

Once we have specified our null hypothesis and chosen our test statistic, the procedure for statistically testing a null hypothesis uses a combination of the approaches of Fisher (1935), termed significance testing, and Neyman & Pearson (1928), termed null hypothesis-testing. Neyman & Pearson (1928) included a specific alternate hypothesis $H_A$ in their scheme, which is the hypothesis that must be true if the $H_0$ is rejected. If our $H_0$ is that there is no difference in algal biomass between locations upstream of the mine and locations downstream of the mine, then the $H_A$ is that there is a difference in algal biomass.

Consider our previous simple example of a monitoring design, modified from Faith et al. (1995), to determine whether there is an effect of a mine (and implied pollution of a river) on algal biomass on snags in a lowland river. We measure the average algal biomass per snag from a number of control locations upstream and supposedly impacted locations downstream of the mine. The logical null hypothesis is that there is no difference in mean algal biomass between locations upstream and downstream of the mine. Although simple in appearance, even this type of design can be difficult in practice and careful thought must be given to the spatial scale of replication within locations (snags within locations?), whether the only difference between locations upstream and downstream of the mine is due to the mine (are there confounding variables?) and whether there will be a temporal component to monitoring (see chapters 5 and 7).

We might fit two competing models to our sample data:

Model 1:    algal biomass = overall mean + upstream
                vs. downstream + unexplained error
Model 2:    algal biomass = overall mean + unexplained error

To test the $H_0$ that there is no difference between locations upstream and downstream, we compare the fit of these two models to our observed data. We can compare the variance explained by the two models with an F-ratio statistic, which is the explained variance of the full model divided by the explained variance of the reduced model. If the null hypothesis is true (and the effect on the explained variance of omitting upstream vs. downstream from our model is zero), we would expect an F-ratio around 1. If the null hypothesis is false, the full model will have a larger explained variance than the reduced model and we would expect an F-ratio greater than 1. Equivalently, our F-ratio could be the residual variance of the reduced model divided by residual variance of the full model.

The comparison of full and reduced models is tedious and the same result can be achieved by making use of our ANOVA table and mean square estimates. Referring to Table 4.1, we can see that if the $H_0$ is true, we would expect that $MS_A$ (i.e. the variance due to the upstream vs. downstream effect) and $MS_{Residual}$ estimate the same variance and therefore their F-ratio should be 1. So the F-ratio test that is standard output from a single-factor ANOVA is identical to that from comparing full and reduced models.

Note that the F-ratio statistic we calculate for tests of null hypotheses in more complex, multifactor linear models depends on whether predictors are fixed or random factors. The general approach is to calculate the F-ratio from the two mean squares that have the same expected value under the $H_0$. When all factors are fixed, we usually test each term against the $MS_{Residual}$. If we are testing the $H_0$ that there is no effect of factor A in a nested model where the nested factor B(A) is random, then the F-ratio will be $MS_A$ divided by $MS_{B(A)}$ because these two mean squares have the same estimated value when the $H_0$ is true (Table 4.2). This makes sense because the replicates at the appropriate scale for testing A are the levels of B (e.g. the replicates for testing upstream vs. downstream of the mine are the locations, not the individual snags).

For two-factor crossed models with one factor fixed and the other random, the test of the $H_0$ of no effect of the fixed factor A uses an $F$-ratio of $MS_A$ divided by $MS_{AB}$, not $MS_{Residual}$ (Table 4.3). Tests become even more complicated in models that include combinations of crossed and nested and fixed and random factors. Indeed, sometimes there are no appropriate $F$-ratios for some null hypotheses if two or more factors are random in three-factor models.

So, the formal steps for a statistical test of a null hypothesis using the $F$-ratio statistic are:

1. Specify $H_0$ that there is no difference in algal biomass between locations upstream and downstream of the mine. This is identical to stating that the two models described above fit the observed data equally well. This implies the $H_A$ that there is a difference in algal biomass between upstream and downstream locations.

2. Choose appropriate test statistic, in this case an $F$-ratio statistic.

3. Specify a priori a significance level (termed – and often set at 0.05), which is our decision criterion used in steps 6 and 7.

4. Calculate the test statistic, in this case the $F$-ratio statistic, from fitting the two models to our sample data. This $F$-ratio statistic is the standard output from a one-factor ANOVA.

5. Compare that value of the statistic to its sampling probability distribution, assuming $H_0$ is true. We would use the central $F$ distribution, which is the probability distribution of $F$ when there is no difference in fit of the two models and therefore no difference between means for locations upstream compared with downstream.

6. If the probability of obtaining our sample value of the test statistic or one more extreme is less than our pre-defined significance level (usually $\alpha = 0.05$), then conclude that the $H_0$ is false and reject it (a statistically 'significant' result).

7. If the probability of obtaining our sample value of the test statistic or one more extreme is greater than or equal to this pre-defined significance level, then conclude there is not enough evidence that $H_0$ is false and do not reject it (a statistically 'non-significant' result).

The probability of obtaining our sample value of the test statistic *or one more extreme* if the $H_0$ is true is termed the $P$-value (see Fig. 4.3). So the $P$-value is the $P(\text{data}|H_0)$. The $P$-value is termed a **tail probability** because we are calculating the probability from the tail of the probability distribution of our test statistic. This means that we are using our data plus data that we did not actually observe to decide whether or not to reject

Table 4.4. *Decision errors in statistical hypothesis testing*

| | | Statistical decision from sample data | |
|---|---|---|---|
| | | Reject $H_0$ Statistically significant result | Do not reject $H_0$ Statistically non-significant result |
| State in population(s) | Effect ('impact') exists $H_0$ false | Correct decision $H_0$ rejected Effect ('impact') detected | Type II error $H_0$ not rejected Effect ('impact') exists but not detected |
| | No effect ('no impact') exists $H_0$ true | Type I error $H_0$ rejected Effect ('impact') detected, none exists | Correct decision $H_0$ not rejected No effect ('no impact') detected, none exists |

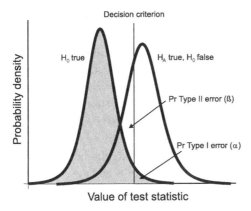

**Fig. 4.3** Probability density functions (pdf) of a test statistic under some $H_0$ (left-hand distribution, shaded) and under some alternative $H_A$ (right-hand distribution). The test statistic would usually be an $F$-ratio, although the pdf for the non-central $F$-ratio is complex and changes shape markedly depending on the degrees of freedom and the non-centrality parameter (a measure of effect size). For illustration, we have used the simpler $t$ distribution. The vertical line represents the critical value of the test statistic for a specified $\alpha$. If the $H_0$ is true, our test statistic should follow the left-hand distribution. The area under the curve to the right of the critical value is the probability of making a Type I error (rejecting $H_0$ because we obtained a value of our test statistic greater than the critical value even though $H_0$ is actually true). We set this probability with our a priori significance level ($\alpha$). If the $H_A$ is true, our test statistic should follow the right hand distribution. The area under the curve to the left of the critical value is the probability of making a Type II error (not rejecting $H_0$ because we obtained a value of our test statistic less than the critical value even though $H_0$ is actually false). We can only determine this probability ($\beta$) for specific alternative hypotheses (i.e. effect sizes).

the $H_0$. This is in contrast to the Bayesian approach that doesn't consider tail probabilities when assessing evidence against $H_0$. One aspect of the Fisherian component of significance testing is the common practice in many disciplines of presenting the $P$-value as a measure of the strength of evidence against the $H_0$ (Fisher 1935). How useful the $P$-value is for this purpose will be considered briefly in section 4.8. Bayesian statisticians would argue that the $P$-value is not usually comparable to $P(H_0|\text{data})$. However, presenting $P$-values does allow a reader to use their own significance levels for making decisions following the steps described above.

The Neyman–Pearson approach closely linked statistical hypothesis-testing to decision making by introducing the a priori chosen significance level, which is our decision criterion for deciding whether to reject the $H_0$ or not. Neyman & Pearson (1928) argued that this significance level must be fixed in advance, although the conventional level of 5% or 0.05 actually has its origins with Fisher (1935). Neyman & Pearson also introduced the concept of errors in decision-making when testing null hypotheses. These errors occur because we are making decisions under uncertainty. This uncertainty results from spatial and temporal variation between sampling or experimental units and also because we are making inferences about unknown population parameters from imperfect samples drawn from the population(s). When we test a null hypothesis from a monitoring program, there are two possible states and two possible outcomes (Table 4.4): four combinations in total. Let's return to our mining example. Our $H_0$ is that there is no difference in algal biomass per snag between locations upstream and downstream of the mine. When we test this $H_0$, the two possible outcomes of our statistical decision-making process are to reject or not reject the $H_0$. If the $H_0$ is actually false (i.e. there really are differences in algal biomass between the populations of possible locations upstream and downstream of the mine) and we reject it, then we have made a correct decision. Similarly, if the $H_0$ is true (i.e. there really are no differences in algal biomass) and we do not reject it, then we also make a correct decision. In contrast, we can also make two sorts of decision errors following this scheme. If the $H_0$ is true but we reject it based on our sample data, then we have made a Type I error because we have incorrectly rejected a true $H_0$. If the $H_0$ is false but we do not reject it based on our sample data, then we have made a Type II error because we have incorrectly retained a false $H_0$.

Classical hypothesis-testing emphasizes the costs or errors associated with decisions about rejecting null hypotheses. Traditionally, scientists have been most concerned with Type I errors. Type I errors have a direct link to our desire for improbabilities. Mayo (1996) presents a simple example, using her terminology of 'severity' to describe tests yielding improbabilities. In her test of hypothesis H, a null hypothesis ($H_0$) is rejected with an observed $P$-value of 0.03. Mayo emphasizes that this value is the probability that the test would pass H when in fact $H_0$, the null hypothesis, was true – the familiar Type I error equal to the probability of wrongly rejecting the null hypothesis. Severity is then calculated as one minus this probability, or 0.97 – the probability of not passing H when H is false ($H_0$ is true). Mayo (1996, p. 193) notes: 'by rejecting the null hypothesis $H_0$ only when the significance level is low, we automatically ensure that any such rejection constitutes a case where the non-chance hypothesis H passes a severe test'. Scientists set the probability of a Type I error with our a priori significance level or decision criterion, which is designated $\alpha$. For example, rejecting a $H_0$ at a significance level of 0.05 sets the probability of a Type I error at 5% if the $H_0$ is actually true. Note that this error probability is a long-run frequency from repeated sampling (i.e. if we repeatedly sample from our populations when $H_0$ is true, we will falsely reject the $H_0$ 5% of the time). It is not strictly the probability of a Type I error in any individual case. In practice, we only need to know the central probability distribution of the relevant test statistic to determine the probability of Type I errors and these central distributions are readily available for test statistics such as $t$, $F$ and $\chi^2$.

The other error, incorrectly retaining a false $H_0$, has until recently received less attention in applied statistics, although its importance was appreciated by Neyman & Pearson (1928) when they first developed the ideas of errors in statistical decision-making. The probability of Type II errors (designated $\beta$) is more difficult to determine because we need to specify the $H_A$ that we believe to be true (or at least that we would like to be able to detect; see discussion of statistical power following) and then use the non-central probability distribution of the test statistic appropriate for that $H_A$ (Fig. 4.3). These non-central distributions have only been easily accessible to practitioners with the advent of powerful desktop computers. The converse of the probability of Type II errors is the power of a test $(1 - \beta)$: the probability of correctly rejecting a false $H_0$.

We can calculate the power of a statistical test to detect a specific alternative hypothesis ($H_A$). In monitoring programs, this alternative hypothesis usually represents an effect of a particular human activity. To determine power, we must specify the size of this effect; the power of a statistical test depends on the size of the effect we wish to be able to detect. For example, consider the example we described above to measure the effect of a mine on algal biomass in a lowland river. Given the sample size (the number of locations) we used and the variation between sampling units (locations) we observed in our sample data, we can determine the power of a statistical test (such as an $F$-ratio test) to detect a difference in algal biomass of a given magnitude between locations upstream and downstream. The power of a statistical test can be represented by:

$$\text{Power } (1 - \beta) \approx \frac{\text{ES } \alpha\sqrt{n}}{s} \qquad (4.5)$$

where:

ES is the effect size as specified in the $H_A$. For example, we might express the ES as a difference in mean algal biomass between locations upstream and downstream

$\alpha$ is the a priori significance level (i.e. the probability of a Type I error we are willing to accept if we reject the $H_0$)

$n$ is the sample size used in the monitoring program, such as the number of locations

$s$ is the standard deviation between sampling units (e.g. locations) used in the monitoring program (and an estimate of $\sigma$) and indicates the degree of variability in the variable of interest.

This relationship is not an equation to be used in practice. The actual equation will depend on the specific analysis and test statistic being used, which will define how the ES is measured. This relationship implies, however, a number of general principles about the power of a statistical test, particularly F-ratio tests based on comparing variances. First, larger effects are easier to detect for a given sample size and level of variation. This means that the probability of detecting a large effect of a particular human activity will be greater than the probability of detecting a smaller effect for a specific sample size and level of variability between sampling units. Note that we may reach a situation where we have nearly 100% power to detect a certain effect size and then, by definition, power cannot increase for even larger effects. Second, effects of a given size for a given level of variation are easier to detect with a larger sample size. This simply means that, for a given level of variability, we can generally increase the probability of detecting, say, a 50% change in some variable, by increasing the number of sampling units. This will be an asymptotic relationship and above a certain sample size, there will be little gain in power as it will be close to 100%. Finally, more variation between sampling units makes effects of a given size harder to detect. We have little control over the unexplained variability between sampling units, although by including additional predictor variables in our models, we may be able to reduce this variability and improve the power of our tests.

Power analysis is most useful for determining an appropriate sample size during the design of a monitoring program. For our monitoring design to compare algal biomass upstream and downstream of a mine, we might ask how many locations upstream and how many locations downstream do we need. First, we must specify the smallest ES (difference in mean algal biomass between upstream and downstream) we wish to be able to detect if it occurs. This may come from previous studies (e.g. a published study elsewhere in Australia might have shown a 35% increase in algal biomass at locations below this type of mine), or a legislative requirement (e.g. a condition of a discharge licence might specify there be no greater than a 20% increase in algal biomass as a result of the discharge), or convention (we might use arbitrary effect sizes of, say, 50% and 100% change). Second, we must specify the power we want to have to detect this particular ES (i.e. the probability of detecting this effect if it exists). We will consider this issue in some detail in chapters 12 and 13 in the context of balancing decision errors and costs. For the moment, let's use 80% power, an arbitrary convention that seems to be developing in the ecological literature. Finally, we need some estimate of $\sigma$ (the variation between sampling units). This is best gained from a pilot study or other data from our system or one similar. Once we have these three components, we can solve the power equation for the appropriate statistical analysis for $n$ to determine the sample size required for a particular monitoring program. For example, if we have an estimate of variation in algal biomass between locations from a quick and dirty pilot study, we can determine the sample size (how many locations) required to be 80% confident (power $= 0.80$) of detecting a 50% change in algal biomass between locations upstream and downstream of the mine.

Power analyses can also be used after we have analysed our monitoring data, particularly if we find no statistically significant effect. In these circumstances, we obviously have $n$ (the sample size we actually used) and $s$ (the variability between our sampling units, such as locations) from our monitoring data, and we can solve the power equation to determine what the minimum detectable effect size (MDES) was for a given power. Again, back to our simple monitoring program. If we found no significant difference in algal biomass between upstream and downstream locations, we might ask a question of the form: what was the smallest difference (MDES) in mean algal biomass between upstream and downstream locations that we could have detected with 80% power, given the sample size we used and the level of variability we had between locations? Ideally, we would want that MDES to be of a magnitude that would allow us to detect ecologically relevant effects with 80% probability (see chapter 12).

It is important to realize that using power analysis to determine design characteristics (especially sample size) of a monitoring program is the priority here. Post-hoc power analyses after a non-significant result may simply indicate that the monitoring program was not sensitive enough to detect effects of biological or management importance.

## 4.8 HYPOTHESIS-TESTING: THE BAYESIAN APPROACH

How would a Bayesian approach hypothesis-testing in a statistical context? Remember that the Bayesian approach involves a more flexible definition of probability, allowing subjective opinion (degrees of belief) to influence the conclusions from an analysis. In particular, prior belief as to whether an hypothesis is likely to be true or not is quantified as a prior probability distribution and incorporated into the analysis. This emphasizes one of the main differences between the two approaches. Conclusions from classical hypothesis-testing are based on comparing our observed sample data, and the test statistic derived from those data, with a theoretical probability distribution for that statistic. No other information except the sample data influence the outcome of the analysis. In contrast, conclusions from the Bayesian approach are influenced not only by the sample data but also by our prior opinion as to whether one or more competing hypotheses are likely to be true; so the sample data are only part of the information used to draw a conclusion.

The aim of a Bayesian analysis is to calculate a probability distribution for a parameter specified in an hypothesis and use this probability distribution to make a probabilistic decision as to whether the hypothesis is true or not. This final (or 'posterior' in Bayesian terminology) probability distribution is interpreted just like probability distributions in classical statistical analyses with one exception – the parameters specified in the hypothesis are treated as random variables, not fixed population values (Ellison 1996). The posterior probability distribution is determined from two other components that are unique to Bayesian analyses. The first is the likelihood that the parameter specified in the relevant hypothesis takes a particular value, given the observed sample data. This is the opposite of the $P$-value in classical hypothesis-testing, which is the probability of observing the sample data or data more extreme, given the parameter value specified in the $H_0$. Just like in ML estimation (see Box 4.7), we can calculate a likelihood function, which indicates the likelihood of different possible values of the parameter. Although likelihood functions don't have to be symmetrical (Hilborn & Mangel 1997), analyses of linear models usually constrain likelihood functions to standard symmetrical distributions such as a normal or a $t$ distribution. So the Bayesian approach has the same distributional assumptions as the classical approach for analysing linear models. Much statistical software now includes algorithms for calculating likelihood functions.

The second unique component of a Bayesian analysis is the prior probability distribution of the parameter/hypothesis, which summarizes our a priori information, both objective and subjective, about the probability that the parameter(s) will take different values or that a particular hypothesis is true. If we have a lot of previous information, or strong subjective opinions, about whether a particular hypothesis is true or not, then this prior distribution may be symmetrical (such as normal or $t$) or may be skewed. If we have no prior information, we consider all values of the parameter equally likely and have a non-informative, rectangular prior distribution. Quantifying our prior opinions and incorporating them into the analysis as prior probabilities is the most controversial aspect of Bayesian analyses. Edwards (1996) showed that, under some circumstances, different prior probability distributions can greatly influence posterior probabilities and therefore our conclusions from the analysis. So two scientists, with exactly the same sample data, may reach very different conclusions about the same hypotheses depending on the strength of their prior opinions about those hypotheses.

The Bayesian approach to hypothesis-testing uses a modification of Bayes' original theorem for conditional probabilities:

$$P(H_i|\text{data}) = \frac{P(\text{data}|H_i)P(H_i)}{P(\text{data})} \qquad (4.6)$$

where

$P(H_i)$ is the prior probability of hypothesis $H_i$
$P(\text{data}|H_i)/P(\text{data})$ is the standardized likelihood function for hypothesis $H_i$
$P(H_i|\text{data})$ is the posterior probability of hypothesis $H_i$.

The determination of the posterior probability distribution may also be summarized as:

$$\text{posterior probability} \propto \text{likelihood} * \text{prior probability} \qquad (4.7)$$

This equation emphasizes that the Bayesian analysis modifies prior information via a likelihood function to produce a final, posterior probability distribution for a parameter or hypothesis (Berry & Stangl 1996; Ellison 1996). Note that the actual sample data enter the process via the likelihood function and only the actual data are relevant. Sample data more extreme than ours are not considered.

The mean or expected value of the posterior distribution for a parameter is our Bayesian estimate of that parameter. We can also use the variance of each posterior distribution to determine the equivalent of confidence intervals (sometimes termed a 'credible' or probability interval in Bayesian literature). If we have used a non-informative prior, then this Bayesian credible interval (such as 95%) will be the same as a classical 95% confidence interval. Note that because Bayesians consider the parameter to be a random variable, not a fixed, unknown quantity, the interpretation of these intervals is different. The Bayesian interval is the range that covers 95% of all possible values of the parameter. The 95% confidence interval is a range that will include the unknown parameter in 95% of repeated samples.

The Bayesian analogue of classical hypothesis-testing would be to establish two hypotheses, one representing the null and the other the alternative. Each hypothesis might be that one or more parameters equal certain values. For example, the null hypothesis might be that the ratio of explained variance for two models (full and reduced) is 1, whereas the alternative hypothesis might be that the ratio is greater than 1. We determine the posterior probability distributions of the two hypotheses and use those distributions to determine the final probabilities of the two competing hypotheses. The hypothesis with the greater posterior probability is the one we would accept. A more sophisticated decision system is to calculate a ratio of the posterior probabilities of two competing hypotheses that is usually expressed as a Bayes factor. Ellison (1996) summarized published guidelines for interpreting Bayes factors as evidence for or against a particular hypothesis compared with an alternative hypothesis. These guidelines can be used as decision criteria for deciding whether to accept or reject a particular (e.g. null) hypothesis. Note that, in contrast to classical hypothesis-testing, the Bayesian approach explicitly compares two or more competing hypotheses or models.

We recommend Berry & Stangl (1996), Gelman *et al.* (1995) and Lee (1997) as good introductions to Bayesian statistics, with Hilborn & Mangel (1997) and Ellison (1996) providing an ecological perspective (but see also Edwards (1996) and Dennis (1996) for cautionary articles in response to Ellison (1996)). Box & Tiao (1972) is the classic source for computational formulae and discussion of assumptions. Bayesian analyses have been used recently by freshwater scientists, particularly to determine conditional probabilities in a modelling framework. For example, Carpenter *et al.* (1998a) have used the Bayesian approach to calculate the probability of an algal bloom given a land-use policy by linking the probabilities of algal blooms in lakes given a lake P concentration and of lake phosphorus concentrations given a land-use policy. Bayesian analyses of time series and forecasting models are also common (e.g. Lamon *et al.* 1998). In contrast, formal Bayesian analyses and decision-making for the types of monitoring programs we recommend with spatial (i.e. control vs. impact) and temporal (i.e. before vs. after) components have rarely been adopted in freshwater systems. This is probably due to scientists still being uncomfortable with considering probability subjectively and therefore being reluctant to incorporate prior subjective probabilities (opinions) into the process for drawing statistical conclusions. In particular, incorporation of prior beliefs into decision-making has the potential to be abused, especially in environmental monitoring when vested interests dominate (Dennis 1996; Stewart-Oaten *et al.* 1992). Also, the complex linear models we use to analyse monitoring programs with both temporal and spatial predictor variables, and our preference for designing such programs based on careful consideration of effect sizes and power to determine sample size, would be difficult for all but experienced statisticians to put into a Bayesian framework. Finally, the Bayesian approach is much more suited to estimation rather than hypothesis-testing (Dennis 1996); some well-known Bayesian texts (see Gelman *et al.* 1995) imply that hypothesis-testing has little relevance in the Bayesian world. In contrast, we support the role of statistical hypothesis-testing in science, and in environmental monitoring. It allows clearly defined questions to be addressed with explicit considerations of the costs involved in decision errors (chapter 12).

In the end, however, the designs we recommend have a strong logical and statistical basis, and good experimental and sampling design is really independent of whether a frequentist or Bayesian approach is adopted. The models we discuss for detecting human impacts in flowing waters can be evaluated from either viewpoint. For example, Crome *et al.* (1996) used a BACI (Before–After–Control–Impact; see chapter 5) design to assess the effects of logging on rainforest birds and small mammals and applied both a classical and a Bayesian analysis.

## 4.9 ASSUMPTIONS OF STATISTICAL ANALYSES OF MONITORING PROGRAMS

The reliability of any decisions based on statistical analyses depends on whether the assumptions behind those analyses are met. All analyses based on linear models, both estimation and hypothesis-testing, have assumptions about the probability distribution of the error terms from the model. Generally, this also means that the assumptions apply to the response variable being analysed. These assumptions are dealt with in detail in many texts (for a biostatistical perspective, see Sokal & Rohlf 1995; Underwood 1997) and we will provide only a brief overview here, with a discussion of alternatives. An important concept is robustness, the ability of the estimator or hypothesis test to produce reliable results even when assumptions are not met. There has been extensive statistical research developing methods that are more robust to underlying assumptions. In general, however, the cost of this robustness is that these methods are only applicable to linear models with few parameters or else are complex to apply. An overriding assumption for the analyses we will describe in later chapters is that of random sampling. We assume that our sample data represent random samples from clearly defined statistical populations. This assumption applies to analyses of linear models, whether using OLS or ML; the only exception is when we use randomization procedures (see below).

Most aspects of the analyses of linear models rely on the underlying distribution of the variable, or the distribution of the error terms from the model, being of a specific form. These analyses are termed parametric. While point estimation of parameters in general linear models, which include regression and analyses of variance, based on OLS has no distributional assumptions, interval estimation (i.e. confidence intervals) and hypothesis tests assume that the variable(s) being analysed is/are normally distributed. The normal, or Gaussian, distribution is a symmetrical, continuous probability distribution defined by the mean and the standard deviation, which are independent of each other. Normality is often not met because we know from theoretical considerations and empirical evidence that many continuous variables, especially measurement variables, tend to have positively skewed distributions (i.e. the distributions are not symmetrical but have a long right-hand tail; Fig. 4.1). In such distributions, the variance is positively related to the mean.

Another important assumption when analysing linear models is that the population variance of the response variable (and error terms) is equal across the range of values of the independent variable(s). This is termed the homogeneity of variance assumption. For example, an ANOVA comparing the mean species richness at a number of locations must assume that the true variance in species richness is the same at all locations; a linear regression of chlorophyll concentration against nutrient load in wetlands must assume that the variance in chlorophyll concentration between wetlands is the same for all nutrient loads.

The assumption of homogeneity of variance in linear models can be untenable if the underlying probability distribution of the variable (and the error terms) is positively skewed so that the variance is related to the mean. Sampling from such distributions will result in samples with larger means having larger variances. Also, outliers (extreme values for the response variable very different from the rest of a sample) may result in a much larger variance at one level of the predictor variable. Such outliers may represent errors (malfunctioning equipment, mistakes in data transcription etc.) which can be corrected or deleted, or may suggest that the population being sampled is very heterogeneous. For example, an unusually small animal in a sample may indicate the presence of a cohort of recently recruited animals that might be better analysed as a separate population. Outliers can result in serious problems. They will often produce large differences in variances and can also have considerable influence on the fit of linear models, resulting in models that don't fit the majority of data very well because the fit is influenced by one or two very unusual values.

OLS interval estimation and hypothesis-testing is robust to the assumption of normality unless the distribution is very skewed or multimodal. In contrast, the assumption of equal variances is more important and differing variances can have severe effects on the reliability of the analyses. Hypothesis tests are much more sensitive to these assumptions if sample sizes vary; this applies mainly to situations where the predictor variable is categorical (i.e. classical ANOVA). For this reason (and others; see Underwood 1997), similar sample sizes should be a priority of monitoring designs, although small differences in sample sizes produce few problems.

There are two (non-exclusive) approaches for determining whether the distributional assumptions of parametric analyses are met:

- We can do formal hypothesis tests for specific underlying distributions ($H_0$: sample came from a population with a particular probability distribution) and homogeneity of variances ($H_0$: samples came from populations with identical variances). Unfortunately, such tests are often more sensitive to non-normality than the linear model analyses themselves and, depending on sample size and power, these tests might not detect violations of assumptions that are serious enough to affect the subsequent linear model analysis. So such tests should only be part of the approach for checking assumptions.
- We can use methods of exploratory data analysis (EDA) to check the assumptions of statistical analyses. EDA uses mainly graphical techniques such as boxplots, probability plots etc. to check for normality and homogeneity of variances. Because the assumptions behind analyses of linear models strictly relate to the error

terms of the model, examinations of residuals (the differences between observed values and those predicted by the estimated model) are very important. Along with statistics measuring the influence of particular observations on the outcome of the analysis (such as leverage and Cook's $D$ and the size of standardized residuals), plots of residuals provide diagnostic information about the adequacy of the model as a fit to the observed data. Hoaglin *et al.* (1993) is an excellent overview, Neter *et al.* (1996) provide detailed statistical background and Ellison (1993) illustrates the methods for ecological data.

There are a number of alternative methods for analysing linear models based on OLS if the assumptions of normality and homogeneity of variance are not met. One approach is to transform the data to a different scale of measurement, which is an important tool for dealing with skewed distributions and will often correct both non-normality and heterogeneous variances. Many variables have right-skewed (heavy-tailed) distributions: measurement variables often have a log-normal distribution, where the mean is proportional to the standard deviation; and counts often have a Poisson distribution, where the mean is proportional to the variance. Transformations that may be suitable for different types of data are presented in Table 4.5. Transforming a response variable to better match the distributional assumptions of a statistical procedure is common practice but the choices of whether to transform, and which transformation to use, are not always straightforward. The appropriateness of transformations depends on the nature of the ecological process we assume is operating. For example, reconsider our monitoring design from above to examine the possible impact of a mine on the biomass of algae on snags in a floodplain river. To keep this example simple, we will use control (upstream) vs. impact (downstream) as one factor and two times (before mining and after mining) as a second factor. Say we measure the biomass of algae at replicate locations in each combination of the two factors and the mean abundances are those presented in Table 4.6. A sensible question might be whether the difference between upstream and downstream locations changes after mining commences compared with before mining. If we use the raw data, we might conclude that the control vs. impact difference is greater after than before because the difference between locations upstream and downstream is 90 before and 900 after. However, if we think of this change in percentage terms, the change is the same for both times (a 10-fold increase in algal biomass downstream). A log-transformation of

Table 4.5. *Types of variables and transformations to make their distributions normal. This table is based on theoretical considerations and our own experience analysing biological data from aquatic monitoring programs*

| Variable type | Probability distribution | Characteristics | Transformation to normal |
|---|---|---|---|
| Binary | Binomial | | |
| 'Discrete' counts (small range of values) | Poisson | $\mu = \sigma^2$ | Square root |
| 'Continuous' counts (large range of possible values; e.g. abundances) | Overdispersed Poisson and log-normal | | Fourth root or log (any base)[a] |
| Measurements | Log-normal | $\mu \propto \sigma$ | Log (any base) |
| Proportion or percentages | Symmetrical but truncated at 0 and 1 | | Arcsin square root |

[a] In practice, fourth root or log-transformation of such variables will result in similar distributions close to normal. If data include zeros, use fourth root or log ($y$ + constant).

Table 4.6. *Mean biomass of algae at locations upstream and downstream of a mine before and after mining commenced (log transformed values in parentheses)*

|            | Before mining | After mining |
|------------|:-------------:|:------------:|
| Upstream   | 10 (1)        | 100 (2)      |
| Downstream | 100 (2)       | 1000 (3)     |

the data reflects this identical percentage change. Clearly, the choice of whether to log-transform the data in this case would depend on whether we are happy that equal percentage change represents the ecological process we are interested in. If so, the raw data would be best fit by a non-additive model:

$$\text{algal biomass} = \text{constant} + \text{upstream vs. downstream} \\ + \text{time} + \text{upstream vs. downstream} * \text{time} \qquad (4.8)$$

where the upstream vs. downstream * time interaction measures how much the upstream vs. downstream difference changes between times. The log-transformed data would be best fit by an additive model:

$$\text{algal biomass} = \text{constant} + \text{upstream vs. downstream} + \text{time} \quad (4.9)$$

where the effect of upstream vs. downstream is consistent between times (i.e. an interaction is not anticipated).

The issue of additivity and transformations in environmental monitoring programs has attracted the attention of statisticians (Sampson & Guttorp 1991) and ecologists (Stewart-Oaten *et al.* 1986). It is clear other considerations besides simply meeting distributional assumptions are relevant for any decision on whether to transform data. Whatever our rationale for choosing a transformation, the assumptions and adequacy of the proposed statistical model should always be re-checked post-transformation; it is possible that the transformation has not improved the distributional properties of the data or has even made them worse. In these cases, there is no advantage to using transformations over the raw data. If the transformation has improved the degree to which the data and the model fit the assumptions, the usual parametric analyses can be used. Note, however, that the hypothesis being tested now refers to the transformed scale of measurement, so we might be testing hypotheses about log algal biomass or square root of abundance (of organisms).

A second approach, if we know the underlying distribution of the error terms from our linear model is one of the exponential family, is to use generalized linear model (GLM) fitting procedures (e.g. Dobson 1990; Neter *et al.* 1996). GLMs allow us to specify an exponential-type distribution, including both discrete (such as binomial for binary data, Poisson for count data) and continuous (such as normal, log-normal, exponential, gamma) distributions. Our choice of distribution should be based on our knowledge of the variable being analysed. Estimation of model parameters uses ML, and hypothesis tests are based on comparing the fits of full and reduced models with a log-likelihood ratio statistic (also termed deviance). These analyses are still parametric in that a distribution of the error terms from the model must be specified and this will usually imply a particular pattern for the variances. The common problem is over-dispersion, where the variance is greater than we would expect based on our chosen distribution.

A third approach to dealing with distributional assumptions is to use procedures that do not require any. An extreme transformation is the rank-transformation, where each observation is converted to its rank value, usually across the whole data set being analysed. A group of statistical tests has been developed based on this transformation, sometimes called non-parametric tests because they make no distributional assumptions about the response variable being analysed (Potvin & Roff 1993). Originally, these tests were randomization tests based on the ranks of the data. As such, they test more general hypotheses about distributions, usually in relation to a location parameter (the median, or the mean if the distributions are symmetrical), or about relationships, such as testing for monotonic relationships between variables rather than relationships of a specific structural form, like linear. These tests include the Mann–Whitney–Wilcoxon tests and the Kruskal–Wallis test for comparing groups, and the rank-based correlation and regression procedures (see Sokal & Rohlf 1995; Sprent 1989). The sampling distributions of the test statistics are generated by randomization procedures, which were easier to do in the days before computers if the data were ranked. An extension of these tests is the rank-transform procedures, which simply transform the data to ranks and then use OLS procedures on the rank-transformed data, sometimes called rank-transform (RT) tests.

These rank-based tests do assume equal distributions of the variable (except for the median) when comparing groups (i.e. when the predictor variables are categorical). This implies that the group variances should be equal, the same assumption as for OLS analysis of a linear model with categorical predictors (ANOVA). These rank-based tests are inappropriate solutions to unequal variances. Also, they do not provide any straightforward measure of the fit of a linear model so they are unsuitable for comparing model fits. Finally, the simple rank-transform approach is ineffective at detecting interactions between predictor variables (McKean & Vidmar 1994; Seaman *et al.* 1994), such as before vs. after by control vs. impact interactions, the basic test of a BACI design (see chapter 5).

Finally, we can use randomization (or permutation) procedures (Crowley 1992; Manly 1997). These methods randomly rearrange the sample data many times to generate the sampling distribution of a test statistic based on the assumption that if the $H_0$ is true then any random arrangement of the data is equally likely (Crowley 1992). We can compare the value of the test statistic from our sample data to this randomization distribution to determine the $P$-value. These tests are best illustrated with a simple comparison of two means, such as mean species richness at locations upstream and downstream of a human activity. Say we have data consisting of 20 observations: 10 locations upstream and 10 locations downstream. We calculate our test statistic, such as the difference between the two means. We then use a computer algorithm to randomly rearrange the data many times (say 1000) and calculate the probability distribution of the difference between means under the $H_0$. Finally, we compare our sample difference between means to the probability distribution of differences generated by randomization to determine a $P$-value and interpret this as usual. These randomization procedures have been used in the literature in circumstances when the underlying distribution of the response variable is unknown or when the assumption of random sampling cannot be justified. Their main limitation is the availability of computer algorithms for performing the randomizations for complex sampling designs and linear model analyses. If these analyses are used because the assumption of random sampling cannot be justified, then we also cannot easily extrapolate our conclusions to a population of, for example, control and impact locations (i.e. it is more difficult to generalize our conclusions in a spatial or temporal context). Note that, like rank-based non-parametric tests, randomization tests are not a panacea for problems with underlying assumptions. Only normality is no longer necessary; other assumptions, like homogeneity of variance, can still be relevant for these tests (Manly 1997; Stewart-Oaten *et al.* 1992).

## 4.10 UNIVARIATE AND MULTIVARIATE ANALYSIS

The models we have been describing are sometimes termed univariate because they only consider single response variables at a time. For example, we may have hypotheses about the effects of human activities on single response variables, such as pH, phosphorus concentration, or the abundance of a particular species of fish. In practice, however, we record many variables from each sampling or experimental unit and our interest may be in the collection of variables, rather than (or in addition to) individual variables. For example, a freshwater ecologist may be interested in how the assemblage (or 'community') of fish in a river responds to the discharge of a gold mine. While the response of individual fish species may also be important, an analysis that compares all the variables as a group is more appropriate to the question about fish assemblages. Similar arguments can apply to a collection of physical and/or chemical variables, where patterns in all the variables together might be more important than patterns in any individual variable. Analyses that examine a collection of variables together are termed multivariate analyses.

The terms univariate and multivariate are actually used in a slightly confusing manner in the literature. For example, a two-factor ANOVA model is considered univariate even though there are multiple predictor variables. Generally, univariate refers to a single response variable being analysed, irrespective of the number of predictor variables. Multivariate is used when we are modelling more than one response variable or when we are looking for patterns in a collection of variables that might be considered response or predictor variables in subsequent analyses. There are two broad types of multivariate analyses used in environmental monitoring (Fig. 4.4). Both start with a rectangular data matrix of sampling units (termed objects in the statistical literature) by variables (e.g. species abundances). All these analyses were originally developed as descriptive techniques for representing complex patterns graphically in as few dimensions as possible, but they can also be used for testing relevant (such as control vs. impact, before vs. after) hypotheses.

The first type of multivariate analysis is based on correlations (or covariances) between the variables (e.g. physico-chemical measurements, species abundances). These analyses basically create new variables (sometimes called components) by rearranging the original variables (Table 4.7). Each new variable is a linear (additive) combination of all of the original variables and the number of possible new variables

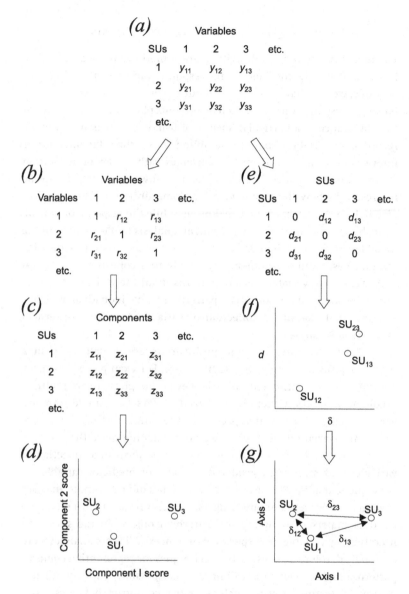

**Fig. 4.4** Diagrammatic representation of the two major forms of multivariate analysis resulting in ordination plots. Both start with a rectangular data matrix of variables by sampling units (SUs) as illustrated in (a). The left-hand side shows methods like principal components analysis which produce a matrix of correlations (or covariances) between variables (b), then decompose that matrix using eigenanalysis into new variables called components (c); a score for each SU for each component is then calculated and the scores plotted in two or higher dimensional space (d). The right-hand

equals the number of original variables. These new variables are extracted from the original data so that they explain successively less of the total variability that was present in the original variables and are also independent of each other. The statistical technique of decomposing a matrix into these components is termed eigenanalysis. If the original variables were correlated with each other, the first few of the new variables (often the first three of four) will explain most of the variance that was present among all of the original variables. Hence, these analyses reduce the number of variables without losing much of the original information (variance).

Two methods, discriminant function analysis (DFA) and multivariate analysis of variance (MANOVA) are designed for situations where we have a priori groupings of sampling units (e.g. upstream vs. downstream or before vs. after). The new variables (each a combination of the original variables) are extracted so that they maximize the difference between these groups. MANOVA then allows testing of hypotheses about these new variables in a similar framework to ANOVA models. DFA works in reverse, finding the new variables that are the best discriminators between groups (Manly 1997; Tabachnick & Fidell 1996).

Some ordination or scaling methods take these new variables a step further. Component scores for each new variable for each sampling unit can be plotted to show the relationship between the sampling units in fewer dimensions than if all the original variables were plotted. Such plots, showing the relationship between sampling units based on a small number of new variables created from a large number of original variables, are termed 'ordination plots' by ecologists, although statisticians use the more general term 'scaling'. The interpretation of these plots is that sampling units further apart on the plot are also more different in their values for all the variables together.

The two most common ordination techniques based on eigenanalysis are principal components analysis (PCA), as described in Fig. 4.4, and correspondence analysis (CA). PCA can be based on covariances or correlations between variables, both of which mean that variables

---

**Fig 4.4** (*cont.*)
side shows methods like multidimensional scaling which produce a matrix of dissimilarities between sampling units (e) and then represent these dissimilarities in two or higher dimensional space (g) so that the distances between SUs on the plot ($\delta$) most closely match the actual dissimilarities (i.e. the relationship in (f) is as good as possible).

Table 4.7. *Overview of multivariate analyses used for environmental monitoring in freshwater habitats. Specialist textbooks should be consulted before using these analyses*

| Analysis | Acronym | Comments based on our experience |
|---|---|---|
| Principal components analysis | PCA | Original variables standardized to zero mean (and unit variance). |
| | | Creates new variables (components), based on correlations between original variables. These new components are independent of (uncorrelated with) each other. |
| | | Components extracted so that they explain successively less of original variance and first few components usually explain most of original variance. |
| | | Contribution of each original variable to each component quantified (loading). |
| | | Components can be used as response or predictor variables in subsequent analyses. |
| | | Component scores for each sampling unit can be plotted, usually for first few components. |
| | | Assumes linear relationship of species abundances to underlying gradients. |
| Multivariate analysis of variance | MANOVA | Multivariate analogue of ANOVA. |
| | | Assumes multivariate normality and homogeneity of variance and covariance. These assumptions are difficult to meet and check. |
| | | Response variables rearranged into (usually one) new variable that maximizes group differences. |
| | | Approximate F-ratio statistics produced. |
| | | Can be applied to robust ordinations enabling tests for interactions supporting inference or impact. |
| Discriminant function analysis | DFA | Variables rearranged into new variables (discriminant functions) that are independent of each other (like PCA) and maximize group differences (like MANOVA). |

| | | |
|---|---|---|
| | | Discriminant functions extracted so that they explain successively less of original variance. Used to predict group membership and classify new observations into groups with estimated probability of success. |
| Correspondence analysis | CA | Used mainly when variables are species abundances. Related to PCA and based on simultaneous eigenanalyses of correlation matrix between species and between sampling units. Inherently standardizes both sampling units and species. Produces joint ordination of sampling units and species. |
| Canonical correspondence analysis | CANOCO | Used mainly when variables are species abundances. Extension of CA where component scores for sampling units are 'constrained' by linear relationships with other environmental variables. Influence of environmental variables on component scores for sampling units can be determined. |
| Multidimensional scaling | MDS | Used mainly when variables are species abundances. |
| Non-metric multidimensional scaling | NMDS | Based on (dis)similarities between sampling units; Bray Curtis dissimilarities allow for unimodal relationship of species abundance to gradients. |
| Hybrid multidimensional scaling | HMDS | Relationship of sampling units based on dissimilarities can be plotted in a small number of dimensions. Dissimilarities can be used to test specific hypotheses about group differences (ANOSIM) and importance of environmental variables. Distance in ordination space can be used for control–impact differences in BACIP designs. |
| Cluster analysis | Cluster | Based on (dis)similarities between sampling units. Relationship of sampling units based on dissimilarities can be represented as tree-like dendrogram. Forces observations into clusters. Many variants of this technique available. |

are standardized to zero mean or zero mean and unit variance, respectively. CA is like a simultaneous eigenanalysis of correlations between variables and correlations between sampling units. The same results can be achieved using a related method called reciprocal averaging. CA inherently standardizes both variables and sampling units, and the final component scores for sampling units are averages of the variable scores and vice versa. This latter feature is probably what has made CA popular among ecologists, producing simultaneous ordination plots of sampling units and variables. There are some extensions of CA, often named after the software designed to implement them. Detrended correspondence analysis (DECORANA) tries to correct an 'arch' effect that sometimes appears in the final ordination plot, where sampling units that are very different in their variables can end up close on the ordination plot. Canonical correspondence analysis (CANOCO) tries to improve the ordination through constraining component scores of, say, sampling units by their correlations with a range of environmental variables.

The second broad type of multivariate analysis is where similarities or dissimilarities are calculated between each pair of sampling units based on all the variables (Fig. 4.4, Table 4.7). This type of analysis overlaps with the first type in that some dissimilarity-based methods can mimic eigenanalysis methods. Similarities measure how similar two sampling units are based on the measured variables; dissimilarities measure how dissimilar they are. There is a very large range of dissimilarity indices. A strong recommendation for Bray Curtis (or related Kulczinski) dissimilarities can be made for multivariate analyses of community data (e.g. macroinvertebrate species found in fresh waters). These dissimilarities have a robust relationship to underlying distances in environmental/gradient space, even under the expected unimodal relationship between species variables and the environmental gradients that may be revealed by the multivariate analysis (Faith *et al.* 1987). The dissimilarities can be graphically represented in a small number of dimensions (Fig. 4.4) so that the distances between sampling units in the ordination plot ($\delta$; which represent underlying distances in environmental/gradient space) closely match their dissimilarities ($d$). Commonly, this is done by a technique known as multidimensional scaling (MDS) that uses a complex algorithm to iteratively move the positions of the sampling units on the ordination plot until their distances apart most closely match their dissimilarities. Metric MDS achieves this match based on a parametric (linear) correlation between distances and dissimilarities (Fig. 4.4) whereas non-metric MDS (NMDS) uses a non-parametric correlation, requiring only a monotonic relationship between distances and dissimilarities. The latter is more robust because the relationship between distance and dissimilarity is often non-linear, particularly for large dissimilarities (Clarke & Warwick 1994; Minchin 1987). A third approach is hybrid multidimensional scaling (HMDS), which takes greater advantage of the expected form of the relationship between dissimilarities and expected gradient distances by using a linear relationship for small dissimilarities and then allows a non-linear relationship for larger dissimilarities (Faith et al. 1987; Minchin 1987). As a final representation strategy, we should also mention that cluster analysis is another method for graphically representing, in a tree-like dendrogram, relationships between objects (e.g. sampling units) in terms of their dissimilarities.

These representations of the robust dissimilarities, or sometimes the dissimilarities themselves, can be used with varying degrees of success in testing hypotheses of impact. Relatively simple hypotheses about a priori group differences in dissimilarities can be tested using the analysis of similarities (ANOSIM) procedure (Clarke 1993) although this procedure does not allow tests of interactions, the most crucial tests in monitoring designs based on the BACI framework. In the multivariate BACIP strategy (Faith et al. 1995), robust dissimilarities between putative impact and control sites form before impact vs. after impact groups. An hypothesis of impact is tested through a t test. Related multivariate approaches may allow hypothesis-testing for a full range of designs and models, including interactions. The redundancy analysis method of Legendre & Anderson (1999) is described by the authors as closely related to application of MANOVA to an ordination based on principal coordinate analysis (PCO; which combines a dissimilarity matrix with eigenanalysis methods) of Bray Curtis dissimilarities. The latter approach is one example of the application of MANOVA to robust ordinations of Bray Curtis dissimilarities described by Faith (1990). An important difference between the original suggestion and the Legendre & Anderson implementation is that Faith (1990) recommends HMDS (Faith et al. 1987) as the ordination method of choice because of its robustness. The advantage of Bray Curtis dissimilarities established by Faith et al. (1987) and acknowledged by Legendre & Anderson (1999) is negated somewhat by an ordination approach like PCO that fails to take into account the non-linear relationship of Bray Curtis dissimilarities to underlying gradient space. Faith et al. (1995) also cautioned that a requirement for applying MANOVA to robust ordinations is that the points in the ordination space truly represent independent replicates. The redundancy analysis approach provides an additional way to test for interactions (preferably using an HMDS ordination rather than PCO) on those occasions when independent replicates are available. Anderson (2001) has also recently described a method for analysing dissimilarities within and between groups in an ANOVA framework using randomization tests.

We need to make a few comments about these multivariate analyses. First, different transformations (log, power) and standardizations (zero means and/or unit variance) can greatly influence the pattern in an ordination plot and the interpretation of these analyses. Some analyses have implicit standardizations. For example, PCA is usually based on correlations between variables standardized to zero mean and unit variance; CA (and CANOCO) standardizes both variables and sampling units. We prefer methods that leave the decision about transformations and standardizations up to the user. Second, we must be careful about automatically attaching biological significance to measures of dissimilarity between sampling or experimental units. Whether a Bray Curtis dissimilarity really measures a relevant difference in ecological community structure depends on the aspects of community structure we are interested in; a simple univariate measure like species richness may be just as suitable. A rationale for the multivariate approaches may be that, when we have no specific aspect of interest (perhaps because the nature of the impact is unknown), these approaches may increase the chance of picking up any change in community composition. Finally, many of these analyses require very specific software. While most general statistical software will do PCA, MANOVA, MDS etc., the range of dissimilarities available is often very limited and hypothesis-testing capabilities (such as ANOSIM) are only available in a few programs. Excellent introductions to these multivariate techniques include Clarke & Warwick (1994), James & McCulloch (1990), Kent & Coker (1992), Legendre & Legendre (1998), Ludwig & Reynolds (1988), Manly (1994), and Tabachnick & Fidell (1996).

Our preferred multivariate analyses are those where the user can choose a dissimilarity measure between sampling or experimental units that can cope with non-linear, often noisy, ecological data. The suitable analyses include non-metric or hybrid multidimensional scaling for graphical representation of the relationship between sampling units, and techniques like ANOSIM for testing simple hypotheses about group differences. Testing of more complex hypotheses about interactions between factors in a multivariate setting is also possible, applying MANOVA tests to robust ordinations like NMDS and HMDS, although their wider use will rely on the availability of suitable software.

## 4.11 IMPORTANT ISSUES

- True random sampling is the most reliable method of sampling from populations, but not necessarily practical or cost-effective. Sampling is often stratified to account for environmental heterogeneity and haphazard techniques are used where random selection of sampling units is not possible.

- The key strategy for inference of impacts is to find some evidence for impact that cannot easily be explained away by various other processes, such as natural variation in the system. Summarizing all those other factors as background knowledge means that we want evidence to be improbable given only that knowledge. Support for an impact hypothesis is only found if the probability of that outcome is small under normal circumstances, in the absence of impact. This pursuit of improbability provides the rationale for specific aspects of monitoring design including controls.

- The statistical analysis of most monitoring programs relies on assessing the fit of linear models to the data. Regression models are those where both the response and the predictor variables are continuous. Analysis of variance (ANOVA) models are those where the predictor variables are categorical. Combinations of both are common in general linear models and categorical response variables can be modelled with generalized linear models.

- The decision of whether a predictor variable (factor) is fixed or random is crucial and can fundamentally change the nature and interpretation of statistical tests of hypotheses about all factors in our model.

- In classical statistical hypothesis-testing, we can make Type I errors (falsely rejecting a true null hypothesis) or Type II errors (falsely retaining a false null hypothesis). An important component of designing a monitoring program is ensuring adequate power to correctly detect an impact (correctly reject a false null hypothesis), usually by modifying sample sizes.

- All statistical procedures have underlying assumptions, especially concerning normality, homogeneity of variance and independence of error terms from the fitted linear model. These assumptions should be carefully checked before analysis and using graphical procedures (exploratory data analysis) and diagnostic tools from the fit of the model are strongly recommended.

- Rank-based statistical procedures are not a panacea for dealing with badly behaved data when assumptions are violated. Transformations, keeping in mind their effects on interaction terms, and generalized linear models, which allow alternative distributions of the response variable besides normal, are preferred approaches.

- Multivariate ordination procedures are important tools for describing the relationship among sampling or experimental units based on many variables. Procedures that are robust and allow the user to choose data standardizations and measures of dissimilarity, such as non-metric multidimensional scaling, are recommended. Statistical tests of hypotheses about group differences and relationships with covariates (environmental variables) are also possible and should be used.

- The logical principles of designing a monitoring program to detect the effects of human activities apply irrespective of whether a frequentist or a Bayesian approach is adopted.

# 5

# The logical bases of monitoring design

In this chapter we discuss the basics of good monitoring design. 'Design' here means the stipulation of where, when and how many observations or sampling units are taken to provide the data from which we will make inferences against some specified objectives. We discuss here the underlying principles that we consider central to good design, and present an ideal case. In the interests of establishing an understanding of why elaborate designs are often presented, we ignore for the moment the ubiquitous compromises that are necessary for logistic, social or economic reasons. We do not focus here on particular variables (chapter 10), what sorts of changes are considered important (chapter 11) or the specifics of natural systems in the interests of presenting the general principles that underlie good monitoring for most variables in almost any system. Nor do we discuss here the analytical tools used to refine or optimize designs or analyse the resultant data (chapters 7–13). This chapter should be read, therefore, as a conceptual overview of the design principles that motivate us and which will be expanded in operational detail throughout later chapters.

   We recognize that 'ideal' designs will rarely, if ever, be feasible (for a variety of reasons) and discuss in later chapters what compromises are most likely to be precipitated by the characteristics of streams (chapter 8) or because of accidents of history, money etc. In beginning here with an outline of the concepts behind an 'ideal' case, we seek to establish the principle that all these inevitable compromises are just that – compromises. They are not equally powerful alternatives to good design, and they bring with them inevitable limitations of inference about the presence and properties of environmental impacts. It is our contention that beginning with an 'ideal target' in each case makes explicit the inferential costs of compromises and where the inevitable compromises impact on our ability to make conclusions about impacts. Establishing a 'best case' also provides a framework from which to explain to stake-holders the genesis of a proposed design and an aid to making decisions about which of the alternative compromises are least costly.

## 5.1 CLASSES OF MONITORING

The concept of 'good' design entails implicitly the notions of efficiency and rigour, meaning that sufficient resources are used and samples taken to answer the questions of interest without being profligate. Thus, what constitutes good design will depend to a large degree on the objectives of monitoring. Our first point here, then, is that it is essential to be very clear about the objectives of a monitoring program before seeking to specify its design.

'Environmental monitoring' can take many forms, for many objectives. For example, monitoring the 'state' or 'health' of the environment (section 3.2.1) basically comprises periodic (usually infrequent) 'status reports' on a range of variables in the interests of establishing a timed reference point on the environment. Such monitoring is not usually targeted at assessing the impacts of specific human activities. Although not always the case, the broad objective of State of the Environment monitoring might be more to do with gaining an approximate indication of state, rather than precisely describing variables in particular places or pinpointing sources and consequences of human impacts on the environment.

Long-term monitoring (LTM) and reference site monitoring (RSM; e.g. AusRivAS and RIVPACS) are other monitoring activities that fall into 'environmental state' monitoring (see section 3.2.1). These are perhaps considered to be most useful for providing a background of the long-term dynamics of natural systems that might be used to indicate apparently systematic, monotonic or cyclical changes in the environment at large scales over long times. Such changes might be 'natural' or anthropogenic, but in general these types of monitoring will not provide the necessary data to distinguish between natural processes or human impacts on the environment (Box 3.2). Both LTM and RSM can provide, however, a background or context against which shorter term or localized changes, possibly arising from anthropogenic impacts, can be interpreted and instances of localized 'degradation' inferred.

Compliance monitoring (section 3.2.2) typically involves taking samples from a specific site (e.g. point of discharge of treated effluent) or prescribed situation (e.g. water supply reservoir) from which to assess whether some mandated standard is being satisfied. The objectives and scope of such monitoring usually are clearly stipulated by regulation or 'health' standards derived without reference to the time or place for which monitoring is required. Accordingly, compliance monitoring resembles a quality-control process. The status of sites outside of those of specific interest is not of interest and the benchmark that would indicate that a violation, or impact, had occurred is not dependent on the contemporary state of notionally undisturbed systems.

Impact monitoring (section 3.2.3) is targeted at assessing human impacts on the 'natural' (non-human) environment. Typically, impact monitoring is underpinned by an expectation that at some level anthropogenic impacts become unacceptable and action will be taken to either prevent further impacts or remediate affected systems (we consider what might be considered 'unacceptable' changes in chapter 11). We are concerned here with impact monitoring, and will not discuss in detail the requirements of 'state' or other generalized monitoring schemes.

## 5.2 MONITORING TO DETECT HUMAN IMPACTS ON THE ENVIRONMENT

Impact monitoring might be considered usefully in two general categories: compliance monitoring and impact assessment monitoring. Compliance monitoring is generally related to specific regulatory standards against which the level of contaminants or pollutants in natural systems are judged. The standards are typically driven by criteria related to human health or points at which substantial impacts on other biota are considered likely. The objective in compliance monitoring is usually to assess whether the levels of particular compounds are below the critical levels stipulated under some regulatory framework. The origins of the contaminants (whether from natural processes or through discharges from human activities) are often either known (in the case of contaminants from point-source discharges) or of secondary importance, and not necessarily the focus of monitoring, at least in regard to detecting their existence at unacceptable levels. For example, in monitoring the suitability of water for drinking, the exact source of potentially harmful pollutants is not of primary concern, although it may become so once the statutory limits for safe consumption have been exceeded. There is an abundant literature on compliance monitoring and the methods for analysing compliance data are well developed in a variety of forums, including environmental monitoring, industrial quality-control procedures, and monitoring of controlled systems. We will not discuss compliance monitoring in detail in this book.

Impact assessment monitoring has as its focus the discrimination of effects of human activities from patterns or changes arising from entirely non-human environmental processes. Impact assessment monitoring, therefore, usually relies on comparisons within the collected data to assess whether an impact has occurred, and how large it was. That is, there often is no preconceived, externally mandated, absolute threshold that defines an important impact, and so an impact is implicitly defined as an unusual or extreme change compared to the usual status of the measured variable. This is the focus of our discussions in this book. We stress later in this book, however, that the absence of a regulatory threshold or trigger for response to impacts does not absolve those involved in impact monitoring of the responsibility to tackle the question of 'when is an impact unacceptable?' Indeed, we argue (chapter 11) that the stipulation of a (probably case-specific) set of triggers for response to impacts is a crucial 'up-front' step in using impact monitoring data in decision making (Keough & Mapstone 1995; Mapstone 1995, 1996). In chapter 12, we extend this argument to the formal integration of such triggers (or 'critical effect sizes') into the statistical analyses of monitoring data and the inferences arising from those analyses.

### 5.2.1 Detecting change

Both compliance and impact assessment monitoring have a key objective of detecting change in selected variables. Further, in most cases variables are measured at a specific location or locations where impacts are most likely (hereafter 'the impact location(s)'). The change of particular interest is from a variable's status prior to the commencement of an activity of concern (the baseline condition) to a different status consistent with the hypothesis that the activity has affected that variable (i.e. had an impact). For a change to be attributed to the impact of an activity, it is a necessary (although not sufficient) condition that the change occurred coincident with the start-up of the activity, or afterwards (Green 1979). It is also important that there is a feasible mechanism by which the change might have been caused by the activity (Green 1979; Keough & Mapstone 1995; Underwood & Peterson 1988). Thus, it is necessary to have samples or observations from the impact locations from before the commencement of an activity (baseline samples) and for some period of routine operations (operations samples), at least until it can be concluded confidently that the activity is environmentally benign (Stewart-Oaten et al. 1986). It might be desirable also to take samples or observations during the commencement of the activity (start-up samples), although the interval over which start-up occurs will usually be very short. This, then, is the first requirement for good design for impact monitoring: sampling from baseline, through start-up and during operational conditions. Leaving aside the start-up period, the baseline period is usually referred to as the Before (start-up) period and the operational period is referred to as the After (start-up) period. A Before–After change in variable status would normally be expected if an impact occurred, although this need not always be the case if natural changes in the variable are suppressed by the impact of the development.

For compliance monitoring, such as monitoring the concentrations of a chemical in the discharge waters of an industry, there may be little concern about concentrations under natural circumstances, but great concern that the concentrations at the end of the discharge pipe, or in receiving waters, remain below the level stipulated by regulations. Thus, the Before–After sequence of monitoring might be restricted to the impact location (end of the discharge pipe, receiving waters within 1 km etc.) alone, and the trigger for action is stipulated independently of the data (i.e. by regulation).

### 5.2.2 Discriminating impacts from natural changes

For impact assessment monitoring, there usually is no independent standard or threshold level of 'contamination' by which an impact is judged. Thus, in impact assessment monitoring impacts tend to be defined relative to natural conditions, rather than with reference to external criteria. Comparisons between the status of the impact location Before and After start-up indicates whether a change in state has occurred (Green 1979; Keough & Mapstone 1995). It does not allow us to distinguish, however, a change caused by an impact from a change that would have occurred even if the activity had not begun, but which just happened to occur coincident with start-up of the activity (Green 1979; Keough & Mapstone 1995; Underwood 1991a). Comparisons between the impact location Before and After start-up indicates whether a change in state has occurred (Stewart-Oaten 1996b). This implies, however, that we would need to have data for the impact location both in the presence and absence of the activity, but over the same time. Clearly this is impossible.

In order to estimate whether a change coincident with start-up would have occurred in the absence of the activity, we collect and analyse data from locations that are considered beyond the influence of the activity and therefore not subject to its impacts. These non-impact locations usually are termed control locations (see Box 5.1 for an explanation of the meaning of the word 'control' in this context). The underlying assumption in this approach is that the impact location would have behaved approximately the same as the control locations in the absence of the impact (Keough & Mapstone 1995; Underwood 1991a). That is, data from the control locations provide a surrogate or proxy for (the impossible) measurements at the impact location during the After period but in the absence of the activity. If similar changes occurred at both the control locations and at the impact locations, then it would be logically inconsistent to infer that the activity had caused such changes, given that the control locations had been selected to be outside the influence of the activity whose impacts were being monitored. If, however, changes at the impact locations differed from those at the control locations, then it would be appropriate to infer that an impact had occurred that had caused the impact locations to depart from 'normal' behaviour.

## 5.3 BACI DESIGNS

From the above argument, then, it is clear that impact assessment monitoring requires that samples or observations are taken from both Impact and Control locations during both the Before and After periods. Only when all these data are in hand can we logically distinguish natural changes at the impact location from those caused by specific human activities. Even so, there will always remain the possibility that the changes at the impact location(s) were natural phenomena that were not present at the control locations, but the likelihood of such an event leading to an incorrect inference of impact is vastly less than the likelihood of an incorrect inference if no control locations are sampled or if there are no data from the Before period.

Thus, to this point we have developed an argument that in order to assess whether an activity has caused an environmental impact at particular locations we need to have:

- Data from the Impact location(s) over some period Before the activity commences
- Data from the Impact location(s) for some period After the activity commences
- Data from Control location(s) over the same period Before the activity commences
- Data from Control location(s) over the same period After the activity commences.

These requirements were summarized by the terminology Before–After, Control–Impact, or BACI designs by Green (1979) and are the key underlying elements of rigorous impact assessment monitoring designs.

### 5.3.1 Natural dynamics and the duration of monitoring

In most natural systems there is considerable change at any given location because of natural processes (Andrew & Mapstone 1987; Sokal & Rohlf 1995; Underwood 1981, 1992). Some of the dynamics of flowing waters (see chapter 2) clearly indicate that streams and rivers are typically highly variable over most time-scales. If such variation is not accounted for in monitoring, there is great risk that either a natural change coincident with start-up of an activity might be incorrectly interpreted as an impact or, conversely, that a real impact might be masked by underlying natural variation (Keough & Mapstone 1995; Underwood 1991a). Where there already is extensive human activity affecting streams, these natural dynamics might be exaggerated or dampened by the overlay of human impacts (Underwood 1992). Indeed, it has been argued that impacts on the frequency and/or magnitude of naturally occurring events (that result in the observed variation in natural systems) may be as important as impacts that simply change the state of a system. That is, impacts that change the underlying regime of natural variability of a system might be particularly important (Underwood 1992, 1994a).

To assess whether natural variation has changed because of impacts and to be able to separate impacts on the average status from the effects of natural variations requires that we know the characteristics of natural variation in both the Before (un-impacted baseline) and After (operational) periods (Stewart-Oaten et al. 1986; Stewart-Oaten 1996b). This means repeatedly sampling over some period both Before and After start-up. Ideally, the duration of sampling within each of the Before and After periods should span several occurrences of the major sources of natural variation that might be expected within the life of an activity (Keough & Mapstone 1995). Thus, for streams with strong seasonality in flow or biota, sampling should cover the full range of seasons and more than one instance of each season should be sampled. In this case, sampling would need to span at least two, and preferably more than three years. The scale at which sampling should be repeated within Before and After periods should be related to the dynamics of both the variables being measured and the environmental processes expected to drive those dynamics. Thus, the need for seasonal sampling might be

**Box 5.1** The difference between control and reference locations

In the environmental monitoring literature, two terms are used to describe locations to which a putative impact location is to be compared: reference and control. 'Reference' locations are used in two different contexts – in some recent literature, they denote undisturbed 'standards' against which impact areas are compared, while in the second context, 'reference' is used by people unhappy with the term control and its associations with laboratory experimental situations. Below we discuss our interpretation of this conundrum and our use of the terms 'control' and 'reference' in this volume.

**Reference** locations are chosen to be as close as possible to the state of an environment undisturbed by human activity. These locations are often not chosen with a particular impact in mind, but to represent what a water body *could* be, or probably would be, in the absence of human disturbance. They are often located at some distance from a putative impact area and remote from centres of population or human activity. Reference locations play a major role in 'status' reporting, and often a large number of variables will be measured to provide as complete a description as feasible of a suite of locations, capturing the overall biological status, biodiversity etc. of a region or habitat type.

**Control** locations are chosen to be as similar as possible in all respects to the impact location, except for the presence of the putative impact. The intention is to use the control locations to *isolate the effect of the particular human activity* from a range of other processes. Under some circumstances, when a human activity is to occur in an otherwise undisturbed area, control and reference locations may have the same attributes. However, when a new activity is contemplated for an area that has already been highly modified, the controls should be locations that are themselves highly modified. That is, control locations should be chosen for their similarity to the putative impact area, and we will often measure a smaller suite of variables that we expect to clearly respond to the impact and distinguish the effects of the particular activity of interest from other processes, natural or human, that affect the suite of control and impact locations.

In some of the stream literature, there is the misconception that control areas resemble controls in laboratory situations, in which attempts are made to tightly regulate environmental variables. This is incorrect, and the extensive literature on experimental design, especially in agriculture, emphasizes that controls do not require that other, extraneous variables be controlled, only that they be matched between impact (or experimental) and control locations. Our aim is not to produce constant conditions, but to set up a sampling scheme under which any observed differences can be attributed to the human activity in question, and not to other processes operating at the impact (and control) locations.

related to climatic events, but it might be necessary also to sample at quite short time intervals within seasons if the variable being measured is the abundance of an organism with a very short life cycle and high population turnover. We will discuss later (chapter 7) the importance of the relationship between sampling frequency and variable dynamics for the analysis of resulting data.

### 5.3.2 Spatial variation and multiple locations

Just as it is to be expected that the impact location might change even without the presence of impacts, it is also to be expected that control locations will have their own natural dynamics. Any two locations that appear very similar at one time will very likely differ in some ways and their similarity will likely change through time. That is, whatever natural changes occur, it is likely that they will be different at different locations, or at least not synchronous. This means that if a single impact location is compared with a single control location, even with multiple sampling during both the Before and After periods, there is a real, non-trivial possibility that their relative status will change near to or after start-up of the activity of interest. If the change in the comparison between the two locations was simply because of natural location-to-location variation, then an impact would be inferred erroneously. If there is only a single impact and single control location, any inferences about impacts are particularly susceptible to 'unusual' events not related to the activity being monitored influencing just one of the locations.

What is needed here is an estimate of the range of dynamics that might be experienced by the impacted location and the control conditions with which it is being compared. Such an estimate can be obtained by sampling several impact and/or control locations concurrently. From such multiple locations it is possible to describe the 'envelope' within which un-impacted locations might be expected to exist at any particular time, the envelope within which such locations might change through time, and the envelope within which the impacted locations actually behave, both before and after start-up of the activity. For the above 'envelopes of normality' to be reliably described, the variables being measured should be behaving effectively independently at each of the locations. That is, the dynamics of a variable at one location should not determine its dynamics at any other locations being sampled. This requirement often is achievable in the selection of control locations, but often is difficult to realize in the selection of impact locations (as we discuss in chapter 8).

Thus, because of the inherent variation in most natural systems, the basic BACI design has to be extended to include sampling at multiple Control and (ideally) Impact locations on multiple occasions during the Before and After period. Keough & Mapstone (1995) referred to this design as MBACI, although they did not expect that multiple impact locations would be available in most cases. Analytical models for MBACI designs are presented in chapter 7 and we discuss the location of suitable controls in chapter 8.

### 5.3.3 Asymmetry in impact assessment monitoring

If the same number of Impact and Control locations were sampled at the same times Before and After start-up, then the design would be said to be completely balanced and symmetrical (Sokal & Rohlf 1995; Underwood 1996; Winer *et al.* 1991). Balance is highly desirable for analytical and inferential convenience, but is rarely achievable in impact monitoring.

In the absence of impacts, it is often straightforward to select several sampling locations that function essentially independently with respect to many local processes. Thus, it is often possible to identify multiple control locations for an impact assessment monitoring program. In stream situations, however, multiple locations on the one stream are unlikely to be independent because of unidirectional flow and multiple controls usually will mean multiple streams. This will present serious obstacles in some cases, especially where large lowland rivers are being monitored (chapter 8).

For many (perhaps most) impacts, however, there is only one source of impact beginning operations at a time, even though there may be other instances of the same activity starting up at other times and in other places. In such cases, there is often only one functional 'location' of impact (Keough & Mapstone 1995). Even though multiple 'sampling locations' might be identified at which an impact is possible, they are all subject to the same cause of impact – the operation of a particular instance of an activity. Data from these multiple sampling locations will be influenced by a common source of disturbance and so cannot be considered to represent separate or independent instances of such disturbances, even though they might be independent with respect to the un-impacted dynamics of certain variables (Hurlbert 1984; Keough & Mapstone 1995). Thus, the multiple (sampling) locations will in fact be subsets of a larger area that is impacted, and cannot be considered to provide independent measures of the impact. In other situations, where the source of impact is diffuse, the same set of circumstances may affect large areas more or less simultaneously. Here too, a series of sampling locations is unlikely to be independent with respect to the cause of impact and no one location is likely to adequately represent the scale of the impact of interest. This means, in practice, that in most impact assessment monitoring designs there can only be a single legitimate impact location for a given instance of an activity (see also Box 5.2).

Hence, most impact monitoring designs become unbalanced, and asymmetrical, with several control locations being compared with a single impact location (Keough & Mapstone 1995; Underwood 1991a). This means that the control locations, and their replication, take on added importance because they are the only source of data from which to estimate the envelope of normal location behaviour with which to compare the dynamics of the (single) impact location.

### 5.4 SCALES OF IMPACT AND MONITORING

So far we have discussed the basic logical necessities of impact assessment monitoring. The definition of what is meant by an impact location, and the choice of locations (whether impact or control locations), however, involves a raft of logistic and functional considerations related to the spatial and temporal scales of impact and the dynamics of measured variables. Perhaps first among these is consideration of the extent – or scale – over which an impact is expected, or over which concern about impacts extends. This issue will essentially define the scale of an impact location, and in turn dictate the appropriate scale of control locations with which the impact location is to be compared (Keough & Mapstone 1995). For example, it may be expected that a sediment plume for a development will extend along a 1 km stretch of stream and that if

**Box 5.2** Gradients and inference of impact

We have noted the design problem arising whenever multiple sampling locations are wrongly considered to be independent measures of impact. One special case of this general problem occurs when the multiple sample sites are found to indicate an apparent gradient of response to impact (say, over sample sites moving downstream from a putative source of impact). A significant, non-random gradient may be taken to imply impact. However, such a design provides only weak inference. The gradient defined by the multiple sample sites provides a valid form of evidence for impact, but there is no form of control to assess whether that apparent 'evidence' could have arisen even without impact (as discussed in chapter 4). Observing such gradients in streams may not be at all improbable even in the absence of impact. We therefore need other control streams, in order to determine the envelope of normal location behaviour (in this case, behaviour with respect to the kinds of gradients that might normally be expected).

That caveat does not mean that gradient analyses are not valuable. Not only may a gradient validly form the evidence as part of an ideal design, but also examination of gradients may help evaluate the utility of more conventional forms of evidence, such as changes in species abundance. Gradient analysis provides ways of learning about the nature and extent of impacts (see section 7.6) and the best variables for detecting those impacts. In chapter 9, we also discuss how gradient analysis may be linked to causality arguments, when an increase in the magnitude of an effect corresponds to increasing intensity or frequency of human impact.

Lastly, we note that ordination methods provide a special form of gradient analysis for multivariate data (see chapter 4), and, when the impact and control sites in such a gradient space are independent, MANOVA and other methods can provide powerful inference of impact.

impacts within this area are monitored, and minimized, then impacts beyond that area are unlikely to be important. The impact location then, would be a 1 km stretch of stream, and control locations of similar scale would also be sought.

It is important also to consider the physical and ecological characteristics of the impact location when selecting control locations and designing the details of sampling. If, for example, the 1 km stretch of stream that constitutes the impact location contains a combination of pools and riffles, with a narrow riparian margin, then it would be important to look for control locations with similar characteristics. It would be important also to stratify sampling within the impact and control locations to take account of the different habitats present.

In some cases the likely scale of impact will be very uncertain and one objective of the monitoring program might be to identify the scale over which an impact occurs. In such cases a range of scales needs to be considered explicitly and systematically in sampling, possibly both temporally and spatially. Two approaches might be considered here, both beginning with an impact 'location' being defined by the largest distance over which an impact might conceivably occur. If it is expected that there might be discrete, stepwise differences in the scale(s) at which an impact might occur, a number of smaller scale locations might be defined within the impact location. These smaller locations would represent impacts over smaller areas. This would result in a spatial hierarchy of 'locations', with each location nested within all those of larger scale. At its simplest this might be viewed as a set of concentric polygons or nested stream sections defining the different distances from a point source of impact over which impacts might be expected. Boundaries between the polygons or stream sections would be defined by the breakpoints in scales of impact that seem most likely. Because many ecological processes may be scale-related, it would be inappropriate to compare the impact data from one scale with data from a different scale at control locations. Thus, the hierarchy of scales should be repeated at the control locations. In designs such as these, sampling would entail the Before–After sampling at multiple Control (and perhaps) Impact locations, as described above, but there would be additional levels of sampling nested within each of the Impact and multiple Control locations from which the scale of impact would be estimated.

If, in addition to uncertainty about the spatial scale of impact, there was uncertainty about the time frame(s) over which impacts might manifest (and disappear, thereby possibly being missed), then a similar hierarchy of sampling might be necessary within the Before and After periods. This might entail, for example, sampling weekly within seasons over several years. Such additions to the basic sampling design rapidly increase the number of times and places sampled. Clearly, then, the greater the uncertainty about the spatial or temporal scales of impact, the greater will be the sampling effort (and cost) needed to resolve that uncertainty, and the greater will be the logistic obstacles to rigorous monitoring.

A second approach to resolving the scale of impact is to establish a regular series of sampling over the gradient of potential impact. The most likely gradient of impact would be related to distance from a point source of impact. For point sources we expect a gradient of disturbance corresponding to, for example, diffusion or transport of waste from the point of discharge. In some cases a formal mixing zone is defined within which a substantial impact is expected, and tolerated, and so monitoring is focused on detecting unacceptable impacts beyond that area. In other cases, however, sampling over a (potential) gradient would be appropriate. Such cases would include those where there was no defined mixing zone, there was no a priori reason to expect stepwise changes in the likelihood of impact and it was considered important to document the way in which a variable changed with distance from a source. In this case, it is possible that the gradient sampling would occur only at the impact location if there was no expectation of non-impact sources of scale-related behaviour or gradients in variables at the control locations. In streams, more than many systems, however, there may be compelling reasons for monitoring longitudinal gradients for variables at both impact and control locations.

### 5.4.1 Sampling within locations – impacts on status

So far we have discussed only the factors influencing the choice of locations to be sampled, but have not considered how each location might be sampled at each time. For most variables, it is not possible or efficient to enumerate or take a census of their status (e.g. the average value of the variable across a whole location) over the entire area of a location, unless it is a very small location and there are relatively few, easily counted cases of the variable (e.g. large fish in a small pond). This necessitates the estimation of variable status from smaller, manageable samples or observations, taken within each location and time, and assumed to be representative of the location's status in general. Such samples are termed representative subsamples (Sokal & Rohlf 1995). Perhaps the key issue to be considered with respect to subsampling is how best to ensure that they do adequately represent the larger area or time that they are expected to represent (Hurlbert 1984).

Most ecological or environmental variables demonstrate considerable variation in space and time at many scales. Moreover, most samples that are logistically feasible are smaller in size and taken over a shorter duration than most of the scales at which such variation is conspicuous. This means that single samples taken on different occasions or in different places within a location may vary greatly simply because of variability in the measured variable at scales between the size of the sample and the size of the location (Sokal & Rohlf 1995). The result of taking only single subsamples would be absolute uncertainty about whether changes through time or differences between control and impact locations represent real impacts or are simply expressions of smaller-scale variation manifest in the small-scale subsample from each location. If the latter were true, then a number of samples taken within a single location might have showed as much variation as single samples taken from different locations. Thus, taking a single sample that is small relative to the location it is expected to represent will be insufficient.

Taking several samples scattered over a location on each occasion will reduce these uncertainties. Indeed, to adequately 'represent' a location with an estimate based on subsampling, it will almost always be essential to take several (perhaps many) samples from within each location at each sampling time (Hurlbert 1984; Underwood 1981). Because it is the entire location that we seek to represent with the collection of subsamples, it is important that the latter are dispersed throughout the location (or those parts of it that are relevant, such as specific habitats), and not all taken from, say, one corner of the location (Hurlbert 1984). It is important to recognize here that if samples were taken from a small part of the proposed location because of strictly random sampling, the result would be effectively a redefinition of the location to that area over which samples were actually taken. Hence, any inferences would fail to relate properly to the location originally defined in the design phase.

Adequate dispersion of observations can be achieved by either collecting subsamples in a very regular pattern over the location or at random from within the location. There are advantages (and disadvantages) to each strategy. Regular sampling will ensure maximum dispersion. Randomization, by definition, might result in clumped sampling or missing large sections of the location simply by chance. This argument would tend to favour regular sampling. However, if the regular sampling happens to coincide with some regular feature of the location (such as undulations in a stream bed), then the samples may be biased because they were mostly taken from, for example, the swales of the bed ripples. Moreover, the coincidence of regular samples with important natural features might vary among locations and/or times (swales in one place, ridges in another), resulting in completely erroneous inferences about location-scale patterns or impacts. Randomization will usually minimize the risk of such coincidence confounding interpretation of the data. This argument, then, would favour strict randomization. Perhaps the safest strategy for subsample allocation is a two-stage process in which each location is first 'gridded' at some fairly coarse scale and then (an equal number of) samples are taken at random within each grid. This will ensure adequate dispersion but avoid systematic relationships among samples over the location.

### 5.4.2 Sampling within locations – impacts on variation

As mentioned earlier, it may be as important to consider the effects of an environmental impact on the variation(s) in a variable as on the overall average status of that variable. Here too, there may be uncertainty about the scale-related nature of effects (on variation) and a desire to resolve whether an impact simply results in a 'blanket' effect over the impact location, or a patchy effect with specific scales of patchiness. In order to assess the latter, Underwood (1991a, 1992, 1994a) has recommended a design framework that he termed 'Beyond BACI'. Beyond BACI designs share many of the features of MBACI designs, such as multiple control (and if possible) impact locations, and multiple sampling occasions before and after start-up of an activity, but in addition include one or more hierarchically arranged scales of sampling within each location (see chapter 7 for more information). These nested scales (which we will term sites) within locations are allocated essentially at random within the locations. Thus, they do not equate to the nested concentric locations, discussed previously, that might be used to define the scale at which an impact occurred. The concentric arrangement of locations of different sizes represents a strictly systematic allocation of smaller scales within larger ones, not a random allocation of smaller-scaled sampling spaces across a larger, conceptually homogeneous location.

Although Keough & Mapstone (1995) and Underwood (1991a, 1992) indicated that the Beyond BACI designs were appropriate for estimating the scale of an impact on the status of a variable, this is not automatically the case. Only if the nested sites are specifically arranged within specified distances from the impact will they systematically address the question of scale of impact. If this is the case, however, they then are unlikely to satisfactorily estimate the scales of variation or patchiness in impact within the impact location, for which a randomized allocation would be required. Again, the specific questions being addressed, and the assumptions underlying decisions about the likely features of an impact need to be clarified before useful choices can be made between these alternatives. Ultimately, if there remains considerable uncertainty about both the scale(s) of impact and the scale(s) of patchiness in the impact, a combined sampling strategy may be required. In this case, sampling spaces (sites), possibly of more than one size, might be allocated at random within each concentric location, thus enabling estimation of whether the impact was defined better at some scales than others and also the characteristics of impact on spatial patchiness within scales. Such a scheme would be extremely expensive.

### 5.4.3 Sampling within periods – duration and fluctuations in impact

Again, similar issues arise with respect to sampling in time as well as space. If a principal interest is to determine the period over which an impact occurred, then regular sampling throughout the potential (largest) period of impact would be preferable. If, however, the major focus is on the effects of an activity on the temporal variation(s) in a variable, then sampling at randomly allocated times when impacts are likely would be required. As before, combinations of both approaches might be considered in some cases, but will dramatically escalate the costs of monitoring. As with the spatial analogue, these options reflect clear uncertainties about the characteristics of an impact. In many cases, references to impacts of similar activities that have been monitored before may clarify the likely properties of an impact such that monitoring designs can be simplified and focused more powerfully on the discrimination of impacts rather than the exploration of their characteristics.

### 5.4.4 Collecting the samples

Underlying much of the preceding argument is the proposition that the putative impact and the control locations are being monitored in parallel. It is particularly important to sample all locations within a small window at each sampling occasion (but see Stewart-Oaten 1996b; Stewart-Oaten *et al.* 1986), although usually it will be impossible to sample at all locations exactly concurrently. The greater the difference in timing of sampling among locations the greater the potential for confounding changes at one or more locations with differences in the timing of sampling at those locations. For example, if an impact location is sampled on one day and the control locations are sampled on subsequent days, but there is substantial rainfall overnight, there is substantial potential for the data from the impact and control locations to be seriously flawed because of differences in stream flow among them.

It may be important also to avoid sampling the locations in the same order on all sampling occasions. Sampling locations in the same order on all occasions increases the possibility that unknown or unexpected biases related to the logistics of sampling might taint inferences of impact. For example, if sampling is condensed into a short period and is arduous, fatigue may result in biases in the collection or treatment of samples. Changing the order in which locations are visited is perhaps the easiest way to insure against the influence of such biases. It is better to have such (usually unknown) biases spread among all locations over the life of a monitoring program than to have them consistently arrayed across locations. Strict randomization of the order in which locations are sampled on each occasion is perhaps most desirable, but often will not be feasible. An acceptable alternative would be to sample the locations in the same sequence, but with a different starting position on each occasion. The starting position might be chosen either at random or sequentially across sampling occasions.

The logistics of sampling may be complicated also when multiple variables are being sampled, as will usually be the case. Although the overall logical structure of a monitoring program usually will be the same for multiple variables (e.g. the need for Before–After and Control–Impact sampling), the details of sampling strategies (e.g. numbers and scale of sites sampled, numbers of observations at each site) optimized for different variables may vary among variables. Clearly this is less of an issue when the data are expected to be analysed together by multivariate methods (e.g. Faith et al. 1995; Humphrey et al. 1995) and must, therefore, be collected from the same sampling units. Even in these cases, however, it is unlikely that all variables will be treated in a single analysis and so there exists the potential for different sampling requirements for different sets of variables.

Usually it will be desirable logistically to collect all the observations (for all variables) from each location on a single visit to that location at each sampling occasion. Often it will be easier to standardize the sampling design across most or all variables, irrespective of the sampling details appropriate for each, rather than trying to implement separate, different sampling regimes for each of the variables. In such cases it is preferable to 'design-up' to the greatest level of detail required by any variable rather than to 'design-down' to the lowest common factor. We recommend this strategy because it is easy to subsample or aggregate data that has been gathered already to fit a desired analytical method, but it is very difficult, and ill advised, to 'fill-in' data that has not been collected for some variables because sampling was geared to the least demanding design. Further, the aggregation of observations often improves the properties of data for statistical analyses (discussed in chapters 4 and 13). Of course, this approach will have to be balanced against the costs of sampling all variables with the greatest intensity, even though some of them may not require such effort for acceptable inferences about impacts.

### 5.4.5 Other considerations

The design of an impact assessment monitoring program should not be considered in isolation each time a new activity is considered in need of assessment. Data from previous studies of similar activities and their impacts may provide valuable insights to the likely scale, duration, manifestations of impact and the variables most sensitive to impact. Clearly, this implies that there is a place for the collective consideration of data from monitoring programs for similar activities at different times. Such synthesis of data from different programs will be enhanced greatly if there is consistency of sampling design, variable selection, sample collection and processing among those programs. Accordingly, there is a place in the design of new programs for considering their consistency with prior work. Where possible, consistency should be maximized – but not at the expense of improvement in basic design and implementation. It is of little consolation that a current monitoring program is the same as previous ones if it is (as they were) inadequate to distinguish the impacts of the driving activity from disturbances caused by other events. We recommend, therefore, that prior experience be considered in the design of new monitoring programs, but that the prior experience is subject to critical review and improvements in design are favoured over adherence to tradition or dogma.

A related issue here is the proper care of data arising from monitoring programs. The existence of prior information is only useful if it is accessible for review and the data are available for (re)analysis. Similarly, the synthesis of information from multiple studies is greatly enhanced if the underlying data are available for analysis. Availability of data across multiple studies will be improved substantially if the data from each are carefully processed, stored on well-constructed databases, and archived for future use. Although more a facet of implementation than of design, we consider it essential that the management of the data from the chosen monitoring program be planned and rigorous. This is perhaps best begun at the early stages of the program, when other aspects of the work are also being designed.

## 5.5 CAREFUL DESIGN IS THE MOST IMPORTANT STEP

We have endeavoured here to explain the issues in monitoring design that we consider most important. We consider these principles to be important because of the logical foundations that they provide for making conclusive and legitimate inferences about the existence or absence of environmental impacts of specific human activities. Inevitably, compromises will be necessary in monitoring programs and we discuss throughout the remainder of this book where, and how, such compromises might be made – and what their consequences are for inferences about impacts. It is critical in each step that choices about compromise are explicit and documented with reference to their impacts on the potential to meet the objectives of the monitoring program. Careful attention to the design of the study will aid in making choices about compromises, and clarify their implications.

It is important to recognize that although data arising from these designs are more easily analysed by some methods than by others, the choice of analytical tool should not dictate the basic design choices. Once the data are in hand, any number of analytical methods can be applied to them. They can be analysed and re-analysed, either as a whole set or in subsets, according to the preferences and skills available. If the sampling design under which the data were collected was deficient, however, no amount of different statistical treatments will fix those deficiencies. We contend, therefore, that thorough attention to the design of monitoring is perhaps the most important step in setting the scene for robust, defensible inferences later in the monitoring–decision–management process.

## 5.6 IMPORTANT ISSUES

- Good monitoring design means having sufficient resources and samples to answer our questions both efficiently and rigorously but without being wasteful.
- The first requirement for good design for impact monitoring is sampling from baseline, through start-up and during operational conditions.

- For impact assessment monitoring, impacts tend to be defined relative to natural conditions, rather than with reference to external criteria. Hence, a second requirement for good design is to collect and analyse data from locations that are considered beyond the influence of the activity and therefore not subject to its impacts. These non-impact locations usually are termed control locations.

- These two requirements are summarized by the terminology of Before–After, Control–Impact, or BACI, designs and are the key underlying elements of rigorous impact assessment monitoring designs.

- To assess whether natural variation has changed because of impacts, and to be able to separate any effects of impacts from natural changes, we need to sample repeatedly during both Before and After periods. Additionally, we need to estimate the range of dynamics that might be experienced by the impacted location and the control conditions with which it is being compared. Such an estimate can be obtained by sampling several impact and control locations concurrently.

- Together, the above samples help describe an envelope of variation among control locations over time, and the envelope within which impact locations actually behave. Replication of controls and samples through time are important because they are the only data from which to estimate the envelope of 'normal' location behaviour, as, in most impact assessment monitoring designs, there will only be a single legitimate impact location for a given instance of an activity.

- One of the most common deficiencies in impact monitoring designs is insufficient data gathered Before a putative impact occurs. Hence, it is critical that monitoring designs are evaluated early in the development process. Formal monitoring of control and future impact locations should commence as far as possible in advance of start-up of the impacting activity.

- The definition of what is meant by an impact location, and choice of control locations, involves many considerations related to the spatial and temporal scales of impact and the dynamics of measured variables. These considerations will affect the spatial and temporal extent of control and impact locations and periods, spatial and temporal placement of subsamples, and the resulting analytical model.

- We will usually use more than one variable to examine any putative impacts. It is often logistically efficient to collect data for multiple variables simultaneously but the optimum design might not be the same for each variable. Hence it is important to consider explicitly the trade-offs between using one sampling design planned around the more logistically demanding variables (e.g. those requiring most frequent sampling) or using independent designs for different variables or groups of variables that have different sampling requirements. In all cases, impact and control locations should be sampled as closely together in time as possible on each sampling occasion.

- We can learn much from previous examples of human impacts, so it is important that monitoring data be archived properly and be accessible by others.

# 6

## Problems in applying designs

Chapter 5 dealt with the ideal design criteria appropriate for the logical detection of impacts. In this chapter, we illustrate why these design criteria are important by outlining the problems that have been faced by ecologists seeking to apply at least some of the principles that were discussed in chapter 5. Many studies were conducted before much was known about good design, but problems in applying monitoring designs have arisen partly from the nature of rivers themselves (chapter 2), partly from the nature of the variables that river ecologists deal with (chapter 10), and partly from institutional and political arrangements that affect the timing and scope of such studies.

Before describing these problems in detail, we first present an historical sketch of the previous attempts at implementing programs for impact assessment and monitoring using biological data in rivers and streams. This brief history lesson illustrates the apparent simplicity and hidden complexities of assessing and monitoring in rivers and streams. We will then elaborate on how each of the three problem areas makes it difficult to implement ideal experimental designs in rivers and streams.

### 6.1 A BRIEF HISTORICAL SKETCH

The vast majority of published field studies of impacts in streams have been directed at investigating pre-existing impacts (see reviews by: Hellawell 1986; Hynes 1960). Many of the earlier studies were aimed at least at corroborating conclusions drawn from observations of large, obvious human impacts and from physicochemical investigations (Fjerdingstad 1964; Forbes & Richardson 1919; Kolkwitz & Marsson 1908; Patrick 1949), and sometimes at establishing that biological variables were detecting changes that had been missed by using physicochemical data alone (reviewed by Wilhm 1975). Much of this research was conducted in Europe and North America, and often concentrated on rivers with conspicuous point sources of pollution, such as untreated sewage outfalls, and untreated effluent from heavy industry and mines. By the late 1940s, the common patterns due to organic pollution were felt to be well known (Bartsch 1948). Hynes (1960) summarized much of this descriptive research on both organic and inorganic pollution and formulated some simplified and frequently cited qualitative models, which we reproduce here in Fig. 6.1.

It would be fair to say that the relationships presented in Fig. 6.1 have strongly influenced many subsequent studies, either directly or indirectly. Three features of these models recur in the design of many survey-based studies. First, there is a reach upstream of the discharge that can act as a 'control' for locations downstream of the potential problem; second, the biota differs dramatically downstream of the discharge, often with a succession of different groups of taxa replacing each other as the dominant organism as one moves further downstream; and, third, eventually the biota returns to a composition similar to that of the 'control' location upstream of the input.

There are surprisingly few empirical studies that conform to this simple scheme, and several reasons for this (besides its obvious oversimplifications) are commonly proffered. Often, it is felt that the biota rarely 'recovers' before another impact is wrought on the river. Either additional outfalls discharge into the river, or land use changes with concomitant alterations in runoff and, hence, the potential for diffuse pollution (e.g. Hawkes & Davies 1971). In a similar vein, the river itself may change downstream, so that the effects of the impact are confounded with natural changes in the underlying habitats in the river; as a result the upstream 'control' no longer provides an appropriate standard with which to compare impacted locations (e.g. Stewart & Loar 1994). Less frequently, unexpected droughts or floods have coincided with the survey so that it is difficult to distinguish 'control' from 'impact' locations because the biota has been severely reduced by the drought or flood. Finally, much is sometimes made of the taxonomic differences between a particular survey and whichever version of Hynes's models is current; at times an obsession with regional faunal differences has obscured other problems with drawing inferences from the data. Similar problems arise when assessing diffuse sources: impacted locations are often downstream of control locations and differences in the biota are confounded with background spatial changes that may be unrelated to the impact concerned.

## (a)

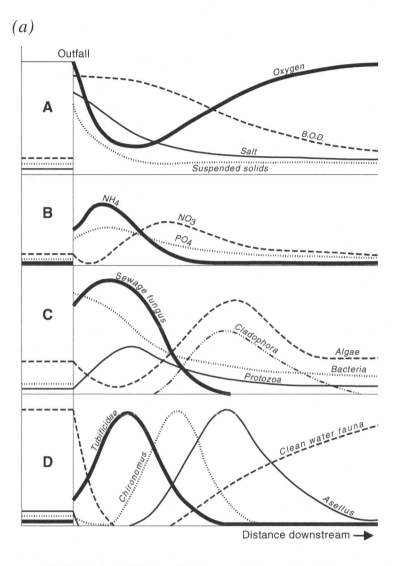

**Fig. 6.1** Hynes's (1960) qualitative, graphical models of the expected changes in rivers downstream of (a) a point-source discharge of an organic effluent (e.g. untreated sewage), showing expected downstream changes in (A) oxygen, salt and suspended solids concentrations and biological oxygen demand (BOD), (B) concentrations of common forms of nitrogen and phosphorus, (C) densities of algal, microbial and protozoan taxa and (D) densities of particular invertebrate taxa; and (b) (*overleaf*) a point-source discharge of a toxic substance (e.g. effluent containing heavy metals) showing downstream changes in algal and animal abundance. (The figures are redrawn from Hynes (1960), Figs. 15 and 16.)

*(b)*

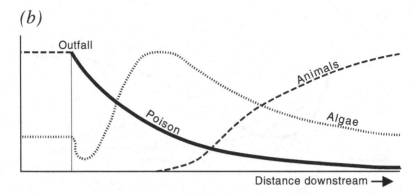

Fig 6.1 *(cont.)*

The notion of upstream control locations was sometimes expanded to include unaffected controls on tributaries and, more rarely, similar rivers in nearby catchments. The major problem with this strategy is with the representativeness or comparability of the tributary locations, because often there are no historical data for comparing the tributaries with the main stem prior to the impact. Arguments about the advantages and disadvantages of this procedure have generated a large literature about 'bioregions' and 'ecoregions', especially in North America (Barbour *et al.* 1992; Hughes *et al.* 1994; Rohm *et al.* 1987). This problem becomes especially pronounced in large rivers, where the tributaries may be very dissimilar from the main river. There may also be natural, location-to-location differences in the biota that are rarely either measured intentionally or discussed.

Although it is often not couched in the terms of formal statistical inference, the issue of confounding natural, longitudinal changes in the river with the effects of impacts (Underwood 1994b) has clearly worried many investigators. Several have noted that collecting baseline data for more than one annual cycle before the impact strengthens conclusions considerably (Humphrey & Dostine 1994; Humphrey *et al.* 1995; Tubbing *et al.* 1994), but there is still a dearth of published studies where adequate pre-impact data have been collected for a sufficiently long time (Rosenberg *et al.* 1981; Warnken & Buckley 1998). The overwhelming impression that we get from the major reviews and texts is that the vast majority of published studies of impact assessment or monitoring of rivers using *in situ* biological variables under field conditions has only been undertaken *after* a putative impact has started.

Inadequate funding, a lack of appreciation of the importance of pre-impact data, insufficient lead times and a continuing emphasis on assessing existing impacts have probably combined to hamper progress in applying good statistical designs with strong inferential bases (Fairweather 1994; Green 1979). In the absence of temporal controls, a cocktail of *ad hoc* strategies has been employed by researchers in an attempt to assemble multiple lines of evidence about the nature and size of an impact (Abel 1989; Chapman 1996; Wilhm 1975). These strategies might be summarized as follows:

1. Collection of supporting physical and chemical data. If these data are consistent with the observed biological patterns (or vice versa) then more confidence is expressed in the conclusion.
2. Use of a system of zonation or bioregionalization to corroborate expectations of the comparability of the biota in the impact and control locations in the absence of the impact. If impact and control locations are within the same zone of a river, or the same biological region, then there is an expectation that the biota should have been similar before the impact occurred.
3. Use of indicator species, where such species are held to be specific to a particular type of impact. Some species are thought to be well adapted to or tolerant of particular environmental conditions and are hence regarded as characteristic of a particular form of impact.
4. Characterizing the tolerances of the constituent species using some form of scoring system. This can be an extension of the indicator species approach, where an attempt is made to summarize the composition of the community in terms of its tolerance to the impact. If the assemblage is dominated by tolerant taxa with few sensitive individuals, then this is deemed to indicate that the biota is, in some way, impacted.
5. Analysis of individuals for deformities, biomarkers or chemicals thought to be specific to the impact. Often these attributes are the principal variables measured in an investigation, but sometimes they are used as additional evidence to back up conclusions drawn from physicochemical, population and community-level data.

All of these strategies have been criticised, and it is probably fair to say that, for biological variables, a disproportionate effort has been devoted to critiques of strategies 3 and 4 above. This has resulted in a large literature on different ways of summarizing and analysing data resulting from field surveys, and a substantial cottage industry in reviewing this literature. From the debate about these issues, some felt that the relatively informal, often qualitative approaches to quantifying biological responses to human disturbance were inadequate, leading to the increased use of statistical methods and multivariate displays of the data (e.g. Green 1979).

Three themes emerge from this historical sketch. The first involves the difficulties of defining a control or set of controls on a linearly connected system such as a river. The nature of rivers seems to offer some advantages (e.g. upstream locations or tributaries to act as controls) but, in practice, have posed some difficulties in implementing reliable experimental designs. The second theme involves the practical difficulties of using river biota in implementing impact and monitoring studies. The third theme concerns political and institutional issues that prevent researchers from implementing good designs. We will now discuss each of these themes in turn, bearing in mind that there is some overlap between these sets of problems.

## 6.2 PROBLEMS INHERENT IN THE NATURE OF RIVERS

### 6.2.1 Interdependence between locations

Sampling locations on the same river or stream presents both an opportunity and some problems in applying conventional statistical models (described in chapters 4 and 7) for making strong inferences about impacts and changes over time. The opportunity exploited by the vast majority of published field studies is that rivers usually flow in only one direction so that one might expect that locations immediately upstream of the impact would represent the biota that should have existed at the impacted locations before the impact started. There are two main problems with this approach. First, the river biota changes in species composition naturally as one proceeds downstream, so that impacts 'detected' by changes in species composition or population densities may be confounded with natural, downstream changes in the biota. Second, intercorrelations between locations closely located to each other may violate the assumptions of independence between locations required of classical statistical designs (section 5.3.2). Upstream controls or those on tributaries may be a net source of colonists for downstream impacted locations leading to underestimates of effects (see also Underwood 1994b).

In formal terms, the lack of independence between sampling locations is termed serial correlation in space (sometimes referred to as autocorrelation). The problems this presents to conventional analytical techniques such as ANOVA have been raised by marine and lentic freshwater scientists (Millard et al. 1985; Reckhow & Stow 1990), but remain largely uninvestigated for biotic variables in rivers and streams when considered in the context of environmental monitoring and impact assessment. Locations that are close together, relative to the mobility of the target organism, are unlikely to be independent, as has been appreciated by some of those appraising the use of fish populations and community structure, even if they do not conduct their discussion in the jargon of formal statistical design (Schlosser 1990). The problem remains that, for some variables, sequential locations along a stream will be spatially correlated, whereas others may not be.

Clearly, using only spatial controls in lieu of temporal controls as a basis for making inferences about impacts in rivers has some problems. Stewart & Loar (1994), for example, noted that the upstream controls used in studies around US Department of Energy facilities near Oak Ridge, Tennessee, were likely to be depauperate in both fish and benthic invertebrate groups because of the lesser habitat diversity in these locations compared with the putatively impacted locations downstream. Similarly, replicating 'impact' locations is also problematic. Many studies include several locations downstream of the impact to determine whether 'recovery' is taking place (e.g. Gaufin & Tarzwell 1952, 1956; Hawkes 1964; Hawkes & Davies 1971; Norris 1986; Norris et al. 1982), or to try to quantify the effects of additional impacts further downstream (e.g. Arthington et al. 1982; Campbell 1978). As has been noted before, the lack of any data prior to the onset of the impacts makes it difficult to judge either the degree of impact or the extent of recovery.

For slow-moving or sessile benthos, investigators often implicitly assume that locations $> 1$ km apart on a river are independent. Resh & McElravy (1993) examined 48 quantitative studies from rivers and streams compiled by Voshell et al. (1989) and found that more than half the lotic studies (63%) used controls solely in the same water body as the impact locations. They also noted, but did not quantify, a large degree of inappropriate replication, where only a single sampling location was used to represent control or impact conditions. Similar patterns are cited by the major reviews of field studies of populations or communities (e.g. Hellawell 1977, 1978, 1986; Hynes 1960; Metcalfe 1989; Rosenberg & Resh 1993). A recent review (J. R. Thomson & B. Downes unpublished data; see Box 6.1) of 140 studies looking at a variety of human impacts on lotic systems found that $\sim 59\%$ used controls. Of these, 57% used locations upstream of impact points to provide controls, and thus have potential difficulties with confounding. Encouragingly, though, 35% had multiple controls located on separate non-impacted rivers; most of these studies were published in the 1990s suggesting that sampling designs in rivers are improving.

Larger, vagile organisms such as fish present more obvious problems. First, control locations may not be truly independent because of movements between them by the animals; spatial autocorrelation and its effects on the width of confidence intervals may therefore be substantial, although if sufficient data are collected to model these autocorrelations, then this problem may be solvable (Conquest 1993). The second potential problem with vagile organisms is difficult to overcome: if organisms can move freely between control and impact locations, these locations become confounded. Fish may, for example, disperse from a control location into the impacted location leading to underestimates of the change, or they may disperse from the impact location into the control location, which would double the effect size (see Smith et al. (1993) for a salutary example and also chapter 11 for discussion about effect sizes). Furthermore, the degree to which component species are resident in a location affects researchers' perceptions of stability in terms of population size or community composition, and there have been vigorous exchanges about how sampling locations might be defined unambiguously for vagile fish (e.g. Grossman et al. 1982; Herbold 1984; Ross et al. 1985). Defining the spatial extent for sampling fish populations or assemblages is, therefore, difficult but the debate and awareness of the issue is more developed than it is amongst benthic biologists (Grossman et al. 1990; Meffe & Sheldon 1990).

More attention has been paid to pre-existing differences in habitats between control locations in benthic sampling programs. Researchers recognize that the benthos is strongly influenced by substrate type, water velocity and the presence of vascular aquatic vegetation (e.g. Allen 1959; Hynes 1960, 1970; Winterbourn 1981). Comparing like habitats is essential, therefore, and the merits of stratifying sampling programs have been promoted heavily (Elliott 1977; Green 1979; Hellawell 1978; Resh 1979), with the practice now being widespread (Resh & McElravy 1993). Confusion still remains, however, about the appropriate level at which sample units should be replicated, with subsampling within a location being mistaken for replication of 'treatment' conditions (Resh & McElravy 1993; and see section 6.3.1 below).

Overall, research in rivers and streams has been hampered by a poor understanding of spatial scales of variation of the variables under examination. Part of this can be attributed to early expositions of the prohibitive sampling efforts required for modestly precise estimates of population means, which did not address the connections between variance and scale (see section 6.3.1 below). However, spatially nested sampling designs seem very uncommon in riverine benthic studies

**Box 6.1** A recent review of human impact studies

J. R. Thomson & B. Downes (unpublished data) examined the methods used to detect and measure the impacts of a variety of human activities on lotic ecosystems in 140 recently published studies. The review focused on papers published from 1994 through 1999 in the following journals: *Freshwater Biology, Journal of Freshwater Ecology, Hydrobiologia, Australian Journal of Ecology, Canadian Journal of Fisheries* and *Aquatic Sciences, (Australian Journal of) Marine and Freshwater Research, Journal of the North American Benthological Society* and *Regulated Rivers.* However relevant papers from other journals (see full list below), and some published prior to 1994, were also included to increase sample size and to ensure a large geographic coverage of studies.

Papers that reported the results of studies designed to examine the impacts of human activities on lotic ecosystems were included in the review. Studies examining both point and non-point source impacts were sampled. Comparative studies with the primary aim of comparing different monitoring methods (e.g. invertebrates vs. diatoms, multivariate vs. multimetric statistics) were included as long as they involved an assessment of an actual human impact on a lotic system. Only studies that included at least some field-based assessment of impacts were included. Thus studies reporting only laboratory-based toxicity tests, for example, were not included; however studies that incorporated laboratory tests with field surveys/experiments aimed at measuring an actual impact were included. Data collected for each paper included: type of impact (e.g. regulation, urban/industrial effluent, agriculture, forestry, power generation, mining etc.), effect variables (e.g. temperature, pH, siltation), response variables (e.g. macroinvertebrate community composition), sampling design, degree of spatial and temporal replication (number of sites, samples per site etc.), spatial and temporal scales of impact, type of statistical analysis, and justifications (or lack thereof) for site selection, sampling designs, response variables etc. Information about sampling designs included whether a full BACI-type design was used or not, whether controls were located on one or multiple rivers, and where they were sited within catchments.

Full list of journals from which papers were sourced:

*Acta Hydrobiologica*
*Archives Hydrobiologica*
*Australian Journal of Ecology*
*Canadian Journal of Fisheries and Aquatic Science*
*Ecological Applications*
*Ecological Monographs*
*Ecology*
*Environmental Management*
*Environmental Monitoring and Assessment*
*Freshwater Biology*
*Hydrobiologia*
*Internationale Revue der Gesamten Hydrobiologie*
*Journal of Environmental Management*
*Journal of Fish Biology*
*Journal of Freshwater Ecology*
*Journal of the North American Benthological Society*
*(Australian Journal of) Marine and Freshwater Research*
*New Zealand Journal of Marine and Freshwater Research*
*Regulated Rivers*
*Transactions of the American Fisheries Society*

(Downes *et al.* 1993). River ecologists have been fixated with 'sites' or locations at the expense of articulating how these correspond with the spatial scales that the target organisms move over during the course of their life history (Downes & Keough 1998). The spatial scale of the sampling program, therefore, needs to take account of serial correlations among sampling locations. If such correlations are evident, then sampling locations either need to be selected so that independence is ensured, or the sampling design needs to be modified so that the correlation structure can be modelled appropriately (Conquest 1993).

### 6.2.2 Variation in time

In addition to any seasonal changes in flow, rivers are also subject to stochastic floods and droughts that can have substantial and persistent effects on the biota (e.g. Boulton & Lake 1992a,b). The pattern of high and low flows can vary considerably even within fairly small geographic regions (e.g. Hughes 1987; Hughes & James 1989), and flow events interact with the type of substrate, the thermal and nutrient regimes, and the heterogeneity of the habitat to influence the sequence of subsequent recolonization events (Biggs 1995; Poff & Ward 1990b; Yount & Niemi 1990). Antecedent flow events can and do affect the populations and the community composition of biotic variables in monitoring and assessment studies (e.g. Gibbs & Penny 1973; McElravy *et al.* 1989; Pearson 1984; Toshach 1977; Winterbourn & Stark 1978). However, obtaining sampling sequences that are both sufficiently long and frequent to be able to account for such natural variation is problematic.

In fresh waters, monitoring and assessment studies using biotic variables have tended to be shorter in running waters than in standing waters (Resh & McElravy 1993; Resh & Rosenberg 1989), although McElravy (summarized by Resh & McElravy 1993) noted that the proportion of published lotic studies lasting longer than one year increased from 20% for the period 1980–84 to 56% over the period 1985–87. Nevertheless, of the 48 quantitative studies from rivers and streams compiled by Voshell *et al.* (1989) one-third of the lotic studies involved a single sampling event and only 11% spanned two years or more. Resh & McElravy (1993) did not comment on the number of studies with pre-impact baseline data.

Thomson & Downes (see Box 6.1) found only 17 out of 140 studies (or ~ 12%) had any data collected before impacts occurred. This almost certainly reflects the preoccupation of many managers and researchers with existing impacts, but may indicate also that the necessity of collecting such data for strong inferences is not broadly understood (we consider this problem specifically in chapter 9). Where there have been some attempts to collect pre-impact data, the length of time before the impact is usually limited. In major studies of two large dams in Victoria, Australia, for example, the pre-impact sampling coincided with dam construction (Blyth *et al.* 1984; Davey *et al.* 1987; Marchant 1988; West *et al.* 1984). In the case of the Mitta Mitta River, subsequent investigation of the impacts of cold, hypolimnetic irrigation releases of water had to rely on spatial controls in a tributary and interpretation of the nature of the fauna that was found downstream prior to the sediment inputs from construction (Doeg 1984). There are some recent examples of high-profile, large projects where some pre-impact data have been collected, but often physicochemical data have been collected for longer periods of time and more frequently than biological data. For example, Smith & Morris (1992) noted that only one 'expeditionary fish sampling' was undertaken prior to the development of the Ok Tedi gold and copper

deposit at Mount Fuliban, Papua New Guinea, whereas the documentation of 'pre-operational chemistry of waters, soils and sediments in the
region' was 'comprehensive'.

The lack of sufficiently long sequences of pre-impact data for
biological variables probably has several causes. Political and institutional issues obviously have a prominent role, especially if the project
has a high profile and is being 'fast-tracked'. These issues are not entirely
limited to political expediency, and are discussed more fully later in this
chapter. We have already noted, from the history sketched earlier in this
chapter, that many published studies concern impacts that already
exist. We note, however, that there are some recent examples of studies
directed at detecting recovery of severely impacted rivers and streams,
but still the frequency of sampling is often limited to that permissible
within the ambit of a PhD project (Likens 1984).

Some data have been collected just to establish baselines (e.g.
Bennison et al. 1989; McCarthy et al. 1997; Tubbing et al. 1994), but few of
these investigations find their way into the refereed literature. The aims
of such investigations include detection of trends and documentation of
'background variation' so that unspecified future impacts might be
detected. Sometimes, where such data have been collected, the design of
a sensible monitoring program has been feasible. For example, Resh and
co-workers collected seven years of baseline data on benthic invertebrates from Big Sulphur Creek, Sonoma County, California (Resh et al.
1988), a stream likely to be affected by future geothermal development.
Sampling was restricted to the end of the wet season, and the chosen
variables regressed against antecedent rainfall to account for variations
due to flow events. This information was then used in a sequential
sampling design (Resh & Price 1984) to provide a cost-effective sample
processing strategy that required processing many fewer sample units
than suggested by conventional analyses of sampling precision (cf. Chutter 1972; Chutter & Noble 1966).

Nevertheless, some data have been collected over multiple time
intervals, but have often been analysed inappropriately. In an extensive
review of studies of aquatic insects, Resh & Rosenberg (1989) found that
many authors ignored variation amongst times or locations by averaging data across quite different events or combinations of location types
and events. They entered a plea for both sampling at multiple scales and
the use of appropriate analyses.

However, we still have a poor understanding of temporal scales of
variation for many biological variables. Apart from the logistic constraints of maintaining long-term programs with sufficient flexibility to

take advantage of high and low flow events, ecologists have lacked a coherent framework for designing such sampling programs in the first place (Poff 1992; Poff & Ward 1990a,b). As for serial correlations in space, serial correlations in time (temporal autocorrelation) also present problems in applying conventional statistical analyses relying on independence between times (Millard *et al.* 1985). Those dealing with physicochemical and hydraulic variables, however, have better appreciated these problems than riverine biologists, albeit relatively recently (Loftis *et al.* 1991). However, as Loftis *et al.* (1991) emphasize, serial correlation depends on the scale of the observations and the questions being asked, and pre-existing 'routine' water-quality data sets are often poorly designed relative to the questions managers and the public ask of them.

Finally, a conceptual problem arises even when we do have long-term data that encompass large natural perturbations such as droughts and floods. The effects of such natural (but infrequent) events can result in massive changes to the variables being measured (Boulton & Lake 1992a,b; Boulton *et al.* 1991, 1992; Hall *et al.* 1978). If an anthropogenic agent produced changes that were inside the envelope of such natural changes, should it still be regarded as an impact? The answer to this question is not straightforward and will depend on the relative frequency of the anthropogenic perturbation and the value judgements inherent in how the monitoring and assessment questions are framed.

### 6.2.3 Logistic and technical issues

The majority of the space allocated in general manuals about sampling procedures is for smaller, 'wadeable' rivers (e.g. Hellawell 1978, 1986; Rosenberg & Resh 1996). Even within this class of river type, there has been further concentration on tractable habitats consisting of riffles, runs and pools for both benthic plants and animals, and fish. Alternatively, large rivers and a number of other habitats, such as logs and macrophyte beds, can pose substantial difficulties in reliable sampling.

Large rivers present a variety of logistic and technical problems for sampling the biota. In deep, swiftly flowing rivers, it is difficult or impossible to set nets to catch fish in the mainstream (e.g. Smith & Morris 1992), and dangerous to operate benthic equipment by scuba. Such rivers may even flow too strongly to deploy surface-operated equipment such as benthic grabs and air-lift samplers. Slow-moving large rivers seem, initially, to be safer. Many, however, are highly turbid, making scuba operations difficult. In tropical areas, the added risks from large, semi-aquatic vertebrates such as hippopotamus and crocodiles provide an additional frisson when operating nets, diving or using small boats. Under such circumstances the majority of the habitats in the reach, in areal terms, may not be available for sampling. Only some of these problems are likely to be overcome by ingenious new sampling methods.

The fine substratum of some large rivers is often all but devoid of attached benthic algae and macroinvertebrates, with the flora and fauna being concentrated in accumulations of woody debris and littoral beds of macrophytes. Woody debris has long been known to be an important habitat for fish, providing shelter, food and, for some species, a spawning location. However, it can be difficult to sample fish in such intricate habitats: nets and traps can become snagged, the fish are hard to see while electrofishing, and even the use of poisons is problematic if the debris is dense enough to trap and hide the fish from observers. Macroinvertebrates and algae also make extensive use of woody debris, but consistent non-destructive sampling of this habitat is problematic (DeLong et al. 1993). Artificial substrates have often been promoted as alternative sampling methods when habitats are either too difficult to access or too complex to quantify using active techniques (Biggs 1988; Flannagan & Rosenberg 1982; Hall 1982). Most of this literature has focused on the representativeness or relative cost-effectiveness of artificial substrates in collecting either benthic algae or macroinvertebrates (Cattaneo & Amireault 1992; Goldsborough & Hickman 1991; Mason et al. 1973; Reynolds & Hunter 1985).

Although methods for mapping vascular macrophytes are now well developed, there is less consensus on methods suitable for the flora and fauna that live on or amongst them (Aloi 1990; Morin & Cattaneo 1992). Although ingenious methods have been employed to quantify the surface area of plants of different shapes, it is unclear whether all of the species that might use the surface of a plant respond to this area in the same way (Downing 1986; Lillie & Budd 1992). Moreover, the abundance of some of the rapidly moving nektonic species is likely to be underestimated by most of the techniques currently used.

Finally, some major projects, such as mines and large hydroelectric dams, are located in remote areas, which results in limited opportunities for the collection of baseline data prior to the construction phase of the project. Recent examples are the Ok Tedi gold and copper mine in Papua New Guinea and hydroelectric proposals for southwest Tasmania.

## 6.3 PROBLEMS ARISING FROM THE TYPES OF VARIABLES USED

### 6.3.1 Variation and imprecision

The necessity for statistical comparisons in benthic studies was tempered by early investigations that emphasized the variability inherent in sampling stream fauna and flora. Needham & Usinger (1956) and Chutter (Chutter 1972; Chutter & Noble 1966) for example noted that a prohibitively large number of square-foot Sürber sample units was necessary to bring estimates of benthic invertebrate abundance to within $\pm$ 20% of the mean, and this observation was reiterated by Elliott's (1977) frequently cited handbook on statistics for freshwater benthic biologists. Hellawell (1977, 1978, 1986) further emphasized the prohibitive sample sizes necessary to achieve population estimates of $\pm$ 10% of the mean. On these grounds he opined that '[f]ully quantitative surveys are very demanding in terms of time and resources, but they are normally neither necessary nor practicable for the conventional routine methods used for assessing environmental quality' (Hellawell 1986, pp. 421–2). These influential authors focused on the *precision* of sampling (e.g. Winterbourn 1985) rather than addressing issues about the *power* of an analysis to detect a given difference (Allan 1984). This undoubtedly led to some confusion, and few studies that have involved statistical comparisons seem to have considered the issue of power explicitly.

All else being equal, a variable that can be measured with high precision is likely to be more useful than one requiring very high sample sizes to gain similar levels of precision. High precision results from low standard errors about means, which yield smaller confidence intervals; thus a given difference between control and impact locations is more likely to be detected. Commonly, highly variable data are attributed to streams being 'complex' or 'heterogeneous', but there is no reason why streams and rivers should be inherently more variable than other sorts of environments. It is more likely that the reason for highly variable samples is that the size of common invertebrate samplers (usually close to 0.1 m$^2$ or 1 square foot) is often badly mismatched with the grain over which many invertebrate species are dispersed (Downes & Keough 1998).

Mismatches between equipment and organisms are common sources of measurement error (see section 2.3 and Schneider (1994) for a general discussion). Organisms are often aggregated spatially and if sizes of aggregations are of the same order as that of sampling quadrats, or if few quadrats are collected relative to the number of aggregations through the sampling location (Andrew & Mapstone 1987; Morin 1985), then problems arise. In this situation, most values will be either zeros or high numbers, resulting in mean abundance estimates with high variance and very low precision; that is, we will have very low confidence that the sample mean is close to the actual population mean (see Andrew & Mapstone 1987 for a comprehensive discussion). The same problem can occur temporally. If organisms are much more abundant at some times than at others, and if the time over which samples are collected is of the same order of magnitude as these fluctuations, or if samples are collected relatively infrequently, then some sampling times will produce zeros and others high abundances, resulting, again, in high variance and very low precision.

High variance of variables can also result when sampling programs are mismatched with the spatial and temporal scales over which the impacts themselves occur (e.g. Underwood 1994a). For example, Mackay & Mackay (1996) found that the distribution of acid-volatile sulphides varied enormously over small spatial scales, with these spatial patterns shifting seasonally. Sampling at the wrong scales would produce imprecise and possibly misleading estimates of the degree of impact. The same difficulties occur with temporal variability as exemplified by Brewin et al. (1996), who found that water samples from acid-sensitive streams with intermediate values of pH had the highest variance. Sampling had to be more frequent at those times of the year when impacts were least and pH values were less extreme. Unfortunately, obtaining good representation across multiple spatial and temporal scales is costly and rarely achieved (Wiley et al. 1997).

### 6.3.2 Physicochemical variables as surrogates for biological variables

Physicochemical variables have a long history of use in monitoring and assessment of rivers and streams. As with biological variables, there has been some confusion about subsampling, sample precision and power to detect an effect. Physicochemical variables are often perceived as being cheaper than biological variables to collect, which has resulted in larger and longer runs of data from many rivers. However, the value of many of these data sets is questionable because, often, they have still not been collected sufficiently frequently to answer the questions of most interest to water managers or the public.

Many chemicals exist in a number of different states, not all of which are biologically active. In addition, there may be considerable interchange between labile and inactive forms resulting in a prohibitive-ly large list of compounds that need to be measured (Chapman 1996). As a result, some chemical variables can become very expensive and time-consuming to measure (e.g. pesticide residues). This can lead to too few true replicates being collected. Alternatively, less expensive, surrogate or summary variables (discussed in chapter 10) are measured instead, with attendant risks of either over- or underestimating the concentra-tion of the chemical of concern.

For long-run data sets, additional problems include changes in analytical procedures and poorly defined monitoring questions that lead to data being collected at inappropriate frequencies. As for biologi-cal variables, the fluctuations in flow (including floods and droughts) result in large changes in concentration of most chemical components. Although the invention of automatic water samplers has enabled the use of event-related sampling protocols, the intensity of sampling re-quired to characterize, for example, water quality during a single flood event can be prohibitively expensive. Additionally, because some physicochemical variables can be collected at high frequencies, serial correlation in the data may be problematic, depending on the question being addressed (Loftis *et al.* 1991). A further problem is that biological responses may occur over quite different temporal scales to the fluctu-ations detected in physicochemical variables so that the two are not statistically correlated – for example, chemicals (e.g. heavy metals) that accumulate only slowly in tissues before causing effects.

### 6.3.3 Univariate biological variables

Variability in variables measured on individuals (e.g. enzyme concentra-tions) or on populations have always posed well-recognized problems for biologists, although as we have seen they have sometimes focused on sampling precision at the expense of other important aspects of the design of their experiments or surveys.

For variables measured on individuals, one of the chief difficulties is in interpreting how important sub-lethal changes inside an animal or plant are to the population dynamics or community structure of the river in which the organism lives. In addition, innate variability both between individuals and in changes in exposure pathways in different ecosystems means that sometimes species respond differently in differ-ent habitats or locations; consequently, it is possible for there to be poor correlations between concentration of the pollutant and effects on indicator or sentinel species (e.g. Bervoets *et al.* 1997; Buikema &

Voshell 1993; Johnson *et al.* 1993; Plénet 1995; Weatherley *et al.* 1997).

Of population variables, large vertebrates such as fish have often been targeted because of their obvious appeal to the public. Although the disappearance of an abundant fish species from a reach or the occurrence of fish kills can indicate serious impacts on a river, the use of less dramatic changes in the population densities of fish or assemblage structure to detect more subtle impacts can be difficult. The high amount of year-to-year variation in fish population sizes is often unrelated even to obvious variables, and it is difficult, therefore, to detect anthropogenic disturbances by population or assemblage structure data alone (Grossman *et al.* 1990). In addition, there may be spatial variation within a catchment. Schlosser (1990), for example, noted that upstream fish assemblages in mid-western USA were more variable in their structure than those in downstream reaches; he also felt that the upstream assemblages recovered from perturbations more quickly than those downstream.

Another difficulty with using populations, which can be exemplified by the studies of fish, occurs when assumptions are made about what limits the distribution of organisms in nature. In some studies, it is assumed not only that the physical environment *per se* determines distribution and abundance, but that the mechanism behind this is 'preferences' by the organisms for 'optimal' conditions. Use of the term 'preference' means that, given an equal choice of locations, organisms will choose to inhabit particular sorts of places (Johnson 1980). Such results can be established only through the use of manipulative experiments. Preferences for particular habitats *cannot* be inferred from sampling data, which reflect the synchronous effects of multiple variables. Nevertheless, sampling data are sometimes used to infer 'preferences' of organisms like fish for particular habitats, as quantified by variables like water velocities and depths. These supposed 'preferences' are then used to predict the amount of 'optimal' habitat available under changed conditions, such as altered discharge regimes caused by water extraction (Petts & Maddock 1994). If such modelling exercises work (i.e. they successfully predict the abundance and/or location of fish in new situations) then use of the term 'preference' is mainly a semantic problem. Another term could be used without affecting the model. However, belief that sampling data really indicate something about preferences can be misleading because it suggests that organisms will inevitably be found in similar sorts of locations under different environmental conditions or at other times, neither of which is likely to be true. Belief that distributional limits reflect 'preferences' by organisms for 'optimal' environments often results from a rather naïve understanding of natural selection, and is one of a group of popular myths about nature (Fairweather 1993).

Benthic invertebrates and plants are, like fish, highly variable in abundance in time and space. However, long-term quantitative studies of these organisms are surprisingly few, and where studies have been carried on for long enough and addressed clear questions (e.g. Resh *et al.* 1988), the outcomes have been much more tractable than suggested by the gloomy predictions of some reviewers (Hellawell 1986).

There are, however, some disadvantages to using the smaller benthic species. Apart from northern Europe, western Russia and the eastern part of North America, the taxonomy of benthic invertebrates and algae is much poorer and has been an impediment to the use of these organisms (Arthington *et al.* 1982). As a result, the background natural history of benthic species is less well known, as are the presumed tolerances of these taxa to various environmental insults. An added difficulty is that often these organisms are being pressed into service as environmental monitors in rivers already subject to a variety of impacts where there are no appropriate spatial controls. The poor stratigraphy of river sediments also militates against routine use of palaeolimnological information from subfossil remains to infer historical changes.

Importantly, however, the relationships between the densities of benthic organisms and public perceptions of environmental quality (however that is defined) are unclear, at least in the public mind. Although recent appeals to conserve biodiversity for its own sake are laudable, it is doubtful that managers and the public will be as worried about a change from one species of diatom to another congener as they would be about a similar change in the species composition of fish assemblages. Thus far, ecologists have been focused on documenting changes in patterns of species density and assemblage structure without expending much effort in documenting how these changes affect ecosystem functioning (Hynes 1994). Conversely, many of the summary variables of ecosystem function or community metabolism may not change substantially while masking quite large alterations to the densities and structure of the constituent populations (Schindler 1990).

At higher levels of organization, alterations to the community in species richness, species diversity or biotic composition are used to detect a variety of impacts. The earliest forms of these were measures (sometimes called 'indices' or 'metrics') where scores were assigned to taxa so as to represent their respective tolerances to organic pollution (Wilhm 1975). There are now many varieties of such indices or metrics, some of which are ratios of one or more taxonomic groups to others as a measure of pollution (Cairns & Pratt 1993; Metcalfe 1989; Metcalfe-Smith 1994). Many indices or metrics are only useful in the regions in which they were developed, do not provide useful information on other sorts of pollution (such as heavy metals) or other sorts of human impacts, and are problematic in that scores are often assigned rather subjectively (Metcalfe-Smith 1994). Indices and metrics have also been criticized because they provide conflicting or erroneous results and lack sensitivity (e.g. Abel 1989). Nevertheless, because of the rapidity with which locations can be assessed, metrics remain an actively researched approach (Chessman 1995; Growns et al. 1995; Karr 1981, 1991; Kerans & Karr 1994). Other measures of community structure include species richness and the many varieties of diversity indices (Magurran 1988; Washington 1984). Species richness is simplistic in that it takes no account of relative abundances of different species, and diversity indices, while able to capture some shifts in relative abundances, present some problems for statistical tests (e.g. Green 1979; Norris & Georges 1993).

### 6.3.4 Multivariate response variables

To circumvent the loss of information inherent in univariate summaries of community structure, it has become more common to use measures of community or assemblage similarity to measure the resemblance of locations. The advantage is that huge location-by-species matrices, which provide the abundances of all species at all locations, can be simplified into triangular matrices of pairwise similarities, which can then be subjected to ordination or clustering techniques to summarize these data into low-dimensional displays. The researcher then searches for patterns in these displays that are consistent with hypothesized human impacts at some locations. This development has been concurrent with several reviewers' recent attempts to draw attention to improvements in conventional statistical designs (Allan 1984; Green 1979; Resh 1994; Resh & McElravy 1993). There has been much activity in applying multivariate pattern analyses for displaying biotic changes (see discussion in section 4.10). The use of clustering and ordination procedures was partly a reaction to the uncritical application of biotic and diversity indices, partly due to a perception about the high variability of stream biota (and hence, implicit low power of conventional analyses), and partly due to the promotion of these techniques as a means of displaying complex, multi-species data (Green 1979).

There have been several drawbacks to the use of these techniques. The first was the plethora of similarity measures and transformations that could potentially be applied. Little real guidance was given by handbooks (Hellawell 1978, 1986; Southwood 1978), and the necessary research to establish which measures yield realistic responses has only been completed relatively recently (Faith *et al.* 1987; Legendre & Legendre 1998). However, the use of inappropriate similarity measures persists because of the popularity of packages based on correspondence analysis and reciprocal averaging, where the similarity measure employed is hidden from the user by virtue of the implicit nature of the way these techniques work. Much has been written about the distortions that result from using these popular procedures (Minchin 1987; Tausch *et al.* 1995; Wartenberg *et al.* 1987).

The second drawback is that many of these techniques do not usually provide a formal hypothesis-test, although some authors opine that inspecting such displays is inherently a more conservative procedure than formal hypothesis-testing (e.g. Green 1979). Some researchers have compared the relative use of different sorts of community-level variables (e.g. Cao *et al.* 1996; Coimbra *et al.* 1996), but this has also sometimes been carried out without the use of objective tests. To be fair, many have used these techniques in 'post-impact impact analyses', not to test hypotheses formally but more to display data and to generate hypotheses about which species might be included in a future sampling program based on either univariate or multivariate analyses of variance.

Nevertheless, hypothesis-tests have been proposed for comparing similarity matrices (e.g. Mantel's tests; Manly 1985), clustering (e.g. Sandland & Young 1979) and ordination techniques (e.g. Pielou 1984). There has also been a recent trend to employ randomization tests that have similarities to familiar univariate techniques such as ANOVA (Clarke 1993). However, these techniques have rarely been used. Partly this may be due to their absence from the popular multivariate analytical packages, and partly because these techniques rarely extend to the more complex survey designs necessary for many studies, although this is slowly improving (e.g. Anderson & Gribble 1998). More importantly, it is difficult to specify what an effect size means in terms of a change in the value of a measure of community similarity, although Faith *et al.* (1991) have made an attempt in one experimental study.

The final drawback to the use of these multivariate techniques concerns our lack of knowledge about the appropriate models of species' responses to environmental gradients and consequently what a change in a multivariate measure means ecologically. Even the so-called 'robust' techniques assume unimodal, approximately normal responses (Minchin 1987), but there is little empirical evidence for or against this assumption for riverine flora and fauna, or even for terrestrial vegetation, which has been used as the paradigm for developing many of these techniques (Austin 1980, 1987). In addition even 'robust' ordination techniques have difficulty in recovering patterns produced by two gradients where the strength of influence of one of the gradients is not uniform over the entire range of values of the other gradient (Minchin 1987). This seems to be a multivariate equivalent of an interaction effect in ANOVA, and would, we presume, be likely for some combinations of environmental variables in rivers and streams. It appears, therefore, that there is some distance to cover before explicit formulations of hypothesis-tests can be framed unambiguously for some forms of multivariate analyses, although some progress is being made (Anderson & Gribble 1998; Legendre & Anderson 1999; see section 4.10).

### 6.4 SOCIAL, INSTITUTIONAL AND POLITICAL ISSUES

A key problem underlying many monitoring and impact assessment studies is the vagueness of the questions posed. Rosenberg *et al.* (1981) noted this flaw nearly 20 years ago, and the problem persists (Cairns & Pratt 1993; Cairns & Smith 1994; Cullen 1990a; Phillips & Rainbow 1993). Public ignorance about the nature of ecological (and hence freshwater variables) is one of these causes, and Fairweather (1993) provides a useful introduction to the mismatch between public expectations, legal frameworks and scientific issues. Rather than pretend to be anything other than armchair ecosociologists, we seek here to examine those domains of public discussion where scientists have had a role in setting the research, monitoring or assessment agenda, and ask why the managerial and administrative structures have continued to exclude clear, quantifiable and unambiguous questions.

#### 6.4.1 Difficulties caused by different backgrounds

Inappropriate views about how ecosystems operate have lead to unrealistic expectations about notions such as 'a balance of nature' and 'optimality', and ecologists themselves have contributed to public confusion on these issues (Peters 1991; Rapport 1991, 1993). At worst this confusion results in untestable hypotheses being foisted upon managers and the public (Fairweather 1993). In the last two decades, ecologists have tended

Table 6.1. *Parodies of the interests of professionals involved in managing freshwater resources*

| |
|---|
| Engineers don't care why it works as long as they think it does |
| Scientists don't care if it works or not as long as they understand why |
| Economists don't care either way if the internal rate of return is OK |
| Managers don't know unless someone bothers to tell them |
| Planners know how it should have turned out |

*Source:* Cullen (1990b).

to move away from using ideas about equilibria or balance to describe ecosystems, and have emphasized the importance of natural perturbations and the consideration of scale in measuring and manipulating ecological phenomena (see chapter 2). Unfortunately, ecologists are divided about how to translate these measures into operational criteria that might be useful to applied environmental scientists.

In the meantime, the public has demanded better environmental management, and legislators and managers have had to respond, often by using outdated notions about how ecosystems might work to provide the framework for laws and regulations. This has culminated in legislative frameworks and government directives that are often vague (Gilpin 1995; Glasson *et al.* 1994). As a result, the interplay between science and management has proved a fertile ground for mutual misunderstanding of each others' disciplines in terms of objectives, roles and outputs, which Cullen (1990b) summarizes and comments on. He parodied the different cultures amongst professionals involved in fresh waters (Table 6.1), and made the serious point that scientists need to understand managerial and public agendas if appropriate questions are to be framed (see also chapter 3). Sometimes this will mean forsaking detailed mechanistic understanding of processes in favour of purely empirical approaches. As an example of this, he cited the success of the OECD–Vollenweider models as a tool for managing eutrophication in lakes in Europe and North America (an argument further developed by Peters (1991), who also used this example). However, Cullen (1990b) does emphasize that management-oriented questions can be addressed in a Popperian scientific framework, and argues that ignorance of the appropriate spatial and temporal scales has limited the utility of scientific studies to water management so far.

Other commentators on the frequently poor relationship between water managers and scientists have blamed the overemphasis on physical, chemical and microbiological measurements at the expense of biological variables (e.g. Hynes 1960). That these physicochemical and microbiological variables receive so much attention is understandable given the public and political pressure for inexpensive, potable water. However, ensuring that water meets a guideline or standard for human use does not guarantee that the river itself meets with public and scientific conceptions of a 'healthy' ecosystem, and regulatory policies often distinguish between a variety of human uses for water as distinct from some form of 'ecosystem protection' (e.g. ANZECC & ARMCANZ 2001). Herein lies a central cause of the vagueness of the questions posed by so many assessment or monitoring studies: how do we define, operationally, a 'healthy' or 'intact' or 'high quality' riverine ecosystem (Suter 1993a; and see Box 3.1 on ecosystem health)? As we have seen, attempts have been made to measure whether the locations of interest have deviated from some reference or control condition. However, this still does not circumvent the vexed issue of defining the magnitude of the change that needs to be detected (see chapter 11). The conceptual basis of defining and measuring environmental 'condition' using criteria that take into account values other than mere potability is a confusing area in which value judgements inherent in the scientific concepts being used also play a role (Cullen 1990b; and chapter 11).

The limitations of the ecological knowledge of politicians, activists, managers and engineers aside, aquatic ecologists and chemists need to accept some responsibility for the poorly focused questions that continue to be asked. Water scientists have tended to focus on ways of measuring things rather than asking what is being measured and for what purpose. As in any human endeavour, fashion plays a dubious role. For example, the preoccupation by ecologists with fairly flimsy analogies between diversity, stability and systems theory during the 1960s and early 1970s resulted in a plethora of diversity indices and measures of community similarity. Little regard was paid to what, if anything, these indices actually measured; instead these measures were treated as 'magic bullets' by scientists and managers alike (Washington 1984). These practices did little to inform lay readers about how, exactly, a location or system had changed relative to some notional control condition. The mania for documenting patterns and generating indices continues, albeit in a multivariate guise, and led Hynes (1994) to observe that biological monitoring has not really advanced our knowledge or understanding of the ecological processes involved much beyond that apparent by the 1950s.

The way that biological water scientists have been educated about their craft has also contributed to the institutional resistance to implementing good, statistically defensible designs. Concomitant with the elaboration of ingenious indices and multivariate techniques was a curious imbalance in the way ecologists were instructed to collect data on multi-species assemblages or communities. Examination of many of the influential handbooks and reviews for aquatic studies reveals much detail about sample precision and the problems with contagious distributions for estimates of univariate variables, but little guidance about adequate sampling designs for measures of diversity, evenness or community similarity (Cassie 1971; Elliott 1977). Furthermore several prominent texts devoted more space to reviews of indices and sampling devices than to all of the aspects of the logic of survey design combined (e.g. Hellawell 1978, 1986). Indeed, some respected authorities seem to actively discourage the use of formal statistical methods (e.g. Hynes 1994). There have been some notable attempts to inform freshwater ecologists about methods and approaches that are available for dealing with some of the problems inherent in using biological variables in running waters (Allan 1984; Hall et al. 1978; Resh 1979; Resh et al. 1988), but researchers or their funding agencies have been slow to adopt their recommendations.

This inertia against adopting more modern approaches to study design is not unique to biologists. Physicochemical monitoring shares similar problems, where the frequency and timing of water sampling tends to follow precedents rather than be geared to the temporal or spatial scales of interest to managers (Loftis et al. 1991). Improvements in the value of both physicochemical and biological variables to monitoring and assessment will rely as much on asking better questions, which take the relevant scales of variation in time and space into account, as it will on novel chemical analyses or analytical techniques.

### 6.4.2 Insufficient lead time for pre-impact monitoring

Monitoring for extended periods of time prior to the commencement of a putative impact is rare. This is often exacerbated by the fact that many impacts are ongoing press disturbances that began decades before anyone bothered to start any monitoring or assessment. Even for new, large, discrete developments, few countries legally require comprehensive monitoring prior to construction (Gilpin 1995; Glasson et al. 1994). The situation is probably worse for less obvious, dispersed perturbations such as changes in land use. Those data collected under environmental impact assessment (EIA)-type legislation are often of such short duration

and of such hopelessly poor quality that the possibility of either detecting or quantifying impacts if they occur is precluded (e.g. Fairweather
1989; Rosenberg *et al.* 1981; Warnken & Buckley 1998). Consequently, we
are often likely to lack the data needed to manage even modern impacts
on the environment properly.

### 6.5  IMPORTANT ISSUES

- There have been several issues that have prevented river biologists
  from implementing designs with the strongest possible inferential base, as identified in chapter 5. Some of these issues have been
  within the control of biologists, and some are external constraints
  imposed either by the geographical peculiarities of the river under
  study, or by socioeconomic factors.
- Of those issues within the control of biologists, the most prevalent
  have been:
  * inadequate spatial and temporal controls
  * poor definition of the spatial extent of sampling locations
  * confusion between sampling precision and the power of a
    design
  * confusion about the choice of variables to be used.
- These can be addressed, in part, by recalling the goals of strong
  designs (chapter 5), choosing suites of variables carefully (chapters
  9 and 10), and applying analytical models (chapter 7) appropriately
  (chapter 8). Nevertheless, there are still some issues, of which
  spatial and temporal autocorrelations are examples, that remain
  poorly investigated for variables commonly used in riverine studies.
- Of the external geographical constraints, large rivers pose the
  most intractable problems because of the limited opportunities
  for appropriate controls. Such rivers usually also have the longest
  history of human activity and are subject to multiple perturbations, making it difficult to identify impacts unambiguously. In
  remote areas, there are additional logistic problems with often
  little or no baseline information to aid in study design.
- The socioeconomic factors which have hampered good study design include:
  * a mismatch between public expectations, legal frameworks and
    scientific issues
  * poor definition of assessment and monitoring questions

⋆ insufficient lead times before the disturbance takes place.

- Scientists themselves have contributed to the vagueness of the questions posed by assessment studies, and there is often confusion between the types of monitoring and assessment questions (clarified in chapter 3). While funding constraints are always likely to hamper good design and implementation, we hope that the advice proffered in chapters 7–13 provides a framework for more effective assessment and monitoring studies.

# 7

## Alternative models for impact assessment

164    In chapter 5, we developed a logical framework for assessing impacts, including the necessity for control or reference areas, and for sampling to occur before and after the putative impact. In this chapter, we consider practical details of the monitoring, focusing on the formal design and statistical analysis appropriate to the detection of impacts, and on the practical details associated with executing these designs in running waters. It is important to realize that we can only translate general design principles into a specific plan to collect data if we specify the statistical model that is to be fitted to the data. Perhaps the most important message of this chapter is that apparently similar monitoring 'questions' can have quite different statistical models behind them. These different models, in turn, can lead to quite different advice about how to optimize a particular data collection program.

If we consider the two major tasks outlined earlier, the formal test for the existence of an unacceptable impact and the characterization of the spatial extent of any impact, the latter procedure is relatively straightforward in terms of the design and underlying statistical models. The detailed design is modified by practical considerations associated with stream environments, and the characteristics of the activity suspected of having an impact.

In contrast, there is a range of design options for detecting an impact. These designs appear to be broadly similar to each other, but have been the subject of considerable dispute (see for example, Schmitt & Osenberg 1996, in particular the chapters by Stewart-Oaten 1996a,b, Underwood 1996 and Stewart-Oaten & Bence 2001). These different approaches reflect important and substantial underlying logical differences in the view of ecological impacts and, more importantly from a practical perspective, improving the power of each of these designs requires different measures. A change in resources that improves the power of one design will not necessarily help with an alternative design.

These conceptual models are all justifiable approaches to the detection of impacts under particular circumstances, but it is *essential* that anyone implementing one of these designs be aware of the differences between them, and of the important characteristics of each of them.

Below, we describe the background of each approach, the formal question (or conceptual model) being used in each case, the statistical models that are fitted to the data, and the steps that should be taken to increase power in each case. One of the models was derived and recommended by two of the authors of this book (Keough & Mapstone 1995, 1997), so our final recommendation should be considered with that in mind, and we encourage readers to consult the references listed earlier in this chapter, for alternatives. We also describe only the most common formulations of each approach. Environmental data are often messy, and testing for impacts requires flexibility in approaches, to deal with data that may show unusual natural fluctuations and with impacts that may take on forms other than a discrete change through time. The need for flexibility has been emphasized by many authors who have written on the subject, including each of the major approaches described below (Keough & Mapstone 1995; Stewart-Oaten & Bence 2001; Underwood 1996), and a more detailed advocacy of such flexibility is provided by Hilborn & Mangel (1997).

In the formal discussion of models, we use some conventions for terminology, because descriptions of components of space and time are used quite variably in the existing literature. We assume that the time is divided into two major **Periods**, Before and After. Within these periods are **Times**, which can themselves be decomposed into intervals at different temporal scales, such as years, seasons etc. Larger spatial areas are divided into two major groups, to indicate impacts, **Control** and **Impact**. Within each of these groups are **Locations**, which are larger spatial units that may be different areas in which the same kind of human activity occurs (i.e. true *replicates* of the impact), or independent areas that serve as comparison to impacted areas (i.e. *replicate* Controls). These areas are presumed to vary independently of each other. Within each Location, there may be subsampling. We refer only to **Sites**, to encompass a range of spatial units within Locations, and, as for time, Locations may be subdivided in a wide range of ways. The scales on which Times and Locations occur will be determined by a combination of the kind of human activity and its expected area of influence (e.g. a mixing zone for discharge of effluents), the scale on which management decisions are to be made, and the characteristics of the biota (their mobility, dispersal etc.).

We have identified four major approaches to the design of impact sampling programs (see Table 7.1). They arose out of vigorous debate about appropriate sampling programs, largely in marine environments. In describing these designs, we follow a sequence of increasing complexity, with the result that they are not presented in any chronological order.

### 7.1.1 BACI

In early discussions of impact monitoring (see Green 1979), samples were taken at single Impact and Control locations, and once Before and After a particular human activity started. The relative states of Impact and Control were compared between these two time periods (i.e. the **interaction** between Before and After, and Control and Impact, hence BACI). There was no replication in time or space at the appropriate scale; any apparent replicates served only to estimate the variation within a location or time, rather than as measures of the variation among repeated Control or Impact locations, or at other times. As discussed in chapter 5, these sampling programs provided only very weak inference about impacts, and we do not describe these designs in detail.

#### BACIP

A number of authors recognized the necessity for at least more detailed sampling through time, in order to demonstrate that any change occurring at the impact location was not simply a random temporal fluctuation, but was associated with the particular human activity. The solution was to provide estimates of background temporal fluctuations both before and after the activity started.

The simplest sampling design used for this purpose was the BACIP, incorporating sampling Before and After, at Control and Impact locations, with samples Paired in time. This design was suggested simultaneously by a range of authors (Bernstein & Zalinski 1983; Green 1979; Stewart-Oaten et al. 1986). Although Green's book preceded the other two publications, the BACIP design was developed for the San Onofre Nuclear Generating Station in the early 1970s, and appears in early reports from that project (see Murdoch et al. 1989 for a complete chronology). The philosophy behind this design is that the putative impacted area is compared to a single reference, or control location, chosen for its similarity to the impacted location. The design was used also because its advocates considered that the control and impact areas were rarely chosen at random from any larger set of locations; instead, the impact location was often chosen for its special properties, with the control then being selected as one that matched the impact area most closely. The two locations were not representative of any wider group of locations. Because the variable(s) of interest could have different average values at these two locations, a BACIP program focuses on any changes at the Impact location, *relative to* the Control (see Box 5.1), and the variable that is analysed is the *difference* between Control and Impact values.

Sampling through time is used to estimate the temporal variation in these differences, and this variation is used to assess the average difference Before and After the activity commences. Times are separated enough to prevent autocorrelation in these differences, and are viewed as a random sample of possible values in each time period, and a part of the formal analysis is a preliminary test for temporal autocorrelation.

*Intervention Analysis*

In his most recent discussion of BACIP designs, Stewart-Oaten (Stewart-Oaten & Bence 2001) regards the BACIP design as a special version of Intervention Analysis (Box & Tiao 1975), in which there is no Control, and a long time-series of data from the Impact location Before and After. In this analysis, the Before data are considered sufficient to characterize the Impact location. Stewart-Oaten & Bence (2001) argue that a Control (or the mean of a set of Control locations) is useful when such time-series are short. They also argue that the Control is more appropriately seen as a covariate, used to eliminate signals from more widespread (natural) environmental fluctuations, rather than a standard against which to assess the impact. This is a very different interpretation of the BACI and BACIP designs from that used in the past, and in wide use today.

A modification of the BACI analysis, termed Randomized Intervention Analysis (RIA), was proposed by Carpenter *et al.* (1989). They determined the distribution of their test statistic, in this case the difference Before and After the impact of the means of the Control–Impact differences, by randomly rearranging the Control–Impact differences along the time sequence. If the null hypothesis is true, then any of these possible arrangements should be equally likely and the value of the test statistic obtained from the actual data can be compared to its randomized distribution just as in a usual statistical hypothesis-test. This is the

Table 7.1. *The four basic designs used to test for discrete impacts. Definitions of fixed versus random factors are provided in chapter 4*

| | BACI | BACIP | MBACI(P) | Beyond-BACI |
|---|---|---|---|---|
| Control and Impact | 1 each | 1 each | Many | Many, hierarchical |
| Times | 1 | Many | Many | Many, hierarchical |
| Comments | Constrained design, giving weakest inference. Used in desperation. Only possible conclusion is that something happened | Impact assessed relative to standard, not population. Locations fixed. Times must be random | Impact relative to population, not reference. Locations random. Time (at highest level) fixed | Impact relative to population, not reference. Locations random (at all levels). Times random (at all levels) |
| Treatment of space | Fixed. Only two locations of interest | Fixed. Only two locations of interest | Random. Locations are sample of available areas | Random. Locations are sample of available areas |
| Treatment of time | Fixed. Only two times of interest | Random. Times are random selection from periods | Fixed. All times (e.g. years) within period sampled | Random. Times are random selection from periods |
| Replication | Only subsamples within each location. Do not strengthen inference greatly. The comparison is still two places at two times | Times within each period. Subsamples at each location useful only for characterizing location better. Minor effect on power in most cases | Locations. Times used to characterize Before and After; do not appear in analysis. Subsamples at each location have minor effect in characterizing location–time better | Locations at each spatial scale. Times at each spatial scale. Subsamples (at each level of hierarchy) have minor effect |
| Underlying logic/question | Was the Impact location the same, relative to the Control on the two times they were sampled? | Did a change occur at Impact location relative to Control location, in a way that was unexpected, given background of temporal change at the two locations? | Did the Impact location(s) change relative to the group of Control locations, in a way that was unexpected, given the background pattern of changes among these control locations? | At a given scale of space or time, did the Impact location(s) change relative to the group of Control locations, in a way that was unexpected, given the background pattern of changes among these control locations? |

| | | | | |
|---|---|---|---|---|
| Design considerations | May change replication within sites, but only a minimal effect on quality of decision | Selection of Control location is critical. More times = more power. Autocorrelation a potential problem: if times too close together. Within-site replication has diffuse effect on power | Population of control locations should show as little variation as possible, and be closely matched to the impact locations. Maximum improvements in power from increasing number of control sites. More times or more within-location sampling have only diffuse effect on power | Costly; collect more data than you needed. Scale/extent of impact uncertain. Maximal improvements in power by increasing controls and number of times. Power increased by more spatial or temporal sampling, but is optimized separately at each level of hierarchy |
| Value/utility | May be used in subsequent meta-analysis to provide information for future management | Appropriate choice when only one control is possible. Most useful when Control vs. Impact difference is well-behaved (i.e. two locations track similarly in the absence of the human activity) | Simply the best when the spatial scale of impact can be defined, either by accurate predictions of the kind of impact or a management decision about kinds of change to be regulated | Provide lots of information at multiple (discrete, possibly arbitrary) scales; most use as guide to better distributing effort for estimation of unknown effects |

randomization test procedure described in chapter 4 (see also Manly 1997). Carpenter *et al.* (1989) argued that by using the randomization approach, they overcame the assumptions of normality, variance homogeneity and uncorrelated error terms through time. However, Stewart-Oaten *et al.* (1992) disagreed that the randomization test avoids these assumptions, presenting evidence that the test proposed by Carpenter *et al.* (1989) is sensitive to variance heterogeneity and correlated errors. RIA actually has its origins in interrupted time-series analysis, which tests whether a trend through time before some intervention or interruption is the same as the trend after the intervention (Box & Tiao 1975; Rasmussen *et al.* 1993). These techniques are primarily designed for unreplicated time-series (e.g. no control) and allow more complex temporal trends that could not easily be summarized by mean values before and after an intervention.

### 7.1.2 MBACI

A central issue that arose in the marine literature concerned the necessity for multiple control or reference areas. This represents a fundamental shift in the way that impacts are defined, as well as a fundamental shift in practical considerations, and is common to the MBACI and Beyond-BACI designs. The difference is whether an impact should be defined as a set of changes that occur when a particular human activity occurs, compared to natural events occurring at similar sites within a region, or whether each case study occurs in a specially chosen location that is not drawn from any population of broader locations, and can only be assessed against a reference standard.

As we have argued in chapter 5, we believe that stronger inferences are possible when multiple locations are used. The question in this design is whether the population of impact locations differs from the population of control locations, specifically whether the two populations changed in the same way before and after the particular human activity commenced.

In the formulation of this design by Keough & Mapstone (1995), the sampling program covers a specific period. In most sampling programs, there is a fixed period before the activity starts, and (ideally, at least) a comparable period following the commencement of the activity. The question is whether the control and impact locations differ over these two periods. Within each period, it is likely that sampling will be annual (if the length of the period permits) and may be seasonal, with possibly more subsampling within these times. The larger time scale (e.g. years) is sampled exhaustively (i.e. all years available are used) so the formal question is whether the control and impact groups changed over those *two* periods, with the periods assessed by sampling all possible times.

There are some variants of this design that pose slightly different questions – the most common variant is if the control and impact locations consist of a series of matched pairs. Such a design might be used when there are multiple impact locations that are physically well separated, and nearby controls are thought to be better bases for comparison. In this case, the question is whether control and impact pairs converge or diverge following the start of the activity.

In this design, we have some idea of the likely spatial extent of any important impact, and/or the spatial scale on which management responses can or will occur. These considerations dictate the location of samples taken at the impact location(s) and the scale at which controls are placed.

### 7.1.3 Beyond-BACI

There are situations in which we have, a priori, no accurate prediction of the scale of impact in space or the rate at which the impact will become apparent, and more extensive sampling may be required. The Beyond-BACI design was developed to deal with this difficulty, as well as two other problems. First, the designs above deal best with a discrete change in the *average* condition of Impact and Control areas following the activity (i.e. a step change). Underwood (1991a, 1992, 1993, 1994a, 1996) has argued that human activities may also modify variances, rather than means, of variables of interest, and that such changes are not detected well by existing designs. Transient (pulse) responses may also require additional designs (Glasby & Underwood 1996). Additionally, our predictions of the extent of impact may be very imprecise, and Underwood has argued that spatially and temporally hierarchical sampling will allow the detection of impacts at a range of scales.

The Beyond-BACI design shares with the MBACI its view that impacts are detected between Impact and Control populations, with samples of those populations providing the actual locations to be followed through time. It differs, however, from the MBACI, in viewing times (at all scales) as samples from a larger population of times. This difference is most apparent at the scale of years – in the MBACI, those years represent the entire Before and After periods, but in the Beyond-BACI, they are only a sample from all the years in much longer periods.

In assessing the impact, changes at a particular spatial and temporal scale are assessed against the background variation seen at the next level down in the hierarchy of time and space.

## 7.2 THESE APPROACHES ARE DIFFERENT!

A reader who has read the above descriptions, and consulted the original papers, may be puzzled that apparently subtle differences in the questions posed by each design are associated with such vigorous debate, and may think that the sampling can proceed, with the appropriate model chosen at the time of analysis. This view is incorrect, and these different questions are associated with fundamentally different statistical models. As a consequence, the steps taken to generate the most reliable decisions, most often via power analysis, are quite different for each of these designs.

As shown in section 7.8, the BACIP design differs fundamentally from MBACI and Beyond-BACI in its insistence that there are no populations from which Control and Impact areas can be selected, just locations chosen for specific reasons. There is less generality to be gained from a study using this design, because, by definition, it is a special case. We do not agree with this view, but do urge interested readers to consult Stewart-Oaten's papers, particularly his two 1996 papers (Stewart-Oaten 1996a,b) and, most importantly, Stewart-Oaten & Bence (2001). An important consequence of this decision is that the reliability of the sampling program depends primarily on the number of times in each period.

MBACI and Beyond-BACI differ primarily in their treatment of time. While the Beyond-BACI design incorporates a range of spatial and temporal scales, at the highest level of the sampling program (i.e. at the largest resolution of time) the two designs involve sampling at longer intervals, from multiple Control and (preferably) Impact locations, and have multiple times Before and After the particular activity commenced. In most biological sampling programs, the largest time-scale will be years, as many biological phenomena are driven by annual events (reproduction, migration etc.). In the case of MBACI, it is assumed that, in practice, sampling will encompass all possible years prior to the activity commencing, and also at least a series of consecutive years following the activity. In this sampling program, those years represent the entire period of interest – we are trying to make a decision about what happened around the time that the activity commenced, based on what happened in the periods immediately before and after that time. The few years before are not intended to represent a longer time period before the activity started. In contrast, in the Beyond-BACI design, the times that are sampled are presumed to be (random) samples from the Before and After periods, and therefore to be representative of longer time periods. This is an ideal that is rarely reached. Note that if the MBACI design were to be expanded to consider multiple temporal scales, we agree with Underwood that smaller time-scales could involve random samples of available times, in which cases the two approaches could converge.

### 7.2.1 Why it matters

The distinction between these different models is not pedantry – these models represent different conceptual views of the sampling program. They may produce different answers if applied to a particular data set, and it would be easy to dismiss variable answers as another case of scientists providing vague or conflicting advice. We emphasize that this is not the case – each design deals with a precise view of how an impact is assessed, and has associated with it an equally precise statistical model, designed to answer that question. It should not be surprising that changing the question changes the answer. Each of these designs tests for impacts in a different way, with different power characteristics.

Our choice of a particular model will depend on how we regard time and the nature of control and impact locations. A decision about times is a difficult one. Our view in this book is that when sampling is done for all possible times (for a given scale, like years), then that temporal factor should be regarded as fixed (see section 4.6.1) in any subsequent analysis. We are convinced by two arguments. First, statistical inferences are made from random samples from populations, and a series of consecutive years, for example, or an exhaustive sampling of any time period is not a random sample, but a systematic sample. Because we cannot identify a 'population' of times from which we have sampled, our data represent only themselves. Second, in more formal linear model treatments, the formal definition of whether a factor is fixed or random depends on the ratio of the number of levels ($p$) of that factor that were used to the number of possible levels ($P$). For a fixed factor, $p/P \approx 1$, whereas for a random factor, $p/P \approx 0$, reflecting an infinite number of levels. When this definition is applied, time should clearly be fixed.

There are two counter-arguments, neither of which we accept. Underwood (1997) has argued that the distinction between a fixed and random factor should be governed by whether the same levels would be used were the study to be done again. Our view is that an impact study concerns only a particular point in time, so that question is moot. We also believe that we would sample all years again in any new study (if years was the appropriate larger time scale). Stewart-Oaten (1996a) has argued that in a monitoring situation, we see particular trajectories through time at Control and Impact locations. While these may represent all observation times, he argues that there is a range of alternative trajectories that might be observed were the events to be repeated. Because we have seen one of a large range of possible trajectories, he treats times as a random sample. We find this argument hard to follow, and, if anything, his argument provides more support for multiple Control locations (at which we could observe alternative trajectories through time), than for a reconsideration of time.

### 7.3 FORMAL SAMPLING AND ANALYTICAL FRAMEWORK

The multifactorial (treatments, locations, sites, times etc.) nature of measurements likely to be taken in monitoring programs, and the likelihood that dichotomous decisions will be expected from such monitoring programs, suggests the use of one or more linear models (often analysis of variance; ANOVA) procedures for the statistical analysis of data. As a statistical decision-making framework, ANOVA perhaps provides the most flexible, robust and powerful set of hypothesis-testing procedures available. Inevitably also, this recommendation is affected by our experience and biases. Using this framework for illustrative purposes requires that a reader be familiar with linear models and with variance components. These ideas were introduced in chapter 4, where you can also find the descriptions of the symbols and conventions used to describe these statistical models (Box 4.5).

ANOVA is a widely used and well-described, but relatively restrictive statistical method. It is particularly useful when the arrangement of sampling units and the distribution of samples through time are under human control – designed experiments – and in its most common implementation requires data that are distributed approximately normally, with consistent variances etc. In practice, monitoring data often follow different statistical distributions, requiring either data transformation or fitting of equivalent statistical models with non-normal distributions. While monitoring may correspond to a designed experiment, a major theme of this book is the constraints and unexpected events that often occur in practice. We emphasize, therefore, that

ANOVA is used here as an illustration. There is a wide range of alternative statistical approaches – GLMs (see section 4.9), Generalized Additive Models (GAMs), log-linear models for frequency data, randomization tests (section 4.9), tests for multivariate data (section 4.10) etc. – that may be more appropriate for particular situations. It is important to use the most appropriate statistical tool for the particular purpose, rather than forcing a data set into one particular statistical framework.

In this chapter, we use the statistical details as a way of illustrating the consequences of differences in logic, and we emphasize that all statistical analysis requires fitting a statistical model to data, and that this step involves taking a relatively informal monitoring question and translating it into formal statistical terms.

### 7.3.1 The sampling program

*BACIP*

Sampling in a BACIP design is relatively simple. There are only two locations, each of which is sampled through time, with the samples being paired (i.e. each location is sampled at the same time intervals). The analytical models described below require that the times be a random sample of the possible times in the Before and After periods. Logically, this means that sampling at regular time intervals should *not* occur.

Although there is a formal requirement for randomly spaced times, in practice sampling is often done at regular intervals, with these samples then being treated as a random set (e.g. Murdoch *et al.* 1989, for the San Onofre Nuclear Generating Station study, when most variables were sampled quarterly).

It is also important that the samples through time be spaced far enough for there to be no autocorrelation, so there will be some minimal interval between sampling times, determined by the dynamics of the particular biological variables under measurement. For example, microbial assemblages may change rapidly on a scale of hours, whereas samples of riparian vegetation may require long time periods for some mortality and recruitment to have occurred.

The sampling program may involve multiple samples taken at each location and used to provide a better estimate of the mean or variance at each location than would be provided by a single small sample. These samples could be combined at the processing stage (i.e. a composite sample), prior to analysis (by numerically averaging values), or as a part of the analysis (again, by averaging or summing values for individual samples). The decision on when to combine the samples will depend on the cost of the different stages.

We do note that an alternative approach to BACIP designs treats times not as a random sample, which, strictly, they are not, but as a time-series (e.g. Stewart-Oaten & Bence 2001). For illustration in this chapter, we use the former approach.

### MBACI

Sampling in the MBACI case consists of replicate Control and Impact locations, sampled at a series of times. We assume that all locations are sampled at the same times. The times can be separated into years, then possibly seasons, or other, shorter time periods, within years. If they are used in the analysis, the same cautions about autocorrelation as for the BACIP case should be considered.

The sampling program, as for BACIP, may involve subsamples.

### Beyond-BACI

The sampling program for Beyond-BACI has been described in considerable detail by Underwood (1991a, 1992, 1993, 1994a, 1996) and we will not repeat it here. Its essence is the sampling of multiple locations, for extended periods before and after the activity commences. There are various levels of subsampling (i.e. at smaller spatial and temporal scales), the levels of which will be set by the characteristics of the particular environment, the degree of certainty with which predictions of extent of impact can be made, and the scales on which management activities will occur. All of these should be clearly specified in any particular sampling program, but there is no overall prescription.

### 7.3.2 The analytical models and formal hypotheses

#### BACIP

The BACIP design is most easily analysed by treating the variable of interest as $d_{pj}$, which is the difference between the observations at the Control and Impact locations at a particular time $j$, within either the Before or After period ($p$).

The model is then:

$$d_{pj} = \mu + \mathbf{B}_{p.} + \varepsilon_{pj}$$

where

$d_{pj}$    is the $j^{th}$ observation of the difference between Control and Impact in the period $p$ Before or After the activity

$\mathbf{B}_{p.}$    is the average **difference** between Control and Impact, Before or After the activity starts

$\varepsilon_{pj}$    is the residual value after accounting for the above effects (see chapter 4).

The same design could be formalized without computing the differences between Control and Impact, and using data from Control and Impact locations at each time as the variable to be analysed. In this case, the formal statistical model must take into account the paired nature of the data, by explicitly including sampling times in the model, which will be:

$$y_{ijpm} = \mu + \mathbf{C}_{i...} + \mathbf{B}_{..p.} + \mathbf{T(B)}_{jp.} + \mathbf{CB}_{i.p.} + \mathbf{CT(B)}_{ijp.} \, ( + \varepsilon_{ijpm})$$

where

$y_{ijpm}$    is the $m^{th}$ observation at Control or Impact sites $i$ at Time $j$ in the Before or After period $p$

$\mu$    is the population grand mean for the measured variable

$\mathbf{C}_{i...}$    is the time-averaged effect of being in the Control or Impact treatment

$\mathbf{B}_{..p.}$    is the spatially averaged effect of the period Before or After the activity starts

$\mathbf{T(B)}_{jp.}$    is the spatially averaged effect of Time $j$ within Before or After period $p$

$\mathbf{CB}_{i.p.}$    is the effect of being in either the Control or Impact treatment either Before or After the commencement of the activity (i.e. an interaction effect)

$\mathbf{CT(B)}_{ijp.}$    is the effect of being in either the Control or Impact treatment at Time $j$ within either the Before or After periods

$\varepsilon_{ijpm}$    is the residual value after accounting for the above effects (and estimable only when multiple subsamples are taken within each location at each time).

In practice, this latter formulation would rarely be used, as it is unnecessarily complex. If replicate samples were taken at each site–time combination, they could be averaged, and then the difference between Control and Impact calculated. The two formulations are computationally identical, but the simpler equation above can be analysed with a $t$ test, rather than requiring a partly nested ANOVA. The full expression of the model makes it explicit, however, that the data can *not* just be analysed with a replicated two-factor ANOVA (Control–Impact × Before–After).

## MBACI

For the MBACI design shown above, the model used for the ANOVA is (following Keough & Mapstone 1995):

$$y_{inmpj} = \mu + \mathbf{C}_{i...} + l(\mathbf{C})_{in...} + \mathbf{B}_{..p.} + \mathbf{T(B)}_{..pj} + \mathbf{CB}_{i..p.} + \mathbf{CT(B)}_{i..pj}$$
$$+ l(\mathbf{C})\mathbf{B}_{in.p.} + l(\mathbf{C})\mathbf{T(B)}_{in.pj} \left( + \varepsilon_{inmpj} \right)$$

where

| | |
|---|---|
| $y_{inmpj}$ | is the $m^{th}$ observation at location $n$ in Control or Impact treatment $i$ at Time $j$ in the Before or After period $p$ |
| $\mu$ | is the population grand mean for the measured variable |
| $\mathbf{C}_{i...}$ | is the average value for Control or Impact, averaged across times and locations |
| $l(\mathbf{C})_{in...}$ | is the time-averaged effect of being at the $n^{th}$ location within treatment $i$ |
| $\mathbf{B}_{..p.}$ | is the spatially-averaged effect of the period Before or After the activity starts |
| $\mathbf{T(B)}_{..pj}$ | is the spatially-averaged effect of Time $j$ within Before or After period $p$ |
| $\mathbf{CB}_{i..p.}$ | is the effect of being in either the Control or Impact treatment either Before or After the commencement of the activity |
| $\mathbf{CT(B)}_{i..pj}$ | is the effect of being in either the Control or Impact treatment at Time $j$ within either the Before or After periods |
| $l(\mathbf{C})\mathbf{B}_{in.p.}$ | is the effect of being at the $n^{th}$ location in either the Control or Impact treatment in either the Before or After period |
| $l(\mathbf{C})\mathbf{T(B)}_{in.pj}$ | is the effect of being at the $n^{th}$ location in either the Control or Impact treatment at Time $j$ within either the Before or After periods |

$\varepsilon_{inmpj}$    is the residual value after accounting for the above effects (and estimable only when multiple subsamples are taken within each location at each time).

### A special case: MBACI with a single Before and single After sample

If there is no repeated sampling of locations within the main periods, the statistical model can be simplified considerably. The greatest simplification is obtained by calculating the change at each location,

$$d_{in} = l(\mathbf{C})\mathbf{B}_{inB.} - l(\mathbf{C})\mathbf{B}_{inA.}$$

where B and A denote samples before and after the activity started.
The model is then:

$$d_{in} = \mu + \mathbf{C}_{i.} + \varepsilon_{in}$$

where

$d_{in}$    is the observed change at location $n$ in Control or Impact treatment $i$

$\mu$     is the population grand mean for the measured variable

$\mathbf{C}_{i.}$    is the spatially averaged effect of being in the Control or Impact treatment

$\varepsilon_{in}$    is the residual value after accounting for the above effects, and estimable only when there are multiple locations.

In theory, this model can be tested with a $t$-test, but only when there are multiple Control *and* Impact locations. In the common case in which there are only single Impact locations, there is no estimate of the variance in $d$ values among Impact locations, and any test of impact can be made only if it is assumed that the variance of $d$ values at Control locations is a good estimate of the overall variance (i.e. it is assumed that variances are homogeneous between Control and Impact locations).

This model for this special case is conceptually very similar to the BACIP design, with in one case replication coming from samples through time (at two locations) and in the other from samples through space (at two times).

### Beyond-BACI

Underwood's Beyond-BACI designs are extremely complex, and the exact form of the model depends on the number of scales of time and space that are incorporated into the design. The model will also vary considerably if all locations are sampled simultaneously, compared to the situation in which sampling times are distributed randomly among locations. Underwood (1991a, 1992, 1993, 1994a, 1996, 1997) has described the analysis of these designs in considerable detail, and here we present only a very simplified model, to illustrate the differences in his approach. Anyone considering one of these designs must consult those papers or a professional statistician, because they are complex, with great potential for mistakes to be made.

We consider the simplest case, with multiple locations, and multiple times (at a single scale) within the Before and After periods. We consider only a single level of subsampling, that of replicate samples within each location–time combination. The model for this situation is the same for the MBACI:

$$y_{inmpj} = \mu + \mathbf{C}_{i...} + l(\mathbf{C})_{in...} + \mathbf{B}_{...p.} + \mathbf{T(B)}_{...pj} + \mathbf{CB}_{i..p.} + \mathbf{CT(B)}_{i..pj}$$
$$+ l(\mathbf{C})\mathbf{B}_{in.p.} + l(\mathbf{C})\mathbf{T(B)}_{in.pj} \, ( + \, \varepsilon_{inmpj})$$

where

| | |
|---|---|
| $y_{inmpj}$ | is the $m^{th}$ observation at location $n$ in Control or Impact treatment $i$ at Time $j$ in the Before or After period $p$ |
| $\mu$ | is the population grand mean for the measured variable |
| $\mathbf{C}_{i...}$ | is the time-averaged effect of being in the Control or Impact treatment |
| $l(\mathbf{C})_{in...}$ | is the time-averaged effect of being at the $n^{th}$ location within treatment $i$ |
| $\mathbf{B}_{..p.}$ | is the spatially averaged effect of the period Before or After the activity starts |
| $\mathbf{T(B)}_{...pj}$ | is the spatially averaged effect of Time $j$ within Before or After period $p$ |
| $\mathbf{CB}_{i..p.}$ | is the effect of being in either the Control or Impact treatment either Before or After the commencement of the activity |
| $\mathbf{CT(B)}_{i..pj}$ | is the effect of being in either the Control or Impact treatment at Time $j$ within either the Before or After periods |
| $l(\mathbf{C})\mathbf{B}_{in.p.}$ | is the effect of being at the $n^{th}$ location in either the Control or Impact treatment in either the |

Before or After period

$l(C)T(B)_{in.pj}$ is the effect of being at the $n^{th}$ location in either the Control or Impact treatment at Time $j$ within either the Before or After periods

$\varepsilon_{inmpj}$ is the residual value after accounting for the above effects (estimable only when multiple sub-samples are taken within each location at each time).

### 7.3.3 Tests for Impact

*BACI and BACIP*

The BACI test is simple, as the design only includes one time Before and After. The test of impact uses variation within locations as an estimate of background variation against which to contrast the two locations before and after the activity. Formally, a replicated two-factor ANOVA would be used, with the test of interest the Before–After × Control–Impact interaction, tested with an F-ratio having 1, 4 $(n-1)$ degrees of freedom, where $n$ represents the number of samples per location (assuming equal sampling effort for each location; Box 7.1). There are no design options.

In the case of BACIP sampling, the simplest test comes from assuming that the data in the analysis are differences between Control and Impact locations at each time, and the model described above can be described as a one-factor ANOVA (Before–After), with the replicates being times within each period. Because there are only two locations, the analysis can, and probably should be, done using a $t$ test (remembering that $F_{1,n} = t^2_n$), which is simpler, and understood by a broader audience. The test is simply whether the difference between Control and Impact locations was the same in both periods, and the degrees of freedom for the $t$ test depend on the number of sampling times.

If the more complex formulation is used, the model can be depic-

---

**Box 7.1**

**ANOVA models**

To illustrate the appropriate tests for impacts, we show the tables of Estimated Mean Squares (chapter 4), labelled 'MS estimates', and the denominator used to construct an F-ratio to test a particular effect. We indicate the appropriate source of variation (Denominator MS) and its degrees of freedom (Denominator df).

Table 7.2. *Components of variation for a three-factor ANOVA (Control–Impact, Before–After and Times), with no replication, in BACI/BACIP designs*

| Source of variation | Designation | Number of levels[a] |
|---|---|---|
| *Spatial variation* | | |
| Impact–Control – (C) | Fixed | 2 |
| *Temporal variation* | | |
| Before–After development – (B) | Fixed | 2 |
| Times within Before–After – T(B) | Random | $t'$ |
| Impact–Control × Before–After – CB | Fixed | 4 |
| Impact–Control × Times(B) – CT(B) | Fixed | $2t'$ |

[a] The number of levels is shown for each of the factors. $t' = t_B + t_A$, where $t_B$ and $t_A$ are the numbers of times sampled Before and After start-up, respectively.

ted as a three-factor ANOVA (Control–Impact, Before–After, and Times), with no replication (Table 7.2). Note that there are some replicates (viz. samples within each location) and it is possible to include them in the model. They do not affect the test of impact, and the test for impact using the means of those subsamples is identical to that which we would get by adding replicates into the model. Converting data to means before analysis is far simpler. The general form of the ANOVA appropriate to the model, as depicted above, is given in Table 7.3.

### MBACI

The model for this particular design presumes that each location was sampled at the same time (although there is an equivalent model for when they were not). The simplest way to do the analysis is to use as the input data the averages of data collected from each location at each time. Then, the model described above can be depicted as a three-factor ANOVA (Treatment × Before–After development × Times), with repeated measurements of replicate locations within each treatment. (Note that the term $\varepsilon_{inmpj}$ could not be estimated in such an analysis, but, as we discuss later, it is not necessary for analyses of the terms of most interest for monitoring impacts.) In practice, however, the model usually becomes unbalanced because there will be only a single impact location. The only available estimate of within-treatment variation at each time, therefore, is the observed variation among control locations. Accordingly, the a posteriori comparisons of interest are derived from the repeated

Table 7.3. *The general form of ANOVA appropriate to a BACI/BACIP model, when analysed using ANOVA, rather than t test*

| Source of variation | df | MS estimates | F-ratio Denominator MS | Denominator df |
|---|---|---|---|---|
| *Spatial variation* | | | | |
| C | 1 | $\sigma_\varepsilon^2 + 2\sigma_{CT}^2 + 2t'\delta_C^2$ | CT(B) | $t' - 2$ |
| *Temporal variation* | | | | |
| B | 1 | $\sigma_\varepsilon^2 + 2\sigma_T^2 + 2t'\delta_B^2$ | T(B) | $t' - 2$ |
| T(B) | $t' - 2$ | $\sigma_\varepsilon^2 + 2\sigma_T^2$ | CT(B) | $t' - 2$ |
| CB | 1 | $\sigma_\varepsilon^2 + 2\sigma_{CT}^2 + t'\delta_{CB}^2$ | CT(B) | $t' - 2$ |
| CT(B) | $t' - 2$ | $\sigma_\varepsilon^2 + 2\sigma_{CT}^2$ | — | — |

*Note:* For the purposes of testing for variation among times within the Before and After periods (which is not of much interest), the $\sigma_{CT}^2$ must be assumed to be 0. If we had taken replicate samples within each location, at each time, we could have included a residual term, $\sigma_\varepsilon^2$, in the model and analysis, enabling us to test the hypothesis that $\sigma_{CT}^2 = 0$.

The test for impact is the CB term, which is tested using the CT(B) term, which has 1, $t' - 2$ (or $t_B + t_A - 2$) degrees of freedom.

As an F-ratio with one degree of freedom in the numerator, this test is also equivalent to a *t*-test, as described in chapter 4.

measurements at a single impact location and averages of repeated measurements at several control locations.

This approach allows the separation of variation into a number of independent components (Table 7.4). The general form of the ANOVA appropriate to the model, as depicted above, is shown in Table 7.5.

In Table 7.5, $\bar{h}_l$ and $\bar{h}_t$ are the harmonic mean numbers of locations within Control and Impact treatments, and times within the Before and After periods, respectively. We use the harmonic mean to allow for the situation in which there are different numbers of times or locations within each group. With equal numbers, arithmetic means can be used.

Tests for differences among locations or the interaction between locations and times would be possible if all data from within locations were not averaged to produce one datum (mean) per location per time. For clarity, we have not presented subsampling terms here, but they would simply appear as additional nested terms under Locations(C) and in interaction with the temporal terms (e.g. see Underwood 1991a, 1993, 1994a).

Table 7.4. *The components of variation in an MBACI model*

| Source of variation | Designation | Number of levels[a] |
|---|---|---|
| *Spatial variation* | | |
| Impact–Control – (C) | Fixed | 2 |
| Locations within Impact–Control – L(C) | Random | $l'$ |
| *Temporal variation* | | |
| Before–After development – B | Fixed | 2 |
| Times within Before–After – T(B) | Fixed | $t'$ |
| Impact–Control × Before–After – CB | Fixed | 4 |
| Impact–Control × Times(B) – CT(B) | Fixed | $2t'$ |
| Locations(C) × Before–After – L(C)B | Random | $2\,l'$ |
| Locations(C) × Times(B) – L(C)T(B) | Random | $t'\,l'$ |

[a] $t' = t_B + t_A$, where $t_B$ and $t_A$ are the numbers of times sampled Before and After start-up, respectively. $l' = l_C + l_I$, where $l_C$ and $l_I$ are the numbers of Control and Impact locations sampled, respectively.

The principal effects of interest for impact assessment are:

1. The interaction between Impact and Control treatments and periods Before and After development (Green 1979, 1993)
2. The interaction between the Impact and Control treatments and Times within the Before and After periods (the terms CB and CT(B), respectively, in the ANOVA table in Table 7.5).

The sources of variation *within* these interactions against which departures from the null hypothesis would be tested (the error variances) are (1) the interactions between Locations(C) and Before–After; and (2) the interaction between Locations(C) and Years(B), respectively. Hence, the key $F$-ratios of interest in an ANOVA of this form are the ratios $MS_{CB}/MS_{L(C)B}$ and $MS_{CT(B)}/MS_{LT(B)}$. Possible longer-term impacts of the impact in question would be indicated by significant CB interactions. Shorter-term impacts from which the measured variables quickly recovered would be expected to cause a significant CT(B) effect, and be manifest as changes in the Time profile of the Impact location between the Before and After periods that were not evident in the Control treatment. As discussed previously, we consider the basic element of temporal sampling most likely to be annual sampling, in which case 'Time' effects in the above model would represent 'years'.

Table 7.5. *The general form of ANOVA appropriate to an MBACI model*

| Source of variation | df[a] | MS estimates[b] | F-ratio Denominator MS | F-ratio Denominator df |
|---|---|---|---|---|
| *Among locations* | | | | |
| C | 1 | $\sigma_\varepsilon^2 + t'\sigma_L^2 + h_L t'\delta_C^2$ | L(C) | $l'-2$ |
| L(C) | $l'-2$ | $\sigma_\varepsilon^2 + t'\sigma_L^2$ | — | — |
| *Within locations – repeated measures* | | | | |
| B | 1 | $\sigma_\varepsilon^2 + \sigma_{LT}^2 + \bar{h}_l\sigma_{LB}^2 + l'\bar{h}_l\delta_B^2$ | L(C)B | $l'-2$ |
| T(B) | $t'-2$ | $\sigma_\varepsilon^2 + \sigma_{LT}^2 + l'\hat{\varepsilon}_T^2$ | L(C)T(B) | $(l'-2)(t'-2)$ |
| CB | 1 | $\sigma_\varepsilon^2 + \sigma_{LT}^2 + \bar{h}_l\sigma_{LB}^2 + \bar{h}_l\bar{h}_t\delta_{CB}^2$ | L(C)B | $l'-2$ |
| CT(B) | $t'-2$ | $\sigma_\varepsilon^2 + \sigma_{LT}^2 + \bar{h}_l\delta_{CT}^2$ | L(C)T(B) | $(l'-2)(t'-2)$ |
| L(C)B | $l'-2$ | $\sigma_\varepsilon^2 + \sigma_{LT}^2 + \bar{h}_l\sigma_{LB}^2$ | L(C)T(B) | $(l'-2)(t'-2)$ |
| L(C)T(B) | $(l'-2)(t'-2)$ | $\sigma_\varepsilon^2 + \sigma_{LT}^2$ | — | — |

[a] $l'$ and $t'$ are as defined in Table 7.4.

[b] $\bar{h}_l$ represents the harmonic mean of the number of locations within Control and Impact categories. With equal numbers of locations, this value will $= \bar{l}$.

Table 7.6. *The components of variation for a Beyond-BACI design*

| Source of variation | Designation | Number of levels[a] |
|---|---|---|
| *Spatial variation* | | |
| Impact–Control – C | Fixed | 2 |
| Locations within Impact–Control – L(C) | Random | $l'$ |
| *Temporal variation* | | |
| Before–After development – B | Fixed | 2 |
| Times within Before–After – T(B) | Random | $t'$ |
| Impact–Control × Before–After – CB | Fixed | 4 |
| Impact–Control × Times(B) – CT(B) | Fixed | $2t'$ |
| Locations(C) × Before–After – L(C)B | Random | $2\,l'$ |
| Locations(C) × Times(B) – L(C)T(B) | Random | $t'\,l'$ |

[a] $t' = t_B + t_A$, where $t_B$ and $t_A$ are the numbers of times sampled Before and After start-up, respectively. $l' = l_C + l_I$, where $l_C$ and $l_I$ are the numbers of Control and Impact locations sampled, respectively.

### Beyond-BACI

This approach allows the separation of variation into the following independent components, and there is a wide range of possible tests, each designed to assess a particular kind of impact. Here, we present only the test of impact at the largest temporal and spatial scales, to illustrate the major differences between the MBACI and Beyond-BACI (Table 7.6). The general form of the ANOVA appropriate to the model, as depicted above, is shown in Table 7.7. In Table 7.7, $\bar{h}_l$ and $\bar{h}_t$ are the harmonic mean numbers of locations within Control and Impact treatments, and times within the Before and After periods, respectively.

Tests for differences among locations or the interaction between locations and times would be possible if all data from within locations were not averaged to produce one datum (mean) per location per time. For clarity, we have not presented subsampling terms here, but they would simply appear as additional nested terms under Locations(C) and in interaction with the temporal terms (e.g. see Underwood 1991a, 1993, 1994a).

It is important to note how the MS estimates are altered by the designation of time as a random factor, with the main tests of interest, particularly the CB term, tested against the residual term in the model. In the absence of subsampling within time–location combinations, the

Table 7.7. *The general form of ANOVA appropriate to a Beyond-BACI model*

| Source of variation | df[a] | MS estimates | F-ratio Denominator MS | F-ratio Denominator df |
|---|---|---|---|---|
| *Among locations* | | | | |
| C | 1 | $\sigma_\varepsilon^2 + t'm\sigma_L^2 + \bar{h}_l m\sigma_{CT}^2 + \bar{h}_l mt'\delta_C^2$ | — | — |
| L(C) | $l'-2$ | $\sigma_\varepsilon^2 + t'\sigma_L^2$ | Residual | — |
| *Within locations –* | | | | |
| *repeated measures* | | | | |
| B | 1 | $\sigma_\varepsilon^2 + \sigma_{LT}^2 + \bar{h}_t\sigma_{LB}^2 + l'\bar{h}_t\delta_B^2$ | — | — |
| T(B) | $t'-2$ | $\sigma_\varepsilon^2 + \bar{h}_l\sigma_T^2$ | Residual | $(l'-2)(t'-2)$ |
| CB | 1 | $\sigma_\varepsilon^2 + \bar{h}_t\sigma_{LB}^2 + \bar{h}_l\sigma_{CT}^2 + m\bar{h}_l\bar{h}_t\delta_{CB}^2$ | — | — |
| CT(B) | $t'-2$ | $\sigma_\varepsilon^2 + m\bar{h}_l\sigma_{CT}^2$ | Residual | $(l'-2)(t'-2)$ |
| L(C)B | $l'-2$ | $\sigma_\varepsilon^2 + m\bar{h}_t\sigma_{LB}^2$ | Residual | $(l'-2)(t'-2)$ |
| L(C)T(B) | $(l'-2)(t'-2)$ | $\sigma_\varepsilon^2 + m\sigma_{LT}^2$ | Residual | $(l'-2)(t'-2)$ |
| Residual | $(l'-2)(t'-2)m-1$ | $\sigma_\varepsilon^2$ | — | — |

[a] $l'$ and $t'$ are as defined in Table 7.6.

L(C)T(B) term must be used as the denominator for the test of impact, and in doing so, it must be assumed that $\sigma^2_{LT} = 0$.

The designation of times as a random factor also has implications for any power calculations.

As for the MBACI, the principal effects of interest for impact assessment are the interaction between Impact and Control treatments and periods Before and After development (Green 1979, 1993), and the interaction between the Impact and Control treatments and Times within the Before and After periods (the terms CB and CT(B), respectively, in the ANOVA table in Table 7.7). The sources of variation *within* these interactions against which departures from the null hypothesis would be tested (the error variances) are not so simple. From the table, there is no simple test for the CB interaction, and we would need to use a Quasi-$F$ test (Winer *et al.* 1991). The CT(B) term can be tested using the interaction between Locations(C) and Years(B). Possible longer-term impacts of the activity in question would be indicated by significant CB interactions. Shorter-term impacts from which the measured variables quickly recovered would be expected to cause a significant CT(B) effect, and be manifest as changes in the Time profile of the Impact location between the Before and After periods that were not evident in the Control treatment. As discussed previously, we consider the basic element of temporal sampling most likely to be annual sampling, in which case 'Time' effects in the above model would represent 'years'.

Underwood has suggested that this design can be used to detect a number of other kinds of impacts, particularly effects on the variance, rather than the mean of variables. He has described appropriate ways to construct the $F$-ratios for these kinds of impacts (Underwood 1991a, 1992, 1993, 1994a, 1996). We wish only to draw attention to these important alternatives, and to emphasize that the principles remain the same – understand (and specify precisely) the conceptual model, specify the linear model, generate the analysis table and use it to identify appropriate tests.

### 7.4 POWER CONSIDERATIONS

#### 7.4.1 BACI and BACIP

The impact in a classical BACI design is tested against variation among subsamples within each of the four location–time combinations, and can be modelled as a fixed-factor, two-way ANOVA. The power is then determined solely by the number of subsamples.

With the BACIP design, the 'replicates' are the Control–Impact pairs, so the power depends on the variation in those differences, and on the number of time periods. The latter is often under direct control of the researchers, and can be increased. However, the constraints of many monitoring situations means that the number of Before times is likely to be fixed (and often small), and only the number of After times can be increased. The returns on this increased sampling are less than when both times can be increased. There is the added problem that many statistical tests are most reliable when samples sizes are equal between groups, so longer time-series after an impact may involve some trade-off.

The other component of power of a BACIP design is the variance of the Control–Impact differences. The observed variation has two components – the variation in those differences through time, and variation due to using subsamples to estimate the state of the locations at a particular time. If there is considerable spatial variation within each location, and we take only a few samples, the variation in Control–Impact differences may reflect this small-scale spatial variation, rather than the temporal variance that is of interest. Increasing the number of subsamples increases the precision with which we estimate the state of each location, and the variance of the $d$s begins to reflect primarily temporal variation. Increasing the number of subsamples has an effect on power primarily when the within-location variance is high, compared to the between-times variance *and* the initial number of subsamples is small. When there is little within-location variation, additional samples are a waste of resources.

### 7.4.2 MBACI

The primary test of impact in the MBACI situation uses variation among locations through time as the background variation. Consequently, the power depends on the observed variation and the number of locations within Control and Impact groups. Because this number is most often small, increasing the number of locations is usually the most efficient way to improve power.

The observed $L(C)B$ variation is composed of a number of components; in estimating this variation, we have averaged the times within each period. In theory, we can not change this component – because we regard times as a fixed factor, we are assuming that those times represent all possible states within the Before and After periods, so we can not include more times. In practice, we will, of course, feel more confident about a result based on more years of sampling. In calculating the $L(C)B$ variance, we have also used (in most cases) subsamples of each location to characterize its state at each time $T$. How accurately we have done so will depend on the within-location variance and the number of sub-samples that have been taken. As for the BACIP design, more subsamples can indirectly increase power by removing small-scale spatial variation as a contributor to the apparent $L(C)B$ variance, and the improvements will be greatest when this variation is large, and we have only a few subsamples. Increasing the number of subsamples (e.g. from two or three to 10) should rapidly and largely remove this effect.

If the monitoring program is financially constrained, any increase in sampling effort may be difficult to justify, and require trade-offs. A good rule of thumb in such a circumstance is that, when necessary, one should maximize the number of locations, and minimize the number of subsamples. The number of subsamples does not affect the degrees of freedom of the crucial test, so time spent processing subsamples must be minimized. Because there are two levels of sampling (locations, and within locations) that are replicated, there will usually be a trade-off between the two, as we rarely have enough resources to increase all replication at will. In these trade-offs, it is almost always preferable to increase the sampling at the higher level, and to take only enough small-scale samples to get a good estimate of the mean. The exceptional situations are likely to be those in which there is massive small-scale variation, with little additional variation among locations. Understanding the pattern of variation in the variables under examination is an important part of the planning of any sampling program, and a part of the initial power analysis should include an examination of the various subsampling options.

We do, however, recommend taking *some* subsamples, to provide good spatial coverage, and hence a representative sample of the location–time combination. In our experience, almost all systems have small-scale spatial variation, the removal of which would help the sampling program.

In a practical sense, it may not be necessary to process all of the subsamples for a given level individually, but options such as compositing can, and probably should, be investigated in many situations. In marine environments, compositing, followed by sample splitting, is used on benthic infaunal samples, which are often characterized by high diversity, great spatial variation and, most importantly, very long sorting time. Similar procedures are routinely used in environmental chemistry. Pooling all subsamples from a location, and then processing the entire sample or a proportion of it, should still yield a good estimate of the location value at a particular time. The proportion of the sample that is processed will depend on the number of individuals, and the target organisms. If the abundance of a common species is to be estimated, only a small subsample of the well-mixed composite may need to be examined. If a rare species is the target, most of the sample may need to be examined, and if the target is a standard estimate of species richness, we may end up counting the first $n$ individuals in the sample, where $n$ is constant across all locations and times (i.e. using a rarefaction estimate of species richness). This procedure works because the test of impact in the MBACI uses locations as the units of replication, so our aim is to get good estimates of those locations.

One final important aspect of the power is the situation when the number of potential control locations is limited. Our aim is to pick control sites that fluctuate through time in the same way as the impact site(s) in the absence of the particular activity, and this will generally mean close matching of the physical and biological environments. If we seek to add additional locations, we may be forced to begin considering areas that are less similar to the original set. If this is the case, we may find that there is more variation among locations among the expanded set, so that increases in power from greater replication are offset at least partly by an increased background variation. If the initial analyses suggest a big increase in the number of locations, it is critical that we not just add the new locations, and proceed with the sampling program. The variation in the expanded set should be examined, and the power calculations redone.

### 7.4.3 Beyond-BACI

The Beyond-BACI design is a more complex version of the above analyses, and is used to test a wide range of effects. Each of those tests has an associated power function, which will depend on the level of replication at the level under consideration. With so many hierarchical levels, the power of a given test will usually depend directly on the number of spatial replicates at the next level down, and on the variation among those replicates. As with the other designs, the observed variation will reflect not only variation at that particular level, but also the variation at all of the lower levels, with each level lower having a smaller influence on the observed variation.

For a given test in Beyond-BACI, the power considerations follow the principles for the MBACI above. The complication is that the power analysis and computation of the number of replicates must be done separately for each hypothesis to be tested. Further, the power calculations are not independent of each other because, in the simplest case, a given variance may be used as the numerator in one test, the denominator in another $F$-ratio, and appear indirectly in other tests.

Formal procedures for power calculations have not been published, because they are very complicated. We believe that the final design will be derived by an iterative procedure, in which a range of different allocations of sampling effort is tried, and the power values for all tests tabulated. The final decision will depend on an overall assessment of the set of power values, most likely with some weighting of the most important tests. For example, a subtle impact over a large spatial scale may be considered more important to detect than a localized, but possibly more severe impact, so we may wish to ensure higher power for the test at the larger scale, even at the possible expense of the small-scale test.

We know of no formal way of integrating all of the power values, and the final decision will no doubt reflect management priorities, as well as scientific criteria.

## 7.5 DETECTING MORE SUBTLE EFFECTS

All of the tests described above for BACI, BACIP and MBACI designs focus on the detection of 'step' changes – a consistent change to the receiving environment, relative to its control(s). Underwood's Beyond-BACI designs were targeted at detecting impacts at different scales, and at impacts that may act by altering variances, rather than means.

In most of these cases, it is possible that the actual impact may take a different form. A short-term monitoring program will detect essentially acute changes in the receiving environment, but it is also possible that with longer time-series, more subtle effects can be examined. Many of the authors who have written on the subject of monitoring have recommended that these tests for step changes be used as the basis for designing the sampling program, but that flexibility be used in looking for these other kinds of impacts. Our aim, in essence is to describe the changes that occur in response to the particular human intervention, not just to test for one kind of change.

For example, most or all of the common designs treat times within the Before and After periods as a random sample, rather than an ordered set. Treating the samples as a time series allows detection of gradual trends, or contrasting trends between Before and After. Keough & Mapstone (1995) also pointed out that, as the number of sampling times in the Before and After become disparate, such as occurs when monitoring continues for a substantial period, then tests of hypothesis become unreliable. They suggested that for long time-series, tests for trends through time in the After period would allow for detection of less acute impacts, tests of transient effects etc.

The monitoring designs detailed above serve as the ideal situation – the design, statistical model and test of hypothesis that are the focus for the sampling program – when confronted by realities such as limited Before sampling periods, bizarre behaviour of particular locations etc. A mixture of common sense and flexibility must be used in considering alternative ways of assessing impacts.

## 7.6 EXTENT OF IMPACTS

In contrast to the controversy about appropriate models for detecting impacts, estimating their extent presents fewer analytical options. The general problem is one of estimating the rate at which a variable changes with 'distance' from the source of the impact, sometimes called gradient analysis. 'Distance' could be measured as the physical distance from the centre of the human activity in question (e.g. a discharge point) or as the intensity of the activity. For example, if a toxicant is discharged, we may wish to understand the relationship between toxicant concentration and some biological variable, and we could do that by measuring the concentration directly, or by modelling the dispersion, and then using physical distance as a simple proxy for toxicant levels.

In either case, our aim is to describe the relationship between two variables, and there are two basic approaches to this problem (see chapter 4) – simple linear (usually least squares) regression, and non-linear approaches. Most practitioners are familiar with linear regression, which has the advantage of being simple to compute, and simple to explain. However, many underlying relationships are likely to be non-linear – the concentration of a toxicant, for example, does not decrease linearly with distance from a source. When faced with a non-linear relationship, we can either transform the variables to make the relationship more linear, or use a non-linear regression approach. The first option is the most often used, for its simplicity.

The modelling approach is uniform – the characterization of the relationship is the aim, and linear and non-linear methods differ only in the complexity of the underlying relationship and the methods used to fit the models to the data (often least squares vs. maximum likelihood). The principles for maximizing power will be the same and, practically, the final approach used in a particular monitoring situation will depend on the characteristics of the data.

## 7.7 FLEXIBLE ANALYSIS/INFLEXIBLE HYPOTHESIS

In this chapter, we have described the formal structures associated with three common approaches to detecting impacts (see summary in Table 7.1). We emphasize that these approaches differ fundamentally in the way that impacts are viewed. Each of the approaches has its advocates, and although we have argued for a view of impacts being distinguished from a population of control locations, but over a fixed period of time, the other views have merits. Any one could be chosen for a particular monitoring situation, depending on the regulatory and socioeconomic frameworks involved. It is, however, crucial that those involved with monitoring programs frame the objectives of the monitoring very precisely, so the appropriate model is identified *at the start of monitoring.* Each of these approaches leads to different ways of maximizing power, so to design a program under one of the three logical frameworks, and to apply the analytical model for another one, will almost always lead to a weak monitoring framework.

While we do argue for a precise, inflexible philosophical approach and initial hypotheses, we do acknowledge that the actual analysis of monitoring data must retain a degree of flexibility. Most monitoring situations change as data are acquired – things go wrong, our estimates of variation change, data may be non-normal, mistakes occur etc. These changes require us to be creative at the data analysis stage, and to be prepared to alter parts of the statistical models that we use, but we should not forget the underlying logic of the particular program. In chapter 9, we discuss how these constraints will affect our inferences of impact.

### 7.8 IMPORTANT ISSUES

- Different BACI designs lie along a gradient of inferential certainty from relatively strong to relatively weak, rather than providing either perfect or zero inference about human impacts.
- There are different types of BACI designs, which result in distinctly different analytical models that address different questions. Consult Table 7.1 for a summary of the important characteristics of each approach. These conceptual models are all justifiable approaches to the detection of impacts under particular circumstances, but it is essential that anyone implementing one of these designs be aware of the differences between them, and of the important characteristics of each of them.

# Part III

# Applying principles of inference and design

# 8

## Applying monitoring designs to flowing waters

In the last chapter, we introduced the different sorts of analytical models we can use to detect impacts. There are distinct and important differences between these models, and yet understanding these differences is only the first step. The next step is to be able to apply these designs sensibly and usefully in real situations. In many cases as we will see, the ecology of streams and rivers (and, indeed, many other environments) is not sufficiently well understood to make perfect or even very good decisions about design in every instance. As we shall argue below, however, the critical issue is to understand *how* to make good design decisions, and, because monitoring designs always involve compromises, to be very clear and explicit about the reasoning upon which such decisions were based. The remaining part of the process, then, is to understand what monitoring designs can tell you definitively and what they cannot – what can be legitimately and logically interpreted from the data versus what will remain unclear.

Below we describe a hierarchy of decisions that will help define the nature of the problems faced in monitoring impacts in flowing waters properly. There are several problems we have to solve to apply good design principles. These problems are the location and character of control locations, and the frequency of sampling through time. In many places, we will suggest that a systematic and well-structured review of the literature will be necessary to solve these problems.

In some cases, it will prove impossible to gain any control locations. In others, sampling before the start of potential human impact will be impossible because that event may have occurred decades or centuries before. In the most difficult case, neither controls nor Before data are available. We consider all of these cases more explicitly in the next chapter, because they require a 'levels of evidence' approach, whereby conclusions are based on several weak, but independent, lines of evidence. This approach also makes great use of the literature, and may be well worth undertaking even when an MBACI design is possible, because it will help with the selection of controls, the selection of monitoring variables (a topic we cover specifically in chapters 10 and 11) and in improving the inferential strength of designs through indirect means.

## 8.1 SPATIAL VARIATION AND THE LOCATION OF CONTROLS

### 8.1.1 The nature of controls

Before we begin discussing how we might locate controls, it is worth reviewing what controls actually are, because there are a number of common misconceptions about them. First, control locations and reference locations are not the same thing (Box 5.1), and we suggest that these terms not be used interchangeably. Reference locations are typically used to provide an indication of what environments might look like when free of any human activities. They are generally located in pristine or as close to pristine places as possible and can be quite remote from particular impact locations. Comparison between a collection of reference locations and impact locations then has potential to detect changes caused by human activities but relatively poor ability to draw inferences about specific causes of differences at individual locations. This is because any particular impact location is likely to differ in many ways from the collection of reference locations aside from the incidence of the human impact of interest. Additionally, any particular location might be changed by more than one human activity; a comparison between such a location and a collection of reference locations will not tell you which, if any, human impact might be responsible for the differences. Control locations on the other hand are designed to be as similar to the impact location(s) as possible, so that the major source of difference between them will be that of the human impact of interest. Consequently, control locations need not be pristine – and, in fact, shouldn't be, if the impact location suffers a number of human insults but our interest is in one of those in particular.

Control locations thus act to isolate an effect of interest – namely, any changes due to a particular human impact. Using controls does not mean that all variables are 'under the control' of the researcher – the term control is used in the context of field experiments (Box 5.1). Finally, control locations do not have to be 'identical'. As a literal requirement, that is certainly impossible, even in laboratory experiments, because all locations differ in some ways from others. Instead, control locations need to be sufficiently similar to each other and to the impact location that an important change at the latter will have a high likelihood of being detected. Indeed, the degree of similarity among control locations required to detect a change of a particular size with particular probabilities can be quantified, as we will discuss further below and in subsequent chapters.

### 8.1.2 Spatial extent and nature of impact

The first step is to determine the potential or likely extent of the impact in space. The spatial extent of impact is important because it will determine where we can begin looking for control locations. It will also influence where we are able to locate subsamples within the putatively impacted area, as will be discussed further below. Control locations must be free of the impact of concern (section 5.4), and the only way to be certain of this is to know the expected spatial extent of the putative impact. One advantage of river systems is that they are organized hierarchically as branching networks, with each river system or tributary within a river system occupying a defined catchment (Fig. 8.1). Catchments (large or small) form natural boundaries to the movement of water (excepting groundwater), chemicals, sediment and pollutants, and they often form boundaries to biota and human activities as well (see chapter 2). Consequently, catchments provide natural spatial units, and this is one advantage that lotic systems have over other habitats where the boundaries and sources of spatial variance can be far less clear (e.g. soft-sediment, marine benthos; Thrush 1991). However, analogies exist in some marine systems, where embayments or headlands may represent natural barriers to water movement. In deciding the spatial extent of impact, we are really deciding over what size of catchment impacts occur. The use of catchments as spatial units applies both to stressors that occur predominantly on the land of the catchment itself (e.g. land-use changes) as well as to those occurring directly on individual stream channels (e.g. on water quality or quantity) because the latter are still associated with catchments of particular sizes. In either case, catchments form appropriate units of replication for the term Locations in the design.

Obviously, human activities may affect a continuous range of catchment sizes, but we will distinguish two major kinds as the ends of a continuum. The first type of activity is where impacts are restricted to individual tributaries that drain relatively small spatial areas (catch-

**Fig. 8.1** A schematic diagram of a dendritic river system, illustrating small sub-catchments nested within larger catchments. Catchment boundaries are marked by dotted lines. The arrow marks the site of a localized point impact. Note that the branching structure of rivers is dependent upon local topography, gradients etc. and may not always form such a clear branching structure.

ments of perhaps 10s to 100s of km$^2$). In this situation, there may be tributaries on the same river system and hence in the same major catchment that are unaffected and could act as controls (Fig. 8.1). Many of these impacts are likely to be direct, point impacts (chapter 3) on water quality or quantity (e.g. dams, many mining operations, pollutant outlets from factories or sewage treatment works). In these activities, the spatial extent of the impact is usually clear and includes the area surrounding the point of impact together with most downstream locations. Tributaries of other rivers within the same major catchment may be demonstrably unaffected and hence offer potential control locations (subject to further conditions of independence, as discussed below).

The second kind of human activity is where impacts are widespread and affect multiple river systems over catchments of perhaps thousands of square kilometres; that is, tributaries in sub-catchments will not offer suitable controls. These are likely to be non-point and indirect impacts, which are diffuse and spread throughout whole catchments rather than being restricted to defined points along channels. In these situations, it is often difficult to find any areas within the catchment that are demonstrably free of impact. For example, nutrient runoff from multiple septic tanks or agricultural fields can mean that most tributaries within a major catchment are probably affected. Large areas will almost certainly be subject to numerous and various non-point impacts (e.g. urban development in conjunction with agricultural impacts). A burgeoning problem of this type is the spread of exotic species. In these situations, the impacts are generally occurring at the scale of major catchments, and it will be necessary to look to other, equally large catchments for controls.

Potentially then, studies of non-point impacts are likely to be carried out over much larger spatial scales, involving multiple major catchments, than point-source impacts restricted to individual tributaries. What absolute sizes of catchments should be considered 'large' and what sizes 'small', however, are likely to vary between regions depending upon local topography, types of drainage networks etc. The distinction is somewhat arbitrary, but the two sizes encompass the majority of examples because each represents a commonly considered endpoint on a continuum. The principles we discuss below will apply to impacts intermediate between these sizes. A second reason we distinguish these two sizes is that, over very large spatial scales, impacts may encompass whole biogeographical, geological or climatic regions, whereas impacts at smaller scales may fall wholly within such regional zones. Consequently, the likelihood of finding multiple control locations decreases as the spatial scale of impact increases (as discussed further below). This means that researchers trying to measure the effects of large-scale impacts face many more design problems than do those examining small-scale impacts. Indeed, for impacts occurring over very large river systems, it is likely there will be no other river systems available that can be considered truly comparable and able to act as controls. Researchers may have to proceed using a reference system approach (Box 5.1) where possible and proceed using a weaker 'levels of evidence' approach (see chapter 9).

Finally, we note that sorting out whether impacts are point or non-point is not always straightforward. Some human activities may appear to be non-point but actually contain a variety of point impacts. For example, forestry operations can be widespread through an area and may appear to be a non-point impact. A more specific examination may show that forestry can create a series of point impacts. Clearing of vegetation along stream banks can cause bank slumping, erosion and high sediment loads. However, if trees are cut within defined coupes it is

Table 8.1. *Types of sampling design for point and non-point impacts; examples of point and non-point impacts are given in chapter 3*

| |
|---|
| *Point impacts* |
| Multiple or single impact location(s) all within the same major catchment – Fig. 8.3[a] |
| Multiple impact locations, with some in spatially separated, major catchments – Fig. 8.5 |
| *Non-point impacts* |
| Non-point impacts spread over one or more whole major catchments – Fig. 8.7 |

[a] The figures are used later to illustrate possible hierarchies of decisions regarding selection of controls.

possible that areas of forest are left such that some stream locations can provide suitable controls, and vegetation removal is effectively a point-source impact on individual tributaries. Another common impact of forestry is sediment runoff caused by the construction of roads. Again, this impact may appear to be non-point, but if such runoff is localized to specific points along streams, it may be possible to find comparable locations without such impacts in neighbouring sub-catchments (see Welsh & Ollivier 1998 for a nice example). Determining whether impacts are point or non-point and the spatial extent of impacts is sometimes difficult, but it has important implications for the selection and defini-tion of control locations (Table 8.1), and is an important and very necessary first step. The forestry example also illustrates a point we have made throughout this book: it is important to set out each of the perturbations associated with any human activity so that we can pose clear hypotheses about their putative effects.

### 8.1.3  Finding control locations

*Criteria for controls*

To locate spatial controls, the first issue is to decide what criteria loca-tions must meet in order to qualify as controls. These criteria ensure that, in the absence of that human impact, the impact and control locations are expected to be similar to each other. If we do not have reasonably well-defined criteria, we risk choosing control locations that are substantially different from the impact location(s) in various ways besides the incidence of impact (for an example and discussion see

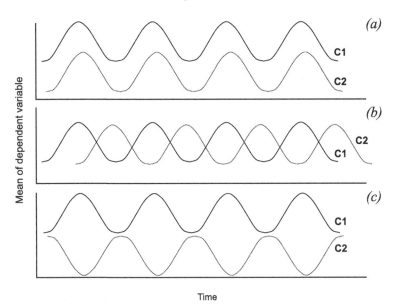

**Fig. 8.2** Stylized graphical illustrations of Location × Time variability among control locations. In each graph, C1 and C2 represent the trajectories of two control locations over time in some dependent variable averaged across any within-location replication. (a) The two locations have a large absolute difference in mean value of the variable, but the difference between the two locations is consistent over time (the two control locations fluctuate in synchrony) so that Location × Time variance is small. (b) There is little absolute difference between the control locations but because they fluctuate completely out of phase, Location × Time variance is high – this will reduce power to detect differences between control and impact locations. (c) There is a large absolute difference between control locations plus they also fluctuate out of phase so that, again, Location × Time variance will be high.

Humphrey *et al.* 1995). Such differences between control and impact locations mean that there will be greater baseline variance among locations, and if the temporal dynamics of these locations are different, then this can lead to survey designs with low power to detect impacts (section 7.4 and Fig. 8.2).

To establish criteria for controls, we need a good understanding of what makes different locations similar, at particular spatial scales, in the absence of human impact. Although there has been substantial research into the relations between catchment characters and type of biota (see chapter 2), the answers we can offer are still very limited because few researchers have explicitly factored spatial scale into their sampling designs. The explicit examination of scale is critical because different factors are liable to be important at different scales. At large spatial scales, factors such as climate or major shifts in bedrock geology might drive lots of biotic differences, whereas these factors may be irrelevant for studies at smaller scales that fall wholly into particular climates or bedrock lithologies. Some studies of the connection between physical environment and biota have incorporated scale, such as hierarchical classification schemes (e.g. Frissell *et al.* 1986; Hawkins *et al.* 1993). While promising, the ability of these schemes to make good predictions and the validity of some critical, internal assumptions still remain largely untested (Cooper *et al.* 1998; Downes *et al.* 1995, 1997). Other schemes that have looked at similarity among relatively 'pristine', reference locations (see Box 5.1), such as RIVPACS-type or IBI studies (section 3.2.1) provide some information about the factors that make locations similar. However, these schemes, again, apply over large spatial areas that may encompass different climatic or other sorts of zones. Their applicability to studies conducted over smaller scales (or targeted to particular species) is unclear (see Marchant *et al.* 1999 for a recent discussion). Additionally, those factors found to create similarity between locations typically vary greatly from region to region and continent to continent, so there are few generalities.

Another likely source of information about suitable controls is studies that have looked specifically at the effects of human impacts. Unfortunately, only a few have recognized the need for controls. In a recent survey of papers examining human impacts on rivers over the last 10 years or so (J. R. Thomson & B. Downes, unpublished data – refer to Box 6.1), relatively few papers recognized the formal need for control locations. Of those that did, only 14 provided any explicit criteria for choice of control locations. These criteria were spatial proximity (especially the need to stay within the same major catchment), similarity in stream size (or stream order or discharge), substrate and local habitat (e.g. riffles and pools were distinguished). Interestingly, however, few of these studies provided any specific justification for using these criteria and most couched selection criteria in relatively vague language. Out of interest, we contacted some freshwater ecologists and asked them to nominate the criteria they would use to select controls for different sorts of rivers (Table 8.2). Although there is some similarity with the criteria identified above, the diversity of replies shows that this is an area in great need of dedicated research and debate.

In the absence of a good body of research with definitive answers

Table 8.2. *Criteria for control locations named by 18 freshwater ecologists in different countries*

| Criterion | Number | % |
|---|---|---|
| *Comparisons among tributaries in the same catchment* | | |
| substrate/geology/suspended load | 12 | 67 |
| discharge regime/stream size/catchment size | 11 | 61 |
| riparian vegetation/zone | 10 | 56 |
| catchment land use/vegetation/soils | 8 | 44 |
| gradient | 8 | 44 |
| altitude/stream order/temperature/distance from source | 7 | 39 |
| current speed/water level/riffle and pool structure | 7 | 39 |
| spatial proximity/external variables (e.g. climate) | 5 | 28 |
| water quality including detritus levels | 4 | 22 |
| channel form/geometry | 4 | 22 |
| algal cover | 1 | 6 |
| where I'm prepared to drink the water! | 1 | 6 |
| logistical practicality | 1 | 6 |
| chance | 1 | 6 |
| presence of fish | 1 | 6 |
| aspect | 1 | 6 |
| | | |
| *Comparisons among catchments* | | |
| substrate | 8 | 44 |
| discharge regime/stream size | 7 | 39 |
| catchment land use/vegetation | 7 | 39 |
| riparian vegetation | 6 | 33 |
| gradient | 5 | 28 |
| geographic proximity | 5 | 28 |
| geomorphic forms/hydraulic habitat | 5 | 28 |
| altitude and aspect/temperature | 4 | 22 |
| water quality/geology | 4 | 22 |
| can't be guessed/not solvable/no answer | 8 | 44 |

*Note:* These data are not from a representative sample of freshwater ecologists, and so should not be viewed as providing definitive information. The ecologists were asked to provide the top five characteristics or features that sites would need to be matched to make them likely to be similar biotically (e.g. faunal or floral composition, species absence/presence, species densities) at two spatial scales of comparison: among sites on different tributaries in the same catchment, and among sites located in different catchments. Reported is the total number of times out of 18 a feature was chosen. Some similarities in selection are present, but replies were often quite idiosyncratic, depending upon taxa and degree of human disturbances in the region. Many individuals found the among-catchment comparison difficult to do and expressed less confidence (in some cases, no confidence) in their answers or refused to answer.

to questions about what, in general, makes locations similar biotically, what suggestions can we offer about developing criteria for controls? It is our belief that we should not provide dogmatic advice here. We suggest that a literature review targeted at studies of the specific impacts, the variables of interest, and the types of rivers and their natural variability in the region will provide more specific and useful criteria than any we can offer here, for several reasons. First, large-scale studies suggest that regional and continental differences can be important. Although we have provided some suggestions above, the variety of impacts, stream and river types, and variables chosen for monitoring means that any criteria we could provide here are too general to be of much use. Second, we must realize that criteria for controls are contingent upon the variables chosen for measurement. The factors that determine fish abundance and diversity, for example, are likely to be quite different and operate over different spatial and temporal scales from those that influence macroinvertebrate or macrophyte diversity and abundance. Consequently, the criteria for choosing control locations are also likely to be different. Third, we should keep in mind that many of our assumptions about the importance of physical habitat types in streams and rivers are untested. Recent work is suggesting that we know a lot less about these things than we may currently realize. For example, Bunn & Hughes (1997) have shown that sites connected by only a few kilometres of stream channel can have populations of the same macroinvertebrate species that are distinct genetically. If such genetic differentiation proves to be generally true, it suggests that locations along rivers may contain individuals that are not part of a larger population of broadly interdispersing organisms, as is commonly assumed (Bunn & Hughes 1997; Downes & Keough 1998). Instead, individuals at separate locations may be members of separate populations (see also Sweeney *et al.* 1992). If so, then such populations may be responding to much more local features of their environment than we have thought up until now, and there may be all sorts of habitat boundaries of which we are unaware.

The lesson for studies of human impacts, then, is that we should not be dogmatic about appropriate criteria for controls. Instead, we should consider each attribute systematically in light of what is known about the type of impact(s), the rivers of interest and the variables to be monitored. We present a step-by-step method for developing criteria for controls in section 9.3.2. Such careful consideration may provide us with better comparisons than we can achieve by clinging to assumptions about rivers. For example, a common assumption is that control locations should remain within the same major catchment to qualify. Locations within the same catchment are often presumed to be similar because of spatial proximity, but some factors can cross catchments. An example is underlying geological rock type, because this can determine substrate type, sediment load and water chemistry (see Huryn *et al.* 1995 for an example). If the latter characteristics are important to the variables of interest, then it may be much more important to keep the bedrock geology of control locations consistent than it is to keep them all in the same major catchment. Setting up this sort of hierarchy of importance for control criteria means each catchment attribute receives careful and measured consideration for the impact at hand.

*The dilemma of the trade-off in similarity and number of controls*

A dilemma that greets anyone who embarks on the above process is that choice of control locations becomes a trade-off between generality and comparability. If we use criteria that define control locations very narrowly and specifically, then we are likely to improve their similarity to each other, but we also reduce the pool of potential locations and the generality of our results. If we define controls much more broadly, then we improve the numbers of potential locations, but we also increase the chance that we are comparing among things that are significantly unlike in important ways. It can be difficult extracting the signal caused by human impacts if the variability among control locations is itself very great (Fig. 8.2). This is particularly important for the MBACI design, which relies on having closely matched control and impact locations. Thus, both the number of locations and their similarity to each other are important, but we cannot usually maximize both of these things at the same time.

We suggest that the criteria for control locations and their selection should proceed by several steps (Table 8.3). First, criteria for controls should be developed based on what is known about the rivers of interest and the variables to be monitored, as discussed above. Next, site visits can determine the number of potential locations available. If the number of locations is too few (or many are unacceptable for reasons discussed below) then the criteria should be revisited to see if they can be broadened to include a greater number of potential locations without compromising the likely power of the sampling design. Finally, a priori decision rules can be set down regarding an agreed level of similarity among locations. A pilot phase, in which some preliminary biological data are collected from potential control locations, allows us

Table 8.3. *Suggested steps in developing criteria for controls and selecting control locations. Note that the process is better if some pilot data can be collected that will assess the degree of similarity among control locations*

---

*Step 1: Conduct a literature review*
This should be a structured examination of the literature examining natural sources of variability in the variables that will be monitored for relevant river types. The review should suggest the biggest sources of variability that therefore ought to be considered for matching control locations. Note that the review can be carried out as part of a larger exercise designed to help determine also the most appropriate variables to monitor (see chapter 9).

*Step 2: Draw up a list of criteria ordered from most important to least*
The review should identify important sources of natural variability. These sources can then be regarded as factors that ought to be matched among control locations as much as possible. For example, if discharge is identified as an important source of variability in the monitoring variables, then it will be important to match locations for discharge regime.

*Step 3: Carry out location visits*
It will then be necessary to visit potential control locations. Initially, possible controls may be identified from maps, but it will be necessary to visit all locations to ensure that they are not different in ways deemed important (such as suffering some other human impact).

*Step 4: Are there sufficient control locations?*
Decisions about what is sufficient need to be made in the context of power analysis, but it may be quickly clear that there are very few potential locations. This is because, of those locations that meet the criteria, some will be rejected because they are likely to be affected by the human impact under consideration or are not statistically independent from other locations (as discussed in section 8.1.3). It is also possible that few or no locations meet the required criteria at all.

*Step 5: Revisit the criteria*
Have the criteria possibly been drawn up too rigidly? Can the necessity for controls to meet one or more of the less important criteria be relaxed?

*Step 6: Does this improve the number of control locations?*
If it seems reasonable to relax one or more criteria, further visits will be necessary to visit other locations that might be suitable as controls. If there are still insufficient numbers of control locations, the criteria can be revisited again, or investigators may have to accept that there are no locations that can act as controls (considered in chapter 9).

---

to examine specifically whether locations meet our criteria for similarity. We can calculate the potential loss of statistical power caused by increased variance among locations versus the potential increased power expected from using higher numbers of locations (Fig. 8.2 and discussed in chapter 12). Hence, decisions to remove particularly variable control locations need not be made in an arbitrary or *ad hoc* way. The advantage of using such a sequence is that we gather a collection of control locations that either form a well-defined population from which we can sample, or at least meet an agreed-upon and rationally argued set of criteria. That is, we are clear about why we chose particular locations and our choices are not arbitrary. Additionally, such a sequence helps us avoid later, *capricious* removal of control locations once monitoring has begun. Because we can examine the variance among locations over time, we can decide ahead of data collection what are the grounds for removing a control location from the data set. We can justify removing control locations, for example, if they are changed by other sorts of human impacts or if there are major natural changes (e.g. landslide causing complete rerouting of the river channel). However, if grounds for removal are not agreed upon prior to monitoring, there may be a high risk that substantial pressure will be brought to bear to remove supposedly 'aberrant' control locations once monitoring has started *because* their removal increases the probability of reaching a particular decision about impacts. This situation is very unfortunate and must be avoided because it has the potential to undermine completely the decision-making process (we discuss a way of preventing such outcomes in chapter 12). It may be particularly likely where insufficient thought was given to the process of selecting controls ahead of time because it is then more likely that some controls will perform differently to others.

As discussed in previous chapters (sections 5.4 and 6.3), locations must meet other conditions to qualify as controls. Control locations must provide statistically independent estimates of the variables, they must be unaffected by the impact in question, and they should not be grouped together in locations that are spatially separated from impact locations or differ systematically from impact locations. Each of these requirements poses further problems for applying BACI-type designs.

*Statistical independence and location of controls*

Statistical independence means that replicate locations must not be highly correlated, an effect often observed when two or more places are in close spatial proximity (section 6.2.1). In rivers, an obvious source of such correlation between locations is water flow. Flow delivers organisms, nutrients, pollutants and sediment as well as extreme hydrologic characteristics like floods. Hence, there is some reason to expect that locations along the same channel may be more similar to each other than they are to locations along different channels (i.e. are spatially autocorrelated). Similarity between locations along the same channel means that the likelihood of detecting impacts is not the same as when independent locations on different channels are used (section 6.2.1; see also Underwood 1994b). This problem is particularly apparent in rivers but is not peculiar to them. Connections through ocean currents or wind are liable to have the same effects in marine and terrestrial environments, respectively, but have received equally little direct attention (e.g. Keough & Black 1996). Overall, we may consider river flow both a boon and a curse. Although it makes locations on the same tributary connected biologically for unknown distances, at least we know the directions of some important and consistent gradients (i.e. upstream and downstream). Many other habitats have fluctuating gradients (e.g. marine intertidal) or substantial gradients not obviously linked to physical features (e.g. marine soft-sediment benthos; Thrush 1991).

Although we can see that flow is likely to make locations along the same channel potentially very similar, we know little about the actual physical distances involved. How far upstream or downstream can we go from any point before other sites on the same river are not connected in an *ecologically* important way? How far apart do we have to keep measurements to ensure they are statistically independent estimates? Again, ecological research into this question is in its infancy (Koenig 1999), and there are few definitive answers. There is some information about the distances nutrients travel downstream before being removed from the water column ('nutrient spiralling'; e.g. Elwood et al. 1983; Newbold 1992), although these values seem specific to stream types. When considering invertebrate dispersal, many stream researchers have been overwhelmed by the high densities of organisms drifting in stream currents, and there has been a presumption that individuals are probably travelling long distances. However, very few dispersal studies in streams have actually attempted to quantify the distances invertebrates travel in the drift, walk over the stream bed or fly along streams as adults (Downes & Keough 1998), so it is possible that many individuals are not travelling very far. For example, Elliott (1971a,b) found many drifting or crawling individuals did not travel more than a few metres in a small (3.5 m wide) stream. The real difficulty, however, is that there is simply not enough information from a variety of rivers for us to know what are typical dispersal distances given particular environmental conditions of discharge and so forth. There is some information about the transport of coarse particulate organic matter (e.g. Raikow et al. 1995) and reasonably numerous studies on fish dispersal (e.g. Gowan & Fausch 1996). Many physical characteristics of streams, such as discharge peaks, sediment transport, temperature and pH are correlated at locations along stream channels as well as between streams in the same catchment (e.g. Walling & Webb 1992; Webb & Walling 1992). In some cases, researchers may have a good grasp of the temporal and spatial scales of autocorrelation among such physical parameters, but, again, the biotic significance of these correlations is not always clear.

In our opinion, the best option, in the absence of any data, is to keep control locations on separate channels as much as possible while keeping parameters thought to be important (e.g. discharge, substrate type) similar. Keeping control locations on separate tributaries is likely to be an acceptable way of ensuring independence for some biota (e.g. macroinvertebrates) that are thought unlikely to travel between them, or are sessile, such as plants. It is much less reasonable for organisms such as some fish, however, that might routinely traverse different tributaries in the same catchment. Additionally, separate tributaries can be connected via movement of groundwater, which can transport water and solutes in a variety of directions (e.g. Jones & Holmes 1996), or floodwaters. Consequently, use of separate tributaries is not a *guarantee* of statistical independence. Again, we advocate researchers examine any available data for the variables concerned, rather than simply assuming that locations on different channels must be independent.

Using more than one location per tributary might be very useful in at least one situation. In point impacts, it may be possible to have a location upstream and one downstream of the point of impact, and similarly separated locations on tributaries free of impact altogether. This design may be especially useful in upland systems where we are confident that other human activities have not caused impacts. If we calculate the differences between pairs of upstream and downstream locations, we remove the problem of autocorrelation, and these differences could be compared over time among impacted and unimpacted tributaries using BACIP or MBACI(P)-type designs. The big advantage is that fewer tributaries overall are required in this design than one where only one location (control or impact) can occur on each channel. However, a potential disadvantage with this approach is that we generally do not yet understand much about longitudinal variation. That is, although we know biotic composition changes along channels, we do not know how biological difference equates with physical distances, nor how biological change versus distance along channel relations varies among rivers. Consequently, we may choose pairs of locations on tributaries that represent quite different levels of difference, resulting in unlike comparisons among control streams. If these differences fluctuate greatly with time, then we have the same problem described above of trying to detect a signal of human impact against a lot of background variability. Additionally, we can only have pairs of impact and control locations on the same tributary if we are confident that impacts are not transferred to upstream locations. An impact zone can represent a barrier to dispersal in both upstream and downstream directions, however, and this may mean that upstream locations are either starved of colonists or accumulate them. In either case, we would not have truly independent estimates of a control state.

*Ensuring control locations are free of the human impact*

Our second characteristic of controls is that they be unaffected by the human impact of interest. Locations on the same channel or channels where the impact occurs are obvious places to seek controls, but flow causes us difficulties here as well. Water flow will often transport the effects of stressors downstream. In many cases, locations downstream in a catchment will be automatically ruled out as possibilities for controls because human impacts such as pollutants, alterations to discharge regimes etc. are transported probably to many, if not all, comparable downstream locations. For the same reasons discussed above, we will not have, in many cases, good estimates of how far downstream one can go and expect rivers to be free of effects from impacts upstream. Consequently, in most cases we expect that we can never site control locations at points downstream of impacts. Usually then, we will be restricted to using locations at upstream points or on tributaries in either the same or other catchments that are free of impact. Again, unidirectional water flow provides both disadvantages and advantages. Although we cannot generally use any downstream locations, we can at least be confident, in some cases, that upstream locations or separate channels are demonstrably free of the impact.

Even when we are confident of an impact's boundaries, we still need to ensure that monitored biota are not travelling routinely between impacted and control locations. Frequent dispersal of organisms between impact and controls means these locations are not independent (section 6.2.1), and that means that comparisons between them will not provide unbiased estimates of the true effect of the impact. Fish, or other highly dispersive animals such as some birds or mammals, are examples of species that may travel upstream or to other tributaries to escape effects of impacts and may end up colonizing, and hence altering, our controls.

Finally even in rivers having neatly defined catchments, there are some impacts whose extents are very difficult to establish and which can be unrelated to catchment boundaries or water flow. Acid rain and other aerial pollutants (e.g. Vitousek *et al.* 1997a) and impacts associated with climate changes are good examples. Such impacts often occur across large areas. The difficulties we may have in establishing the ranges of these impacts add to the problems mentioned above of locating suitable controls over large spatial scales. We expect that these sorts of impacts will often require us to fall back on relatively weaker lines of evidence (discussed in the next chapter) than those we can use when we are able to locate comparable controls.

*Spatial confounding, environmental differences and location of controls*

The final point about controls concerns the need to avoid spatial confounding and/or systematic differences in habitat between control and impact locations. Spatial confounding occurs when all control locations are in one place and all impact locations (or the impact location) are in another (section 6.2.1). In this situation, it is possible for one area containing all the controls to experience a natural change that, by chance, coincides with the onset of human activity at the impact location(s). Likewise, if there are systematic differences in habitat between control and impact locations, it is possible for one habitat type also to experience a natural change coincident with human activity. With either of these designs, we are unable, prima facie, to discount the possibility that differences between control and impact locations are due to natural changes coincident with particular areas or particular habitat types.

Systematic differences in habitats and spatial confounding of control and impact locations often occur together. For example, a common form of spatial confounding occurs where all control locations are in upstream areas and all impact locations are downstream from them (section 6.2). This is a common design because human impacts frequently occur on midland or lowland stretches of rivers, while rocky and hilly upstream areas are often unsuitable for development and remain the only locations free from impact within the catchment. In this case, there are obvious habitat differences related to spatial position, which are systematic changes in channel size, substrate particle size and other physical changes that occur along stream channels. Spatial confounding can occur, however, even when we have carefully matched the habitats of impacts and controls – for example, when we locate controls in a catchment entirely separate from that of the impact catchment. It is possible for there to be natural changes among the upper, rocky locations that do not occur at locations lower down, or for a change to occur in our control catchment that does not occur in the impact catchment. We might be unlucky in that these changes coincide with the onset of human activities, but the real difficulty will be that we simply will not know whether we have been 'unlucky' or not.

To avoid this problem, it is preferable to match the habitat of control and impact locations as much as possible and to seek control locations that are spatially interspersed with, or around, the impact location(s). We then reduce the possibility that systematic differences between impacts and controls will lower or remove our ability to infer differences between them as due to the human activity of interest.

### The relative significance of problems with controls

In the ideal situation, then, we would have many comparable control locations, well separated from each other on different tributaries or in different catchments, but with multiple controls interspersed among one to several impact locations. In reality, control locations are likely to be few in numbers, and little choice may be available. Our control locations may have some of the problems discussed above. However, these problems are not of equal significance (Table 8.4). Some problems with controls might mean that we would need to collect complementary data (as described in the next chapter), whereas others may mean that the sampling design is compromised from the outset. Accordingly, researchers and managers should be aware of the relative significance of violating the principles discussed above. As indicated at the start of this chapter, the priority is recognizing, ahead of time, the sorts of problems that may occur because in many cases we can plan for them.

Perhaps the worse problem is if controls become affected by the impact. In this situation, differences between control and impact locations no longer estimate a treatment effect due to impacts. The fundamental question has changed. Instead of a BACI-type design, at best we have weakly and strongly impacted locations and a design that is analogous to a dose-response experiment in ecotoxicology. Hence, we are no longer testing an hypothesis about the absolute differences between impacts and controls. Instead, we are testing an hypothesis about the effects of different degrees or levels of impact.

When controls become affected, we have a number of possible strategies. First, as discussed above, we can remove impacted control

Table 8.4. *Problems with control locations, in order from worst to least*

| Possible problems with control locations | Remedy |
|---|---|
| 1. Control locations are affected by impact | May require the removal of the impacted controls or the recasting of the hypothesis as a weakly vs. strongly impacted comparison. Situation can be fatal if the controls are affected but to an unknown degree |
| 2. Control locations are not independent of each other (e.g. as can happen if they are on the same tributary) | Can change the tabled probabilities of test statistics (as degrees of freedom are fewer than expected from sample size) and hence increase or decrease the likelihood of detecting impacts but in unpredictable ways. Requires an independent and reliable estimate of the degree of autocorrelation among locations to allow correction, but this is not often available |
| 3. Systematic differences between control and impact locations in, for example, physical features of habitat, and/or control and impact locations are spatially confounded | Extra data needed that support the conclusion that control and impact locations do not differ in features known to influence the variables being used to detect human impacts. Any pre-existing differences need to be factored into the design. Case is particularly strengthened if we have predictions that are unique to the human impact explanation (see chapter 9) |

locations from the survey design if we have sufficient numbers of controls that loss of one or more does not destroy our ability to draw strong inferences. Second, if we have measured the degree to which controls are impacted and if they are only slightly affected, we can recast our question as the aforementioned weakly versus strongly impacted comparison. Indeed, such questions are often of great interest. If considerable data have already been collected then stopping monitoring or removing locations may mean a waste of information and resources. Recasting our question does not mean the statistical analysis of the data must change –

rather, it alters the interpretation we draw from the results because our hypothesis and analysis framework have changed.

The worst case is where control locations are known or thought to have been affected but to an unknown degree because no measurements were taken to assess an impact's intensity, extent or duration. In this case, we do not know whether we have a group of weakly affected locations because some or all of the controls may have been strongly affected and consequently be equivalent to the 'impact' locations. This situation is difficult – probably impossible – to rescue because we have no logical way of recasting a sensible comparison of control and impact locations.

The latter outcome dictates that where uncertainty exists regarding the spatial extent of impact and its likely effects on control locations, prudence should rule. First, we should site control locations as far away from the impact location(s) as is possible without compromising their comparability to the impact location(s). Second, if we have the luxury of having many control locations (and enough money!), we can monitor more controls than are required, so that if we have to later remove some from the data set we can do so without reducing statistical power. Third, we can set up sampling stations at increasing distances from the point(s) of impact, which can provide information about how impacts dissipate over distance (Ellis & Schneider 1997). These strategies reduce the chance that we end up with data with no value for inferring impact.

Our second difficulty is where locations prove to be spatially autocorrelated. It is possible that for some physical variables, such as discharge, sufficient pre-impact data exist that the degree of spatial autocorrelation can be measured or modelled. In this case, a correction can be applied to reduce the possibility that incorrect decisions are reached (see Legendre 1993). For biological variables, it is unlikely that such data will exist. The possibility of estimating independently the degree to which spatial autocorrelation is a problem and applying a correction is thus unlikely. In the future, we should be able to incorporate spatial coordinates of data points routinely into our analyses to estimate and factor out pure spatial variance (e.g. Anderson & Gribble 1998; Borcard et al. 1992; Legendre 1993). Until then, however, conservative solutions that make spatial autocorrelation either unlikely or trivial, such as placing locations on well-separated tributaries, seem the most prudent.

Instances where we know that control locations differ systematically from impact locations in some environmental feature are not fatal in that the estimates of differences between them are, at least, unbiased.

The issue, here, is what may be logically deduced from any differences. Systematic environmental differences between controls and impacts are viable explanations for differences, and it does not suffice for us simply to try to wave these explanations away. Similar efforts are needed when controls and impacts are spatially confounded. Again, some thought ahead of time can help build a much more convincing case, using several lines of weak, but independent, evidence that differences are due to human impacts. We discuss how these lines of evidence may be constructed in the next chapter, when we discuss situations where controls and/or before data are not available. Ultimately however, we should realize that on strictly logical grounds, such a design cannot provide evidence for human impacts as well as a properly applied MBACI design. However, designs to detect human impacts provide *relative* degrees of inference. A design in which control and impact locations differ in some feature or are spatially confounded but where researchers have systematically tackled the problem ahead of time is likely to provide much better inference than one where the problem was ignored and no complementary evidence gathered.

*How many control locations are necessary?*

As emphasized in section 5.2, the more locations there are within the sampling design, the stronger is the inference (especially where we have more than one) and the greater is likely to be our statistical power to detect changes (section 7.4). Nevertheless, there is no 'magic' minimum number of locations that we advocate, because number of locations (and their variation through time) should be examined in the context of the direct effects on statistical power. In some circumstances, locating any controls will prove impossible. We discuss this situation in the next chapter.

### 8.1.4 Subsampling of locations

As explained above, the term Locations can represent anything from a short stretch of a single stream and its surrounding sub-catchment to whole catchments covering thousands of square kilometres. The physical size of locations is driven by the extent of the impact, as described above, but each location, regardless of size, provides one replicate in our survey design, and that equates with, effectively, one value per sampling time for Control vs. Impact hypothesis-tests. Subsamples, which are replicate samples collected from within an individual location, do *not* add replicates to tests of hypotheses about impacts. They serve to improve the precision with which the state of locations are known (or to provide improved estimates of the extent of impact in space where this is not well understood ahead of time as in the Beyond-BACI design). Consequently, it is important that researchers be very clear about what size of area equates with a location to avoid the possibility of confusing true replicates of impacts or controls with subsamples. Consider, for example, when we can use equipment that allows us to quantify variables over large areas, such as satellite imagery that provides a good estimate of the percentage of the whole catchment that is covered by trees (e.g. Wear et al. 1998). Our sampling process generates a single number for each location, and there is no subsampling. However, for most variables we will be using sampling equipment that works over scales far smaller than the likely size of most locations. Macroinvertebrates, plankton, algae, aquatic angiosperms and many fish are all examples of taxa where sampling is likely to use a piece of equipment that examines no more (and in some cases, far less) than a few square metres. We would not be confident that visiting one point in a stretch of stream or a whole catchment and collecting a single sample from that point would provide information that is representative of the rest of the location. One sample is unlikely to represent a whole site well, nor one site to represent a whole sub-catchment, because there is plenty of evidence that site-to-site variation can be considerable in a variety of ecosystems (e.g. Corkum 1991; Morrisey et al. 1992; and many others). As a consequence, we will almost certainly need subsamples in many cases.

A need to subsample immediately poses some problems. First, how do we distribute our subsamples effectively across an entire location? If locations are very large, there may be multiple units within them – river systems, sub-catchments within those systems and multiple sites within those – from which we could take samples. If a Beyond-BACI design is planned, then these nested, spatial units can form natural units for the nested sampling levels of the design (e.g. Corkum 1991). If we wish to take subsamples from only one level below that of locations (i.e. we do not want multiple, nested levels), the important issue is to ensure that subsamples are spatially representative of locations (section 5.4.1). We could locate subsamples entirely at random, but a probably better approach is to ensure a minimum are taken from each major subunit within the location. For example, if our catchment contained three major river systems, we could ensure that one-third of subsamples occurred in each of them, but with allocation within each of the river systems decided entirely at random. This approach ensures that we do not by chance end up with most subsamples from only one river system. Note that there is an acute shortage of information about the spatial scales over which fauna and flora respond to their environment. If we had detailed knowledge about such scaling, it would be much clearer which spatial scales to sample in order to represent well the state of, say, a whole catchment. As it is, we often have to choose scales at random – sizes of individual sites, for example, are often chosen quite arbitrarily, but can have profound effects on diversity and abundance (Downes *et al.* 2000).

The second problem is how we cope with the enormous numbers of subsamples that may be required to represent adequately each location at each time. If each subsample requires a lot of work or time to generate values, then the total effort and cost of the monitoring program could be prohibitive. If the nested subsamples are not being used to look at spatial variation *per se* (as would be the case in a Beyond-BACI or gradients design – section 5.4), then one solution, mentioned in chapter 7, is to composite samples (i.e. take multiple subsamples and combine them) rather than individually enumerate them. The combined sample is then subsampled to generate a single set of values that are likely to be representative of the whole location. Macroinvertebrates are a good example of species that are time-consuming to count and identify. However, techniques are available for combining multiple samples in the laboratory and then subsampling to create a representative sample (e.g. Marchant 1989; Walsh 1997). Compositing is a technique not often used by freshwater ecologists but is a valid way to proceed in these circumstances (see Barbour & Gerritsen 1996; Courtemanch 1996; Growns *et al.* 1997; Somers *et al.* 1998; Vinson & Hawkins 1996, for considerable discussion and examples). We will discuss these issues more in the following chapter on variable selection. The final issue with subsampling is: how many subsamples are needed? As emphasized in chapter 7, we can make that decision only as part of calculations of the statistical power of the design. In this respect, a few pilot data go a very long way. Discussion about how to carry out these sorts of calculations will be addressed in chapters 11 and 12.

### 8.1.5 Examples of decision trees for finding and choosing controls

In Figs. 8.3–8.7 and the accompanying tables (Tables 8.5–8.7), we have provided examples of the sorts of questions and decisions that can improve the selection of control locations. Figures 8.4 and 8.6 also illustrate how differences in the spatial location of controls can result in

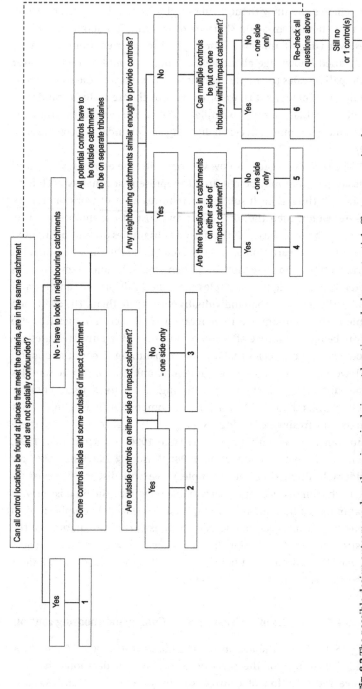

**Fig. 8.3** The possible design outcomes when there is a single impact location and where potential effects are restricted to a sub-catchment on one tributary (usually a point impact). Consult Table 8.5 for notes on each numbered outcome.

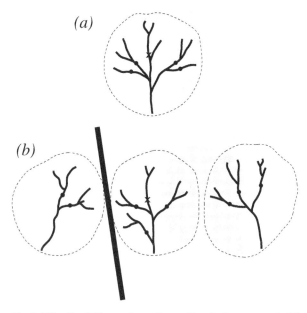

**Fig. 8.4** Idealized illustrations of sampling designs suggested from Fig. 8.3. In each case, the cross marks the impact location and filled dots potential control locations; catchment boundaries are marked by the dotted lines. (a) An ideal MBACI design, where all controls and the impact location occur within the same overall catchment (design 1, Table 8.5). There are no major gradients within the catchment likely to cause high Location × Time variance, and controls are spatially dispersed around the impact location and are not all at upland locales, likely to be very different from the impact location. (b) Designs where large-scale gradients may cause problems. The thick dark line marks a boundary that, unbeknownst to the investigator, causes controls in the catchment on the left to perform differently from those in the two catchments on the right. The light shading marks a further physical gradient that spans parts of the two right-hand catchments and that, also unbeknownst to the investigator, causes these parts of the catchments to perform differently from the rest. Design 1 (see Table 8.5) (which uses only the central catchment) might have higher Location × Time variability than the catchment illustrated in (a). Design 2 (which uses locations in all three catchments) has good replication but may suffer high Location × Time variance among controls caused by the various physical differences both between and within catchments. Design 3, in which only one other catchment can offer controls (beside the impact catchment), may also suffer high Location × Time variance, particularly if the left-hand catchment is the only one available to provide controls. These problems with large-scale physical gradients are likely to be exacerbated in situations where no controls can be gained within the impact catchment (designs 4 and 5).

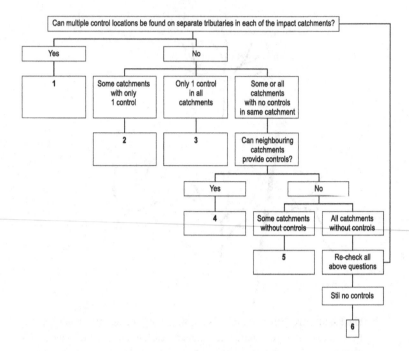

**Fig. 8.5** Some possible design outcomes when there are multiple impact locations, each a point source where potential effects are confined to a tributary. Consult Table 8.6 for notes on each numbered outcome.

different BACI-type designs or raise different potential problems with detecting any signal caused by human impacts.

We have used spatial proximity of suitable control locations to impact locations as the main way of making selections. However, we do not consider that spatial proximity should be used as a fixed rule. Spatial proximity is simply a surrogate for suites of environmental variables whose conditions we hope to keep similar by keeping the distances between locations as small as we can. However, there may be instances where we understand what these suites of environmental variables comprise and how they operate. In that case, spatial proximity could give way to a more reasoned use of these other variables.

The charts assume that the spatial extent (in terms of catchment size) of the impact and whether it is a point or non-point impact have been identified. They also assume that criteria for controls have been developed, the potential pool of locations identified, and the locations likely to have problems with independence (e.g. on same tributary) or with a high chance of being affected by the impact eliminated from the

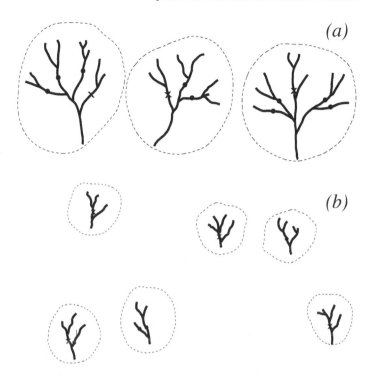

**Fig. 8.6** Idealized illustrations of sampling designs suggested from Fig. 8.5. In each case, dotted lines represent catchment boundaries, crosses mark impact locations and filled dots are potential control locations. (a) An ideal MBACI design, where multiple impact locations are accompanied by replicate control locations within each impact catchment (design 1, see Table 8.6). The catchments do not necessarily have to be in a row or close together. In this example, there are no major gradients within each catchment that are likely to cause high Location × Time variance. Replication of locations within catchments means the latter can be incorporated as a factor in the design, in which case any large variation among catchments will be removed from estimates of variance among Locations. (b) A more likely outcome for multiple point-source impacts, where some catchments have no nearby controls at all, some can gain only one in a nearby catchment, and some can gain both within-catchment controls and controls in neighbouring catchments. In some situations, we can anticipate that all three scenarios will occur within the same design. The degree to which this will create problems is dependent upon whether variation between locations within the same catchment vs. between locations in different catchments causes large Location × Time variance. Such variance can be caused by the sorts of physical gradients and scenarios described in Fig. 8.4.

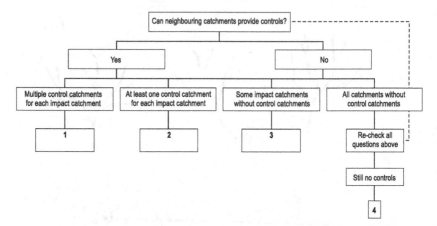

**Fig. 8.7** The possible design outcomes when there are multiple or single impact locations, each non-point and spread over a whole catchment – that is, effects are not restricted to individual tributaries. Consult Table 8.7 for notes on each numbered outcome.

pool. We separate point and non-point impacts because the former can be restricted to individual tributaries, which raises the prospect that controls can be sought within the same major catchment. If so, the unit of replication (i.e. the term 'Location' in our designs) may be small sub-catchments associated with individual tributaries. In the case of non-point impacts, the unit of replication for the design will usually be larger catchments (perhaps even whole major catchments) and will not usually be associated with individual tributaries.

### 8.2 TEMPORAL VARIATION, AND BEFORE AND AFTER SAMPLING

Natural variation through time and the need to sample both Before and After putative impacts raise problems analogous to all those discussed above for spatial variability and the need for Control locations. We need to know the likely temporal extent of impacts because this will define the length of Before and After sampling periods. We need to consider how we replicate times within these periods and, therefore how we will deal with the issues of subsampling over time. This section will be far shorter than that given to spatial replication. This is because we will not, in virtually all cases, have any control over the selection of Before or After periods, in the way that we can select locations to act as Controls. We will be forced to use whatever years (or other units of time) that we can sample prior to start-up and those years following.

### 8.2.1 Temporal extent and nature of impact

In chapter 3, we discussed the different temporal scales over which perturbations can occur. Perturbations may be relatively short-lived pulses, longer-term press disturbances or many other sorts of things. The response of the system may be an abrupt 'step' change, perhaps followed by a recovery, or a shift detectable only as a long-term trend. These different sorts of perturbations pose different problems for detection. Most of our statistical models are designed to detect step changes (section 7.5) but have some capacity to detect more subtle changes depending on how long monitoring continues.

Like the spatial extents of impacts described above (section 8.1.2), temporal extents also have profound implications for monitoring designs. First, we need to know over what time period stressors will be applied. Will stressors be applied only once over a relatively short period of time (say a few months) or will there be a series of pulse disturbances? Will any of the likely stressors act over a much longer period, more akin to a press disturbance? Such information should be part of any development proposal, but may have to be gleaned by looking at published studies of similar developments. Second, we need to know whether ecosystem responses to putative impacts are likely to occur over months, years or decades, because the length of monitoring before and after the onset of human activity needs to be tailored specifically to the appropriate temporal scale. Thus, if we have an impact that causes changes after two to three years, then three years should be the absolute minimum length of monitoring periods.

We should search systematically through the literature for as much information as we can about the likely duration and frequency of perturbations and about what we know of the temporal scales of any ecosystem responses. Review papers do provide such information (e.g. Niemi *et al.* 1990; Yount & Niemi 1990). For example, Niemi *et al.* (1990) surveyed 139 publications, each reporting responses and recovery times of freshwater ecosystems after press or pulse disturbances from stressors like DDT application, logging, dredging, flooding etc. They examined minimum and maximum recovery periods of different sorts of taxa (e.g. densities or species richness of macrophytes, micro-organisms, periphyton, macroinvertebrates, fish) and separated small streams from larger rivers (they also examined lakes). Recovery times varied greatly among taxa and stressors, with minimum times being one to two months and maximum times in excess of several decades. We do not provide more specific details here, because we believe researchers will

Table 8.5. *The diagnosis of design types illustrated in Figs. 8.3 and 8.4, which deal with single point-source impacts confined to individual tributaries in sub-catchments. Locations are sub-catchments, but because some sub-catchments may be in different major catchments, both catchment-level variation should be considered, either as a formal factor within the design or by checking the size of Location × Time variation*

| Number | BACI design type | Level(s) of replication (each level not necessarily replicated formally but may be a source of variation among Locations) | Comments |
|---|---|---|---|
| 1 | MBACI | Sub-catchment | Check that all available controls are not spatially confounded with impacts (e.g. all controls at upstream locales) |
| 2 | MBACI and, potentially, Beyond-BACI | Sub-catchment and possibly catchment | Large-scale spatial variation among catchments (e.g. if there are large-scale physical gradients) may mean high variation between Locations in different catchments. If there are sufficient replicates, it may be possible to incorporate a new term (Catchment), in which Locations are nested and where systematic variation among catchments can be factored out in a Beyond-BACI type design |
| 3 | MBACI | Sub-catchment and possibly catchment | As for 2 above. The possibility of variation between the control Locations within impact catchment and those in the neighbouring catchment could be high, so worth examining with some pilot data |
| 4 | MBACI | Sub-catchment and catchment | There may be significant variation between control and impact Location(s) caused by catchment-level differences, and also significant variation among control Locations driven by catchment differences. Again, pilot data would be valuable |

| | | | |
|---|---|---|---|
| 5 | MBACI | Sub-catchment and catchment | Control and impact Locations are spatially confounded, meaning that differences between them could be due to something other than potential human impacts. Extra data are required to discount the latter possibility (see chapter 9) |
| 6 | MBACI | Sub-catchment | Control Locations are potentially not independent of each other. This arrangement of locations should *only* be considered if it is possible to show, with some surety, that lack of statistical independence will not be a problem |
| 7 | BACIP | Sub-catchment | If one control location, then many observations through time required for BACIP design. If no control Locations available, then data not amenable to BACI-type models (see chapter 9) |

Table 8.6. The diagnosis of design types illustrated in Figs. 8.5 and 8.6: multiple, point-source impacts confined to individual tributaries in sub-catchments and with some in spatially separated catchments. Locations are sub-catchments, but because some sub-catchments may be in different major catchments, both catchment and sub-catchment sources of variation need to be considered

| Number | BACI design type | Level(s) of replication (each level not necessarily replicated formally but may be a source of variation among Locations) | Comments |
|---|---|---|---|
| 1 | MBACI and potentially Beyond-BACI | Sub-catchment and catchment | Catchment can be factored in as another level of spatial replication in which Locations are nested. This allows variation among catchments to be estimated separately from variation among Locations |
| 2 | MBACI | Sub-catchment and catchment | Different catchments may perform differently but are less easily factored out than in design 1, above. May be necessary to carry out preliminary tests to see if variation among catchments is likely to be a significant source of variance between control Locations |
| 3 | MBACIP | Sub-catchment | Each impact Location in one catchment paired with one control Location in a neighbouring catchment. Could be a problem if there are large differences even between neighbouring catchments because this may then create greater variance of differences over time |

| | | | |
|---|---|---|---|
| 4 | MBACI | Sub-catchment and catchment | Control Locations contain a mix of sub-catchments within impact catchments and locations in neighbouring catchments. Variation among catchments could cause high variation among control Locations over time |
| 5 | MBACI (unbalanced) | Catchment | Again, this design is dependent upon the relative importance of catchment-to-catchment variation – if it is high, then having some catchments without controls in neighbouring catchments may mean that there is substantial systematic differences among control and impact Locations over time |
| 6 | none | N/A | If no control Locations available, then data not amenable to BACI-type models (see chapter 9) |

Table 8.7. *The diagnosis of design types illustrated in Fig. 8.7: multiple or single non-point source impacts spread over whole catchments*

| Number | BACI design type | Level(s) of replication | Comments |
|---|---|---|---|
| 1 | MBACI | Catchment | Because we know little about what drives catchment-to-catchment variation, there could be high variation among some controls that may produce relatively high Location × Time variance. Additionally, the level of subsampling (see section 8.1.4) required will need careful consideration for all designs |
| 2 | MBACIP or BACIP if only one impact catchment | Catchment | Any large-scale variation among whole catchments across regions is addressed, at least partly, by pairing impact and control catchments. However, as indicated above, large differences among catchments, independent of impacts, could still cause substantial noise. If we have only one impact catchment, then one control only means we have a BACIP design |
| 3 | MBACI (unbalanced) | Catchment | As for 1 and 2 above, but the situation is worse in that some impact Locations do not have neighbouring control catchments to factor in some regional variation (assuming that this is a significant source of variability). The level of variation among Locations could be very high |
| 4 | none | N/A | If no control Locations available, then data not amenable to BACI-type models (see chapter 9) |

need to conduct their own review, targeted at particular taxa, stressors, and types of river. Niemi *et al.*'s (1990) study provides a nice example of how such a review could be approached.

The more information we have about the duration of perturbations and the likely response of the system, the better informed we will be about the length of time needed for monitoring both before and after the human activity has started. For example, if we know that perturbations will be short-lived and that the system response is likely to be a step change, then we may be confident of detecting such changes within a few years. If changes are likely to be over the long term, then we should plan as much as possible for sufficiently lengthy Before and After periods (see Hershey *et al.* 1998).

In the ideal case, our information about the temporal extents of perturbations would set the overall time frame and inform funding decisions for the monitoring program. More realistically, both of these things are usually set by the time frames over which management gets notice of impending human activities, has funding and must make decisions, even when these are all seriously mismatched with those of any putative impacts. Consequently, we will usually have to be pragmatic about the length of monitoring periods. For example, if the total time frame available is too short to detect important, long-term trends, then it may be wise to plan detection around other, short-term effects, even if these are considered less environmentally important (see chapter 11). The design can still incorporate measures for detecting longer-term changes, should further funds be made available.

### 8.2.2 Frequency of sampling within Periods

The next issue regarding time is the frequency with which we sample within Periods, or what the term 'Times' will represent. In chapter 7, we suggested that Times will often be represented by each year before and after start-up of possible human impact. Certainly some impacts (and the responses of the biota) may occur over periods of a year or more, which makes Times as years appropriate. However, you may wonder whether or not a year may be too long for some circumstances. For example, it is possible for some invertebrates to respond quickly to disturbance and return to pre-disturbance densities within one to two months (e.g. Niemi *et al.* 1990). However, we must keep in mind that typically there is a lot of temporal variability in rivers, much of it associated with seasonal changes. For example, seasonal changes in water temperature and light can mean these factors fluctuate throughout the year. Seasonal changes in rainfall can result in large changes in discharge and hence frequency of flooding or drying. Many lotic species show seasonally related changes in the densities of individuals, what life-cycle stages that are present, when reproduction occurs, and so forth. As a result, when we have a short-term change in something like invertebrate density that might be caused by human impacts, it is likely to be confounded with changes that are related to the time of year when human activities happen to start (see Linke *et al.* 1999 for an example). The only way we will have confidence that we can disentangle any human impacts from natural variability is if we have monitored natural fluctuation through multiple samples of that time of the year prior to start-up. This latter information will give us a good representation of the 'envelope' of natural variability, at different times of the year, among Control locations in the absence of human impact. Hence, even when impacts and responses to impacts are fairly short term, it makes sense to conceive of Times as years, and to expect (or hope!) to monitor for multiple years prior to human activity.

### 8.2.3 Subsamples within Times

Even when we have settled upon the temporal scale that will represent Times, we will probably need to subsample, for the same reasons discussed for subsampling of Locations (section 8.1.4). One sample per year (or whatever period we settle upon for Times) is unlikely to represent one Time particularly well for many variables.

The frequency of any temporal subsampling should be set by the temporal scales over which the variables of interest are likely to change. Densities of macroinvertebrates and algae, for example, might fluctuate markedly over a period of two to three weeks or less, suggesting that subsampling would be needed at that frequency to capture average population densities over the Time period adequately. However, densities of other organisms might be fairly consistent over a few weeks, which would suggest that sampling over that time frame would simply be wasted effort. If we are monitoring reproductive output or recruitment of juveniles into the population, these events may occur only during a particular time of year, in which case temporal subsampling might be concentrated at that time rather than spread throughout the year. There is no fixed regimen of subsampling that we can recommend, because it depends greatly upon the variables selected for monitoring. Again, the literature is the best source of information about likely temporal scales of variability.

By subsampling, we try to make sure that our estimates are precise; that is, that the average of those numbers will represent well the state of a Location during a particular Time. What we do with the subsamples then depends upon the design we have settled upon. If we have planned a Beyond-BACI design, then different sampling frequencies are nested within Times. If we use an MBACI design, then the subsamples serve simply to estimate precisely the state of each location at each Time. In the latter case, we might choose to composite these subsamples (as described above) and create good representative samples from them to keep down costs.

### 8.2.4 Statistical independence and sampling through time

In section 8.1.2 we described the problems created when data from different Locations are not independent because the Locations are close together in space. The same problem occurs with samples through time. Samples that are close together in time may be more similar (or dissimilar, although this is less likely) to each other than they are to data collected much later or much earlier. This is called temporal autocorrelation, and it can have the same effect on statistical tests we described above (Table 8.4). Temporal autocorrelation can occur where the frequency of sampling is high enough that the same individuals are sampled repeatedly because they are relatively long-lived. However, temporal autocorrelation can also occur if populations get a sudden influx of juveniles that greatly increase densities for a time. Samples within this time might be much more similar to each other than to other samples collected at other times.

As for spatial autocorrelation, there are few studies of how ecological variables can be temporally autocorrelated, and therefore few guidelines about how one can reduce its size or incidence. Autocorrelation between temporal subsamples is not an issue if the latter are to be composited or otherwise averaged to generate a single value for each Time period. However, this sort of autocorrelation can be a problem for Beyond-BACI designs, because those nested subsamples are retained as factors within the design. Autocorrelation among Times, especially when Times are represented by years, is much less likely but can occur with particular long-lived taxa. For example, many trees and even some fish can live for many years, and frequent, non-destructive sampling might collect the same individuals repeatedly at the same locations. The best strategy is to use the life cycle of species and the longevity of individuals as a guide to whether temporal autocorrelation is likely to be a problem.

## 8.3 DOING THE SAMPLING

There are just a couple of things to re-emphasize here from issues raised in chapter 5. First, we are very likely to be monitoring more than one variable to examine any putative impacts. Different variables may pose different requirements for the sampling design, such as different sampling frequencies through time. It is important that the final monitoring design not be geared to the sampling requirements of variables requiring the least sampling effort (in either space or time). Second, different variables may create different logistical problems in sampling (i.e. types and numbers of people required to carry out sampling, types and areal extents of sampling gear). Nevertheless, it is logistically efficient to collect all data simultaneously, or as much as possible. Additionally, we must keep in mind that Impact and Control locations should be sampled at the same time or as close in time as is practically possible. If it is not possible to obtain all samples in a short window of time, then it is important to consider ways of randomizing or stratifying the order of sampling of locations to avoid any potential biases (section 5.4.4).

Finally, we re-emphasize that it is important that the outcomes of and data collected from monitoring programs be published in places with wide accessibility to others. We can learn much from previous examples of human impacts, so it is important that monitoring data be archived properly and made accessible to others. Additionally, the results of the monitoring program should be published in journals or other publications with wide distribution. Some examples of published BACI-type studies are given in Table 8.8. The designs are used in both experimental manipulations of stressors in mesocosm-type studies, as well as for assessing human impacts in real settings, at small or large spatial and temporal scales and in a variety of environments. Such studies provide very useful information about human impacts.

## 8.4 A WORKED EXAMPLE – EFFECTS OF LIMING TO DECREASE ACIDITY

In this section we provide a worked example using some real data from a long-term experiment that is examining the effects of acidification of streams and whether such effects can be reversed by adding lime to stream waters. The design is an MBACI, and we examine here each step of the process: the background to the problem and preliminary data collection, the selection of control (and in this case, impact) locations, predictions that are based upon preliminary data, data collection, analysis and interpretation.

### 8.4.1 Background to the problem and preliminary data

Acid deposition into fresh waters – often from acid rain – is a world-wide problem. In Wales, researchers have been studying multiple streams draining into the catchment of the upper River Tywi, a river having a catchment area of 1108 km$^2$ in mid Wales. In 1969 a regulating reservoir (Llyn Brianne) was constructed in the upper reaches of the catchment. After a few years, it became apparent that salmonid fish populations were declining, despite efforts to help fish reach upper areas of the catchment by trapping and transporting them there by truck. It was suspected that juvenile fish were not surviving in streams above the reservoir, and a broad study was begun into the water chemistry of streams draining into the catchment (Stoner et al. 1984). Although some streams were approximately circumneutral in pH, others were quite acidic. Acidity was generated by occasional bouts of acid rain, together with acids generated from soils within the catchment. The effects of this acidity differed between streams however because of variability in buffering capacity and land use. Some streams had relatively high concentrations ( > 10 mg L$^{-1}$) of CaCO$_3$, which buffers acids; these streams had an average pH of ∼ 6.0. Other streams had lower concentrations of CaCO$_3$, lower mean pH and relatively high concentrations of soluble aluminium, which is toxic under acid conditions. Additionally, stream acidity was higher in catchments where the vegetation had been converted to spruce and lodgepole pine forests, probably because of increased evapo-transpiration from these catchments relative to those covered by moorland vegetation. Acidic streams contained no fish (and fish died when transplanted into them) and had a depauperate invertebrate fauna. Further studies (Rutt et al. 1989) established that macroinvertebrates showed shifts in community structure associated with stream acidity, and the researchers were able to identify species that were either sensitive to acidity or were relatively unresponsive to alterations in acidity. They also established that invertebrate densities varied among microhabitats within streams, suggesting that samples from both riffles and marginal areas were necessary to provide a reliable picture of species density or diversity at a location.

Having established that the likely cause of fishless streams was stream acidity, researchers proposed adding lime to these streams to create a circumneutral pH and improved conditions for acid-sensitive biota. To their credit, they set up liming additions as an experiment in which they also monitored streams that did not have lime added to them (details in Rundle et al. 1995). The experiment is important, because liming is a technique used widely in Europe to counter increasing

Table 8.8. *Some recent examples of studies using or examining BACI-type approaches to assess human impacts. The list is not intended as either comprehensive or a representative sample; it serves to illustrate simply the variety of impacts, study types and ecosystems where BACI approaches have been used*

| Environment | Type of impact | Dependent variables | BACI details | Reference |
|---|---|---|---|---|
| Whole river systems | Habitat alteration via flow releases | Numbers and other measures of populations of chinook salmon | One Control catchment, one Impact catchment; multiple years + time-series simulations | Korman & Higgins (1997) |
| Multiple wetlands | Application of mosquito larvicides | Abundances of benthic invertebrates | Nine Control locations, nine locations each for two types of Impact; three years Before, three years After | Hershey et al. (1998) |
| Second-order river | Thermal pollution from power plant | Abundance of phytoplankton, benthic invertebrates and fish taxa | One Control location, one Impact location; seven years Before, seven years After | Smith et al. (1993) |
| River | Exposure to low-frequency electromagnetic fields | Densities and diversity of macroinvertebrates | One Control location, one Impact location; two + years Before, four years After | Stout & Rondinelli (1995) |
| Single creek | Point source pollution from mine | Densities and diversities of macroinvertebrates | One Control location, one Impact location, eight samples Before, seven samples After | Faith et al. (1995) |

| Location | Impact | Variables measured | Design | Reference |
|---|---|---|---|---|
| Multiple streams | Infusion of fine sediments from road construction | Densities of several species of amphibians | Five Control locations, five Impact locations; no Before data possible, one set of After data | Welsh & Ollivier (1998) |
| First-order stream | Experimentally increased water temperatures | Densities of selected species of invertebrates | One Control location, one Impact location; one year Before, two years After | Hogg & Williams (1996) |
| Freshwater ditches | Experimentally applied insecticide and nutrient additions | Number of oligochaetes | Four Control ditches, two Impact ditches per insecticide concentration; three times Before, four times After | Verdonschot & Ter Braak (1994) |
| Upland, second-order streams | Application of a pesticide used to control terrestrial insect pests | Density and diversity of stream macroinvertebrates | Two Control catchments, two Impact catchments; three years Before, one year After | Hurd et al. (1996) |
| Marine near-shore subtidal | Discharge of heated water from nuclear power plant | Underwater light levels; numbers of selected invertebrate species | One Control location, one Impact location; three years Before, three years After | Reitzel et al. (1994); Schroeter et al. (1993) |
| Marine rocky intertidal | Human harvesting of shellfish | Numbers and shell sizes of collected and uncollected mollusc species | Two Control locations, six Impact locations; three years Before, five years After | Keough & Quinn (2000) |
| Marine rocky subtidal | Discharge of sewage effluent | Density and diversity of sessile fauna and flora | One Impact location, two Control locations; no Before data possible, one time After | Chapman et al. (1995) |

Table 8.8. (*cont.*)

| Environment | Type of impact | Dependent variables | BACI details | Reference |
|---|---|---|---|---|
| Marine subtidal rocky reefs | Increases in physical disturbance by divers, brought about by installation of a new diving buoy | Density, size and distribution of colonies of a bryozoan, *Pentapora fascialis* | One Impact, four Control locations; one year Before, three years After | Garrabou *et al.* (1998) |
| Marine subtidal soft sediments | Thermal pollution caused by heated effluent from a coastal power station | Abundances of macro- and meiobenthic fauna, converted to similarity indices | One Impact, three Control locations; no Before data possible, one set of After data | Lardicci *et al.* (1999) |
| Marine coastal mangroves | Removal of a berm | Abundances of fish and macrobenthic invertebrates | One Control location, one Impact location; 13 months Before, 22 months After | Vose & Bell (1994) |
| Marine subtidal soft sediments | Physical disturbance to the benthos by trawling | Abundances of benthic infauna | One Control location, one Impact location; three Before samples, six After | Currie & Parry (1996) |
| Terrestrial rainforest | Effects of logging | Numbers of birds and small mammals | One Control location, one Impact location; three years Before data, one and a half years After | Crome *et al.* (1996) |

acidity, even though the ecological benefits are unclear. By monitoring limed and unmanipulated streams both before and after liming, researchers could be reasonably confident of distinguishing any benefits of liming from background variation in space and time. As we shall see, the ability to do this turned out to be very important. The study is thus one of a restoration technique rather than of a human impact, *per se*. However, the underlying logic of the design, which is an MBACI, is the same (a theme we return to in chapter 14). We note though that their experiment has some added advantages not usually present in human impact studies, which increases confidence in the conclusions: replicate impact locations, plus control over the timing and placement of 'impacts' (i.e. liming). In the following discussion, 'impact' will refer to the addition of lime. To illustrate how MBACI designs and their analysis work, we will discuss the effects of liming as if this were a human impact being compared to otherwise undisturbed streams.

### 8.4.2 Selection of control and impact locations

Six locations were chosen, three of which were selected to be limed, and hence are termed 'impact' locations, and three of which were left unchanged (controls; Table 8.9). From the preliminary work, it was clear that control and impact locations could be on separate tributaries within the upper catchment of the River Tywi. It was also clear that control and impact locations should be matched closely for water quality and for catchment vegetation – these characteristics thus formed the selection criteria. Prior to liming, impact and control locations had similar mean pH, and $CaCO_3$ and filterable aluminium concentrations. Impact and control locations were also matched overall for catchment vegetation, with the inclusion of one coniferous catchment in both the impact and control groups (Table 8.9; see Rutt *et al.* 1989 and Rundle *et al.* 1995 for further details).

### 8.4.3 Predictions and data collection

From the preliminary data, it was possible to make reasonably detailed predictions about the effects of liming. Limed streams were predicted to show increased numbers of acid-sensitive macroinvertebrate taxa and increased overall diversity of such taxa relative to control streams, which themselves should remain relatively constant through time. Taxa not sensitive to changes in acidity should show no systematic differences between impact and control locations.

Table 8.9. Characteristics of control and impact locations in the Welsh liming study, including relevant measures of water chemistry

| Code | Locations | Catchment area (ha) | Slope | Mean pH | CaCO$_3$ (mg L$^{-1}$) | Filterable Al (mg L$^{-1}$) |
|------|-----------|---------------------|-------|---------|------------------------|------------------------------|
| *Control locations* | | | | | | |
| C1 | Moorland | 15 | 0.124 | 5.2 | 3.9 | 0.10 |
| C4 | Moorland | 71 | 0.063 | 5.3 | 4.5 | 0.12 |
| L1 | Coniferous afforestation (~ 30 years old) | 264 | 0.096 | 4.8 | 5.3 | 0.36 |
| *Impact locations* | | | | | | |
| C2 | Moorland | 77 | 0.072 | 5.0 | 3.9 | 0.16 |
| C5 | Moorland | 33 | 0.085 | 5.2 | 4.2 | 0.15 |
| L4 | Coniferous afforestation (~ 30 years old) | 32 | 0.113 | 5.0 | 7.1 | 0.17 |

Source: Rutt et al. (1989).

Macroinvertebrates were sampled at all locations once a year, each year, during spring using a kick sample technique applied to both riffles and marginal habitats. Water quality was measured from spot samples collected at one to two week intervals. These sampling techniques and temporal scales had all been established as appropriate from initial data collection, which had established that April was the best time to detect acidification effect following wet weather flows. Impact locations were limed in late 1987 (C5, L4) or mid-1988 (C2).

### 8.4.4 Results and analysis

We should examine first whether liming the streams produced changes in water chemistry. The liming of the impact locations caused a clear increase in mean pH relative to control locations (Fig. 8.8a). This change in pH was accompanied by increases in calcium concentration at all three impact locations and decreases in aluminium concentration at two of them. The third impact location (C2) did not show a significant drop in aluminium concentrations, but its pH was sufficiently high that the aluminium was thought unlikely to be toxic. Hence, we can conclude that the liming treatment was effective in bringing about the desired changes in water chemistry.

We present the data and details of the analysis for just one of the dependent variables considered by Rundle *et al.* (1995), which is the total number of individuals of acid-sensitive taxa present at locations. The data are provided in Table 8.10, and their analysis by the MBACI model that was described in chapter 7 appears in Table 8.11, with effects of liming illustrated in Fig. 8.8. We have also graphed the responses of individual taxa to consider the overall responses of the fauna to liming (Fig. 8.9).

The two tests for impact are the two interaction terms: *Impact vs. Control × Before vs. After* term and the *Impact vs. Control × Times*$_{within B vs. A}$ term. The former tells us whether any differences between impact and control locations change when we compare the Before period to the After period. The second term tells us whether any differences between impact and control locations alter when we compare among times, within periods. In the analysis (Table 8.11), both of these terms have F-ratios with probabilities of less than 0.05, which in traditional hypothesis-testing (section 4.7) allows us to reject the $H_0$ of no changes. Examination of the means of control and impact locations through time (Fig. 8.8b) shows the source of the interactions. Numbers of acid-sensitive taxa increased the first year after liming at impact, but not control, locations,

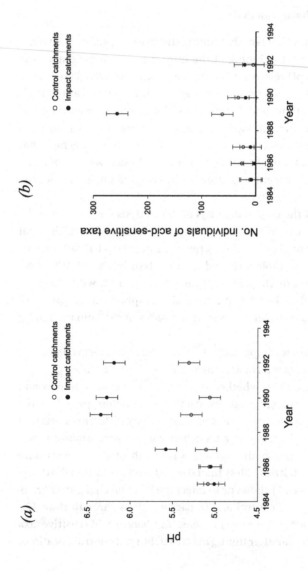

**Fig. 8.8** Results of the liming experiment in the River Tywi catchment. Liming occurred in late 1987 and early 1988. (a) The average, annual mean pH at control and impact locations in different years. (b) The mean total number of individuals of acid-sensitive taxa present in control and impact locations in different years. In each graph, the error bars are standard errors derived from the appropriate MS term in the analysis.

Table 8.10. *Data from the River Tywi catchment liming experiment*

| Year | Before or After liming | Control or Impact location | Location | Number of individuals of all acid-sensitive taxa |
|------|------------------------|----------------------------|----------|--------------------------------------------------|
| 1985 | Before | Control | C1 | 18 |
| 1985 | Before | Control | C4 | 5 |
| 1985 | Before | Control | L1 | 6 |
| 1985 | Before | Impact | C2 | 1 |
| 1985 | Before | Impact | C5 | 7 |
| 1985 | Before | Impact | L4 | 14 |
| 1986 | Before | Control | C1 | 72 |
| 1986 | Before | Control | C4 | 2 |
| 1986 | Before | Control | L1 | 3 |
| 1986 | Before | Impact | C2 | 0 |
| 1986 | Before | Impact | C5 | 1 |
| 1986 | Before | Impact | L4 | 5 |
| 1987 | Before | Control | C1 | 40 |
| 1987 | Before | Control | C4 | 26 |
| 1987 | Before | Control | L1 | 4 |
| 1987 | Before | Impact | C2 | 3 |
| 1987 | Before | Impact | C5 | 21 |
| 1987 | Before | Impact | L4 | 3 |
| 1989 | After | Control | C1 | 134 |
| 1989 | After | Control | C4 | 54 |
| 1989 | After | Control | L1 | 0 |
| 1989 | After | Impact | C2 | 265 |
| 1989 | After | Impact | C5 | 167 |
| 1989 | After | Impact | L4 | 340 |
| 1990 | After | Control | C1 | 72 |
| 1990 | After | Control | C4 | 26 |
| 1990 | After | Control | L1 | 0 |
| 1990 | After | Impact | C2 | 47 |
| 1990 | After | Impact | C5 | 8 |
| 1990 | After | Impact | L4 | 2 |
| 1992 | After | Control | C1 | 8 |
| 1992 | After | Control | C4 | 6 |
| 1992 | After | Control | L1 | 1 |
| 1992 | After | Impact | C2 | 27 |
| 1992 | After | Impact | C5 | 20 |
| 1992 | After | Impact | L4 | 15 |

*Note:* There are six locations: three controls (C1, C4, L1) and three impact streams (C2, C5, L4). Macroinvertebrates were sampled over 3 years (1985–1987) prior to liming of the impact streams, which occurred in late 1987 and in 1988. Shown here as a dependent variable are the total numbers of individuals collected per sample of 18 species of taxa known from previous work to be sensitive to acidity. Note that we have omitted data collected during 1988, as this represents a Before time for one impact catchment and an After time for the others.
*Source:* Rundle *et al.* (1995).

Table 8.11. *Repeated measures analysis of variance of the total numbers of individuals of all 18 acid-sensitive taxa within samples*

| Source | SS | df | MS | F-ratio | P |
|---|---|---|---|---|---|
| *Among locations* | | | | | |
| Impact vs. Control | 6110.03 | 1 | 6110.03 | 2.095 | 0.221 |
| Locations$_{within I vs. C}$ | 11668.44 | 4 | 2917.11 | | |
| *Within locations – repeated measures* | | | | | |
| Before vs. After | 25653.36 | 1 | 25653.36 | 25.824 | 0.007 |
| Times$_{within B vs. A}$ | 79840.78 | 4 | 19960.19 | 19.065 | < 0.001 |
| Impact vs. Control × Before vs. After | 14042.25 | 1 | 14042.25 | 14.136 | 0.020 |
| Impact vs. Control × Times$_{within B vs. A}$ | 38495.22 | 4 | 9623.81 | 9.192 | < 0.001 |
| Locations$_{within I vs. C}$ × Before vs. After | 3973.56 | 4 | 993.39 | 0.949 | 0.462 |
| Locations$_{within I vs. C}$ × Times$_{within B vs. A}$/Error | 16751.33 | 16 | 1046.96 | | |

*Notes:* The two tests of impact, Impact vs. Control × Before vs. After and Impact vs. Control × Times$_{within B vs. A}$, are both associated with relatively small probabilities. The relatively high probability of the Locations$_{within I vs. C}$ × Before vs. After term demonstrates that catchments performed similarly within the categories of impacts and controls over time.

Data were analysed using SYSTAT for Windows Version 8.03. Two outliers were detected but tests revealed they had no effect on the probabilities of the F-tests. Note that use of analyses of variance should always be accompanied by detailed examination of assumptions including that of compound symmetry. We have not reported on whether the data meet these assumptions for the sake of brevity. Readers may wish to check these assumptions themselves and observe the effects that different transformations can have on the F-tests!

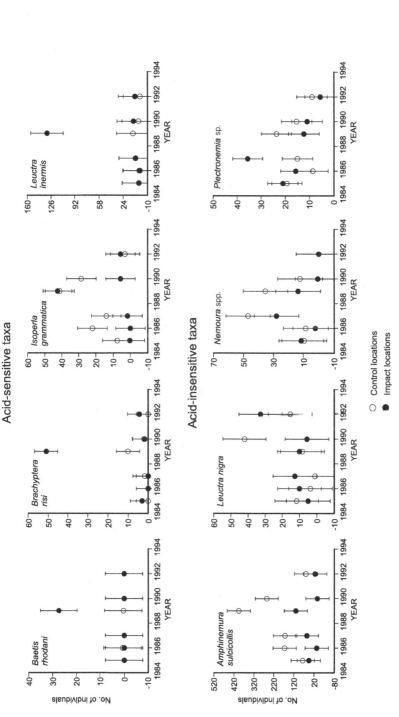

**Fig. 8.9** Mean numbers of each of eight macroinvertebrate taxa, four of which are acid-sensitive (*Baetis rhodani, Brachyptera risi, Isoperla grammatica, Leuctra inermis*) and four acid-insensitive (*Amphinemura sulcicollis, Leuctra nigra, Nemoura* spp., *Plectrocnemia* sp.), at control and impact locations in different years in the liming experiment. *Baetis rhodani* is a mayfly and *Plectrocnemia* sp. is a caddisfly; the rest are stoneflies.

but then dropped back to pre-liming levels in subsequent years. Our confidence that we are examining a real response to changes in water chemistry is also increased by examining the responses of individual taxa of differing acid-sensitivity (Fig. 8.9). Four acid sensitive species all showed peaks in abundance the year following liming, whereas four species not sensitive to acidity show no systematic changes from pre- to post-liming periods, as expected.

It is worth noting that had the liming resulted in a sustained increase in acid-sensitive taxa, with numbers remaining high at impact locations during 1990 and 1992, then the significance of terms in the model may have changed. The *Impact vs. Control × Before vs. After* term would remain significant, but if numbers were relatively similar among times within periods, then the *Impact vs. Control × Times*$_{within\,B\,vs.\,A}$ test may not have been associated with a significantly large $F$ value.

The other term of interest in the analysis is that of *Locations*$_{within\,I\,vs.\,C}$ × *Before vs. After*, which shows whether locations within categories (i.e. either impact or control) performed similarly between the periods. The relatively small value for the $F$-test suggests this was the case. Recall that the similarity of control locations to each other over time (assuming that most impacts will not have replicate locations) is particularly important for the MBACI model presented in chapter 7 (see also Fig. 8.2). The MS for *Locations*$_{within\,I\,vs.\,C}$ × *Before vs. After* (or *L(C)B*) provides the denominator for the $F$-test for *Impact vs. Control × Before vs. After*. Consequently a large MS for *L(C)B* results in a small value for this $F$-test and a likely conclusion of no change, unless the effect caused by the impact is particularly large. Likewise, if control locations differ from each other from time to time within periods, the *Locations*$_{within\,I\,vs.\,C}$ × *Times*$_{within\,B\,vs.\,A}$ MS will also be high. We cannot estimate this term directly for the data shown in Table 8.10 (because we have no way of separating it from the residual error – see chapter 7), but if it is large then the MS error is also large, again resulting in relatively small values for the $F$-test for *Impact vs. Control × Times*$_{within\,B\,vs.\,A}$.

Were this a test of a human impact, our conclusion would be that the impact caused a short-term increase in invertebrate numbers, which then returned to pre-disturbance levels in a classic pulse response (chapter 3). We would conclude that the design was successful at detecting a change caused by the impact, and that the impact caused no sustained alterations. As this was actually an experiment in restoration, the interpretation is that liming caused only a brief and unsustained improvement in invertebrate densities, which did not reach those seen in circumneutral streams in the area (see Fig. 14.2 for further information).

The conclusion was that restoring acid-affected streams is not a simple matter of restoring circumneutral water chemistry.

Finally, we can use the data illustrated in Figs. 8.8 and 8.9 to show how we may be misled when elements of BACI designs are missing. Consider the response of *Isoperla grammatica*, for example. This acid-sensitive species showed an increase in numbers in limed streams in the year immediately following liming – but it also showed a very similar increase in control streams, and this increase was also sustained for longer (Fig. 8.9). Were we monitoring just impact locations with this taxon, we might think we had good evidence of an impact. The control locations however suggest that another explanation – perhaps relating to environmental changes independently of or together with water chemistry – might be just as likely. Our confidence that the responses of the other three acid-sensitive taxa are due to liming is comparatively high because their numbers are otherwise relatively consistent over time for the controls. We can also see why BACI-type designs are useful by examining the changes in numbers of acid-insensitive taxa. Acid-insensitive taxa sometimes showed increases, but different taxa increased in different years. Some increased in control locations and others at impact locations. Such shifts in densities in time or space are *not* unusual; they are why we cannot use change *per se* at a location as evidence of an impact with much confidence. We examine ways of increasing our confidence in our conclusions further in chapter 9.

8.5 IMPORTANT ISSUES

- Overall, to be able to apply monitoring designs well requires a good understanding of the nature of the particular ecosystems that will provide control and impact locations.
- Applying BACI-type designs requires us to have a good understanding of the spatial extent of the putative impact because this allows us to know where we may locate controls. Additionally, impacts that are spread over large spatial scales present many more design problems than those restricted to small scales.
- The need to find controls requires us to develop criteria that help ensure the comparability of control locations to each other and to the impact location(s). There is a dilemma in that the more narrowly we define characteristics of control locations the more similar they are likely to be to each other but the fewer will be the number of places likely to meet our criteria.
- We need to ensure that control locations will be free of the human

impact under consideration, that they are relatively independent of each other, and that they are not located in spatially separated or different types of environments from the impact location(s). These requirements are not equally important however. Some breaches may require us to collect complementary data, whereas others will mean the design can only ever produce ambiguous answers about changes caused by human impacts.

- There are no fixed recommendations for a minimum number of controls, because numbers can only be decided in the context of the final design and effects on statistical power.

- Subsampling of controls will often be required where locations are large spatial areas and/or where sampling equipment works over small areas. Subsamples may be incorporated as factors into the design, but it is important to realize that subsamples do not add replicates to the hypothesis-tests of interest.

- It will be important to know or estimate well the likely temporal extent of any putative impact because this will determine the minimum length of Before and After monitoring periods.

- Likewise, it is useful to know the temporal scale of response by the biota, if any, to determine what the factor Times should represent. In many cases we expect that monitoring will continue for multiple years before and after start-up of human activities even when biota respond quickly, as we will need to disentangle any human-caused changes from seasonally related effects.

- Samples over time should be independent of each other unless autocorrelation can be accounted for in the analysis. Independent sampling can be problematic with long-lived taxa.

- Subsamples within Times can be collected and either composited or incorporated as factors within the design, in the same way that subsamples can be taken within Locations.

# 9
## Inferential uncertainty and multiple lines of evidence

Chapter 5 described the logic behind the BACI approach and chapter 7 described the four basic BACI-type analytical models and the relative strengths of inference each provides (Table 7.1). We can make strong inferences (those with least uncertainty) about the effects of human impacts by examining differences between control and impact locations before and after the onset of human activity, most especially when we have replication of these design elements. However, what happens when one or more BACI elements are entirely missing or when we have no replication? Perhaps the most common problem facing environmental managers is where putative impacts have already occurred, tens or even hundreds of years before, and there is no scope for planning a Before period. There may be no control locations because all suitable locations have suffered the same human activity in question. The latter problem is particularly common when modern human activities are spread over large spatial scales because, as indicated earlier, it reduces the potential numbers of places we can search for controls. How should we proceed in these circumstances?

We must recognize first that the difficulties created here are ones of increased inferential uncertainty, *not* which analytical model to apply. When one or more of the four elements are missing, we lack the information that would otherwise allow us to distinguish, with some confidence, those changes caused by human impacts from those caused by alternative (natural) phenomena (Table 9.1; and see chapter 5). We cannot rectify these situations by simply employing fancier or more complicated statistical models because the latter, no matter how sophisticated, cannot replace information missing from the foundations of the logic we are using.

What can we do in these cases? As we have emphasized earlier, the first step is always the recognition of such difficulties, because this opens

Table 9.1. *Alternative explanations, besides the effects of human impact, for differences seen when using different monitoring designs*

| Design | Alternative explanations for differences |
| --- | --- |
| MBACI or Beyond-BACI | Differences are caused by another change[a] that was coincident with the start of human activity and occurred only at the impact location(s) but not at most of the $n$ control locations. (We have argued that as $n$ increases, our confidence that changes are due to human impacts also increases.) |
| BACIP | Differences are caused by another change that was coincident with the start of human activity and occurred at the impact location but not at the control location. (As the number of time periods before the impact increases, our confidence that changes are due to human impacts also increases.) |
| Control vs. Impact locations – no before data | Differences are caused by important differences unrelated to impact between control and impact locations |
| | Differences are caused by another change that was coincident with the start of human activity at the impact location |
| Before vs. After at impact location – no controls | Differences are caused by another change that was coincident with the start of human activity at the impact location |
| Impact location, after impact only | No change has actually occurred at the impact location since human activity started |

[a] The words 'another change' refer to a change, either natural or human-caused, that is unrelated to the human impact under consideration.

our minds to considering what other sorts of data could be collected that will bear on our problem. In general, these data will provide weaker evidence, but if there are several sets, then collectively we can build a stronger case than we would have otherwise. We detail this 'levels of evidence' approach below (section 9.2).

This chapter largely uses arguments developed over the past 35 years in epidemiology, a use of reasoning that mirrors early discussion in epistemology and logic by Mill (1884). Epidemiology is a discipline where the dire subject matter precludes experiments on people for ethical reasons. Impact monitoring is probably not quite so constrained because human development will continue and we can plan our studies by anticipating them. Also, we have the advantage that experiments are often possible and, with an altered focus about their importance, large-scale experiments are becoming more common.

## 9.1 A BRIEF REVISIT OF INFERENTIAL UNCERTAINTY AND PROBABILITY

Before discussing how we might proceed, we should revisit briefly some ideas developed in chapter 4 about uncertainty and probability (section 4.2), and in chapter 5 about design (section 5.3). Recall that we use probabilities – particularly improbabilities – to help us make a decision about whether we have 'detected' a human impact. We regard the human impacts hypothesis ($h$) as being supported (corroborated) if it is improbable that we will observe certain sorts of data (evidence, $e$) in our monitoring program in the absence of human impact. That is, we observe evidence for impact that cannot be explained away by other processes, such as natural variation (or 'background knowledge', $b$). We can express this as low $P(e,b)$ – a low probability of observing $e$ given only the knowledge $b$. Low $P(e,b)$ is given by a small tail probability in the conventional test of a null hypothesis (section 4.7), in which case the hypothesis $h$ has passed a severe test.

The BACI-type designs described in chapter 5 (and developed analytically in chapter 7) allow us to find evidence for impact (if it occurs) that is improbable under natural conditions (section 5.3). Each of the design elements (Controls, Before data, replication) contributes to this ability to detect variation that is not part of normal variation – each contributes to reducing inferential uncertainty. Hence, designs lie on a gradient of inferential uncertainty. Replicated BACI designs provide the least uncertainty. If we lack proper replication then this increases our uncertainty (Table 7.1). If our design has no control locations and/or no before data, then this also increases our uncertainty. Specifically, we lack information that will help us distinguish changes at the impact location(s) caused by human impacts from those caused by some natural change coincident with the start of human activity. We want to be able to find a low $P(e,b)$.

We need a process that will help us better distinguish between the hypothesis and background knowledge as explanations, and therefore reduce the level of uncertainty we will have in detecting human impacts with these designs. We advocate, therefore, a levels-of-evidence approach as appropriate to our philosophy of monitoring. We have highlighted the problem of weak designs and inferential uncertainty, but there are three other reasons why multiple lines of evidence are desirable. The first, and we argue most compelling, argument is that our best designs (an MBACI or Beyond-BACI design) would seem to be capable of providing us with definitive evidence, but we have good reason not to trust them completely. Even MBACI and Beyond-BACI designs will not have, as a rule, random allocation of locations to treatments (i.e., as either an impact or a control), a process that in experiments helps us decouple treatment effects from other, background variation. So, for all monitoring designs, there are alternative explanations for any differences that are detected besides that of the hypothesis of human impacts (Table 9.1). We can argue that these alternative explanations are less probable when we have MBACI or Beyond-BACI designs (especially as the number of locations goes up), but we cannot discard them out of hand. These lessons about 'errors' associated with different designs are akin to Mayo's (1996) error 'repertoires' – the general reasons we have found for how evidence can turn out to be not so improbable. In Popper's terms (Popper 1983), they reflect awareness of weaknesses in background knowledge.

The other side of the coin, providing the second reason for using multiple lines of evidence, is one where the initial improbability is not low, but we suspect we have been misled. This is a different kind of design failure from that shown in Table 9.1 – the intent may have been there, but the good design was not ultimately realized. Perhaps the replication was not as great as we had planned, so the improbability is merely suggestive rather than leading to rejection of the null hypothesis. In this case, it may be possible for other (independent) lines of evidence to produce improbabilities that, when multiplied together with the first, might be able to give a multiple-tests assessment (see, for example, 'Combining probabilities from independent tests of significance' in Box 18.1 in Sokal & Rohlf 1995). Popper (1983, p. 247) argued for multiple lines of evidence in order to obtain more severe tests overall: 'a statement describing many tests (especially if they are independent of one another) will be less probable than a statement describing only some of these tests.'

The third reason for multiple lines of evidence is simply that any good design should have several different lines of evidence. Our important 'logical elements' include the need to consider different kinds of evidence. The philosophical basis for testing for impacts makes this clear. The hypothesis is that an impact has occurred at some location. Corroboration for that hypothesis is achieved when the observed evidence is judged improbable given only our 'background knowledge' about what is normal. Nevertheless, even when corroboration is achieved, the hypothesis of impact is not proven, and other lines of evidence are expected to provide other tests and potentially further corroboration. Further, absence of corroboration from one kind of evidence naturally does not imply absence of impact (a false hypothesis) and may encourage examination of other lines of evidence.

There is a strong contrast with **verificationist** approaches. Even 'multiple lines of evidence' approaches may take a verificationist stance – that is, accumulating bits of evidence that have some known association with impact (more akin to a 'weight-of-evidence' approach; see section 9.2 below). We caution that the key is that the evidence, even if it looks favourable, must be improbable without $h$, the human impact hypothesis. Further, we look for evidence that supports alternative hypotheses, which ultimately might be corroborated.

Before continuing, we emphasize several things. First, this chapter is not about the sorts of analytical methods available (such as meta-analysis) that allow us to combine data from different studies. This chapter is about the reasoning we can use to assemble such data in a logical manner in the first place, a process that, in our experience, has had far less attention than statistical analysis yet is equally (if not more) difficult to navigate. Second, we do not see a levels-of-evidence approach as an alternative to carrying out good monitoring designs but as a complement to them. The approach can supply a way forward when survey data will be ambiguous from the outset, but can also aid in developing criteria for control locations and in the selection of variables to use.

Finally, a levels-of-evidence approach can be used to answer questions about whether, overall, a particular human activity results in particular changes to the environment. Although reviews of the literature are carried out all the time, they are rarely done in ways allowing a balanced assessment of the evidence for and against particular hypotheses. We suspect that a series of such reviews could be very useful to managers trying to assess current levels of degradation caused by human activities, independently of any specific monitoring program or locale.

## 9.2 A LEVELS-OF-EVIDENCE APPROACH

In order to develop this process, we should turn to other disciplines that have analogous problems with inferential uncertainty. In epidemiology, for example, a common goal is to determine how exposure to something (e.g. a toxic chemical) increases the risk of people eventually developing a particular disease. Typically the relationship between exposure and onset of disease is a complicated one. There are often long time lags and there are usually many other factors that may affect whether a person will develop the disease (such as diet, genetic background etc.). Surveys of people who have had an increased chance of exposure to the toxic chemical (say, through their workplace) versus those who probably do not may show a higher frequency of the disease in the former group. Nevertheless, we will find often that there are other, systematic differences between our two groups (such as genetic differences, diet, a higher frequency of drinking alcohol etc.) that may also explain the higher frequency of the disease. We cannot isolate, clearly, the effect of one potential cause of disease (such as exposure to a toxic chemical) and separate its effects from any other. Moreover, we cannot (and ethically should not!) carry out the definitive experiment that would allow strong inference about whether exposure to the risk factor does indeed cause the disease. Such an experiment would involve selecting a group of people from the population at random and then randomly allocating half of them to be exposed to the risk factor while ensuring that the other half are not. The random selection of people and random allocation of individuals to treatments means we could then disentangle the effects of exposure from all other complicating or associated factors (see chapter 4). The ethical problems here are obvious, and these and logistical problems are the main reasons why disciplines like epidemiology must continually rely upon alternative methods for drawing inferences about cause and effect.

These methods use large amounts of correlative data and try to make a case for causal inference using **causal criteria**. Use of these criteria can be traced back to postulates first made by Jakob Henle and Robert Koch some 150 years ago (Weed & Hursting 1998). Systematic use of the criteria was cemented in 1964, when the US Surgeon General proclaimed that smoking does cause lung cancer by accepting that various sorts of correlative evidence can *collectively* build a sufficiently strong case to infer causality. Hill (1965) formalized these types of evidence into nine criteria (Table 9.2) and these have since formed the basis for building inferential cases in epidemiology (Joellenbeck *et al.* 1998; Potischman & Weed 1999; Weed 1997) as well as in the social sciences (e.g. Reynolds 1998). These criteria are considered particularly important for decisions regarding public health announcements – that is, when it is considered there is sufficient evidence to advise the general public about factors that increase or decrease the likelihood of getting a particular disease.

It is important to note that although Hill (1965) accorded the criteria different levels of importance, he argued strongly against demanding that any particular criterion be fulfilled or that any hard-and-fast rules be developed that must be obeyed in any epidemiological study to infer causality. Because none of the criteria by themselves can establish definite causality, he argued there was no strong argument to weigh some criteria more heavily than others in a formal way.

Before we go on to consider how these criteria might be used in human impact studies, it is worth examining opinions about how their use has fared since 1965. Reviews of the literature suggest that only six or seven of Hill's criteria have played a role in establishing causality (Joellenbeck *et al.* 1998; Potischman & Weed 1999; Reynolds 1998): strength, consistency, specificity, temporality, biological gradient, biological plausibility and coherence. Another has been added: that incidence of the disease decreases when exposure is eliminated (Joellenbeck *et al.* 1998) but this actually falls within Hill's description of experimental evidence. Other sorts of experimental evidence have tended to be rare, and analogy difficult to apply. Coherence, it has been argued, is a 'meta' criterion that is applicable to not only the evidence but also to the criteria themselves (Potischman & Weed 1999). Overall, criteria have been accorded different degrees of importance in different studies, and there is often only approximate consensus on how individual criteria should be interpreted (Potischman & Weed 1999). For example, since 1965 'biological plausibility' has been interpreted in three increasingly rigorous ways, with each subsequent interpretation demanding greater amounts of evidence (Box 9.1). People reviewing an entire literature almost always set down some minimum criteria for individual study designs – those studies deemed 'too flawed' (such that the conclusions within them were likely to be incorrect) were removed from consideration. However, for remaining studies, reviewers rarely if ever specified in advance what specific rules they used overall in judging causality, nor how much evidence was enough, nor how to count it (Weed & Hursting 1998).

Another issue with the use of causal criteria has been 'wish bias', the tendency for investigators to draw conclusions primarily on the basis of their own published results (Wynder 1996). 'Wish bias' was much less likely if inferences about causality were left to those reviewing the entire literature on a topic, especially if the reviewer had not published in the specific area (Weed 1997; see also Loehle (1987) for a general

Table 9.2. *The causal criteria of Hill (1965) in the decreasing order of importance he gave them. The criteria can be used when a correlation is observed between exposure to certain risk factors (such as toxins) and developing a particular disease, to help decide whether there is a case for inferring that exposure causes the disease*

| Causal criterion | Description | How measured and example, where relevant |
|---|---|---|
| Strength of association | Relative to other diseases, there is a particularly large increase in disease incidence associated with exposure to the risk factor | Relative increase in risk of developing the disease when exposed to the risk factor (e.g. chimney sweeps were 200 × more likely to develop scrotal cancer than men not exposed to soot) |
| Consistency of association | Whether the association has been observed repeatedly in different places, circumstances and times | Proportion of studies showing the effect (e.g. by 1965, 36 separate studies had all shown an association between smoking and lung cancer – although Hill did not reveal how many had failed to find such an association) |
| Specificity of association | Whether or not the association is commonly limited to a very particular group of workers, people undertaking particular activities, or a particular locale | N/A[a] |
| Temporality | Whether or not onset of disease always follows exposure to the risk factor | N/A |
| Biological gradient (dose–response relation) | Whether there is higher incidence of or death rate from the disease when there is exposure to increasing amounts or levels of the risk factor | A dose–response curve shows a strong positive relation (e.g. death rate from lung cancer increases linearly with number of cigarettes smoked daily) |

Table 9.2. (*cont.*)

| Causal criterion | Description | How measured and example, where relevant |
|---|---|---|
| Biological plausibility | There is a biologically plausible explanation for causality, even if there is no current evidence for the mechanism | Hypothetical, new mechanisms of causation should not be dismissed out of hand simply because they seem odd |
| Coherence | A cause-and-effect interpretation should not seriously conflict with known history or biology of the disease | All or most of the evidence should support the same explanation (e.g. the evidence that smoking causes lung cancer came from population surveys, histopathological evidence, isolation of carcinogens in cigarette smoke etc.) |
| Experimental evidence | An experiment where exposure to the risk factor is manipulated shows evidence of changed rates of disease | For example, if action is taken to reduce exposure to the risk factor (e.g. number of cigarettes smoked declines) is a drop in disease incidence seen? |
| Analogy | In some cases, effects of risk factors may be argued by analogy because their actions could be similar | Birth defects that are associated with exposure to a drug during pregnancy – we could use well-documented examples of same (e.g. thalidomide) for argument by analogy |

[a] N/A means Hill did not provide a specific example.

discussion of this problem, also called 'confirmation bias', in ecology). Moral stances and political positions also played roles in influencing judgements about which criteria should be considered most important, suggesting that an ethical framework was needed because of the public-

> **Box 9.1** Interpretations of biological plausibility in epidemiology
>
> Weed & Hursting (1998) traced how the criterion of biological plausibility has been interpreted in epidemiology. Up until about 1994, a biologically plausible association was one where a reasonable mechanism for how the risk factor caused the disease could be hypothesized, including those mechanisms for which no biological evidence existed, as Hill (1965) had originally suggested. Subsequently, it was considered that this was insufficient, and that evidence supporting the proposed mechanism was also necessary. More recently, an association has been considered biologically plausible only if there was also sufficient evidence to show how the risk factor itself influences a known disease mechanism.

health implications (Weed 1997). Additionally, it was not possible to define a single set of rules for public-health decision-making from the criteria alone (Potischman & Weed 1999).

From this brief overview, we might draw several lessons for developing the use of causal criteria in human impact studies.

First, causal criteria in human impact studies must be explicitly defined. We anticipate that an ongoing debate among those conducting human impact studies is needed to decide what those definitions should be, how criteria should be measured, and what is 'reasonably' required to infer human impact. At present, use of causal criteria and multiple lines of evidence have only been adopted in a few areas, mainly in the assessment of risks of contamination of fresh water by toxicants (e.g. Beyers 1998; Cook *et al.* 1999; Humphrey *et al.* 1995; Menzie *et al.* 1996; Suter 1993b). Causal criteria need to be applied rigorously, but we may expect that interpretations of what is 'good enough' to fulfil criteria will change as we gain more experience in applying them. This sort of evolution is also inevitable because our state of knowledge will improve.

Second, it seems unlikely that there is any way to formalize, across all possible human impacts, how criteria should be weighted in terms of importance and how inference should be collectively drawn. Inevitably the research base for any particular type of human impact will vary such that one criterion might be weighted more heavily in one review than in another. It is important to remember that no criterion (with some rare exceptions discussed below) can provide information that is definitive.

No criterion will always be absolutely essential to our argument nor any absolutely irrelevant. Instead, the key is to be clear and explicit a priori about how inference for a particular human impact study will be reached. Other researchers may eventually disagree with the conclusion, but at least it will be quite clear how and why an inference of human impact was or was not reached.

We may also wish to differentiate, as epidemiologists do, between inferences that are of interest primarily to researchers and those that might serve the public interest. In epidemiology, early warnings about a risk factor may save lives even if the balance of evidence is not yet sufficient to convince the medical community that exposure to the risk factor increases incidence of the disease. In human impact studies, we face similar decisions. A body of evidence might be insufficient to convince the scientific community that human impact has occurred, but the evidence may be deemed sufficient to warrant modification of human activities associated with those impacts.

Finally, we should keep in mind that the key to inference is to *try to rule out alternative hypotheses* (Reynolds 1998; and chapter 4). Remember that even when we carry out well-designed experiments, we do not prove an hypothesis, because it is logically impossible to do so. Instead, we may be able to show that an alternative hypothesis is unlikely, according to an a priori decided criterion of probability. Consequently, we should not view a process that uses causal criteria as one of trying to 'prove' an hypothesis. Instead, we should proceed by collecting evidence that bears directly on the hypothesis of interest (here, that human impacts have caused some change) and also evidence that bears upon other plausible, alternative explanations. We make our case as much (if not more) by disproving plausible alternatives as we do by showing that the data are consistent with an hypothesis. It is for this reason that, following McArdle (1996), we have called this a 'levels-of-evidence' approach, rather than a 'weight-of-evidence' approach, because the latter usually means simply the number of pieces of evidence supporting the hypothesis of interest ('verificationism').

In sum, it is important to define criteria and spell out how they will be examined or measured very explicitly. It is also important to decide ahead of time how criteria will be used to make inferences about human impact. Experiences in other disciplines suggest against developing fixed rules, for all human impact studies, about how many and which causal criteria must be fulfilled to infer that human impact has occurred. As we shall emphasize in later chapters (chapters 12 and 13), the key is to be clear about how decisions will be made and why, rather

Table 9.3. *Suggested steps in the levels of evidence approach in human impact studies. The bracketed numbers refer to the section where that step is discussed*

1. Define each causal criterion and decide how each will be examined and measured (9.3.1)
2. Use the literature to review all the effects of the human activity and to extract information required to evaluate each effect on response variables, using each of the causal criteria (9.3.2)
3. For each response variable identified in step 2, conduct a separate literature review examining the main natural sources of variability in the absence of the human activity (9.3.3)
4. Put together a list of effects associated with the human activity and evaluate the amount and kind of evidence supporting each (9.3.4)
5. Consider whether the monitoring design could be improved by factoring in natural influences on monitoring variables into the design and removing these as potential explanations (9.3.5)
6. Decide how evidence will be used to draw inferences about human impacts (9.3.6)

than using fixed recipes that 'must' be obeyed. The bottom line is still best expressed by Hill (1965, p. 299):

> What [the criteria] can do, with greater or less strength, is to help us to make up our minds on the fundamental question – is there any other way of explaining the set of facts before us, is there any other answer equally, or more, likely than cause and effect?

## 9.3 A SUGGESTED STEP-BY-STEP GUIDE TO USING A LEVELS-OF-EVIDENCE APPROACH

We can use causal criteria in studies of human impacts by collecting evidence from the literature about the putative effects of the human impact to hand and setting these effects down as a priori hypotheses that can then be subject to test by our monitoring program. We collect information also on natural sources of variability in these same response variables in the absence of human impacts, setting these down also as a priori hypotheses. When a given effect is probable given the impact, and improbable otherwise, we can attain a degree of corroboration of an hypothesis of impact at the location of interest if the effect is actually observed. Our overall case for inference comes not only from the number of predictions consistent with the human impacts explanation, but also the number of predictions that negate alternative hypotheses.

Use of causal criteria should proceed through a number of steps (Table 9.3). These steps allow us to examine critically the possible effects of human impacts and the possible alternative explanations and to look at the overall balance of evidence. It is important that a systematic approach such as this be taken for using causal criteria, because lack of rigour can seriously weaken confidence in the conclusions (Cook *et al.* 1999).

It is also important, before we start, to ensure that we have clarified the human activity as specifically as possible. For example, querying the effects of agriculture upon rivers and streams is not sufficiently specific. Agriculture can include stressors as diverse as toxic pollution (caused by pesticides), eutrophication (nutrient runoff), increased sediment transport or load (erosion), and changed flooding or discharge regimes (water extraction for irrigation). We should be clear whether some or all of these are relevant to the human activity at hand because each sort of impact will require a separately targeted review. We will illustrate the levels-of-evidence approach, where relevant, with the example first mentioned earlier (see introduction to chapter 4 and Box 4.1), of mining in the Alligator Rivers Region.

### 9.3.1 Defining and quantifying causal criteria

To use causal criteria, the first step is to define each of the causal criteria and to decide how each should be measured. We suggest some definitions and methods for measuring them below, but all of this should be viewed as very much open to debate. There are few studies that have specifically attempted this kind of approach (Beyers 1998; Cook *et al.* 1999) and therefore only a small amount of experience at applying them.

There is at least one obvious difference between the task of epidemiologists and those of researchers examining effects of human activities that will affect our use of causal criteria. In epidemiology, the outcome that is to be avoided or reduced in incidence or magnitude is clear – it is the frequency or occurrence of a particular disease. When we examine the effects of human activities on the environment, however, the outcomes we wish to reduce or avoid are not always clear. Sometimes we may wish to reduce specific problems like algal blooms, but in other cases we may be uncertain what the outcomes of any human activity might be. This means we might use the literature and a levels-of-evidence approach to answer different sorts of questions. Where little is known about a particular human activity, we can explore what is known about the changes caused by a particular human activity, as well as the specific mechanisms linking that activity to the changes. We can use the causal criteria described below to help decide which variables to monitor, and this process may ultimately influence decisions about what environmental changes are considered to be 'important' changes (chapter 11). On the other hand, where there is some understanding already of changes related to human activities, we might be able to nominate, a priori, an environmental change considered to be 'important'. For example, an unacceptable algal bloom may be deemed to have occurred when densities of algal cells reach a level at which sheep die if they drink the water or there are large fish kills. In this case, we can look directly at evidence of association between human activities and incidences of this unacceptable change – analogous to the way epidemiologists can look at associations between a risk factor and incidence of disease.

We consider each of the causal criteria in approximately the same order given by Hill (1965), but we are not according them, therefore, the same order of importance. We expect criteria will have different degrees of importance in different studies (as has occurred in epidemiology). Additionally, experimental evidence may be more available in ecological than human studies, which means we might consider it a rather more important criterion than do epidemiologists. Also, considering the problems encountered by epidemiologists with applying 'coherence' as a criterion (section 9.2), we do not consider it in the discussion below.

We relate each criterion to the two sorts of arguments described above in section 9.1 (Table 9.4). One type of argument is that there is a low probability of observing some data in the *absence* of human impact, i.e. low $P(e,b)$. The other type of argument is where the evidence is something that usually (in some cases, inevitably) *follows* from impact, i.e. high $P(e,hb)$.

### Strength of association

Strength of association measures the size of the change associated with incidence of human impact. Where human impacts are associated with particularly large changes in a variable, then we may have higher confidence of some causality than when changes are small, particularly when such a large change is almost never otherwise observed. We are thus making an argument that such a large change is improbable unless a human impact has occurred.

What does a 'particularly large' change mean? We can compare the percentage difference in average value of response variables at locations having the human impact to those that do not (although unfortunately this information is not always available for individual studies). This gives us a measure of effect size (see section 4.7 and chapter 11). Comparing average effect sizes among different variables may provide some clues about those that are causally related to the impact – there may be some changes that stand out as large effects (like Hill's example of the frequency of scrotal cancer in chimney sweeps; Table 9.2). This approach is probably more useful than one where we try to set some fixed effect size as 'large' (e.g. Cohen 1988). We do need, however, to use the literature carefully. Effect sizes are not routinely reported so will often have to be gleaned from individual papers. We need to be sure that means reported in any study are not unreliable estimates (i.e. estimated with poor precision; section 4.4).

*Consistency of association*

Where an association between a response variable and a particular human impact has been observed many times before by other investigators at different times and places, then we will have higher confidence of inferring human impact than if no such consistency is observed. Potischman & Weed (1999) suggest that, rather than demanding every study show the observed association, we require a majority to do so for this criterion to be considered fulfilled. This seems a reasonable suggestion. We can also use meta-analysis to combine the results of different studies and examine the consistency among them with a formal analytical model (although researchers need to be aware of the stringent conditions under which this is possible; see Scheiner & Gurevitch 1993). Again, we need to be sure that any reported differences and correlations are reliable and precise (i.e. come from well-designed studies) before we include them in our sample. We also need to decide the minimum effect sizes and maximum levels of $\alpha$ and $\beta$ that we will accept to regard an ecological significant effect as having been 'detected', an issue we discuss further below (section 9.3.2, step 4).

*Specificity of association*

In some cases, variables can be considered virtually diagnostic for human impact – this is another low $P(e,b)$ argument. Human-made toxins and other substances (like some heavy metals) that are otherwise extremely rare in nature mean that their presence at a location can be a strong signal of potential human impact. Some of these substances cause strange morphological deformities or behavioural abnormalities

Table 9.4. *A description of the type of argument (sensu Popper 1968) made for each causal criterion, the type of evidence collected during the literature review stage and the expected outcomes in the human impact study if the human impact hypothesis is or is not correct*

| Causal criteria | Type of argument[a] | Description of data obtained from literature | Outcome of human impact study | |
|---|---|---|---|---|
| | | | Result consistent with human impacts explanation | Result inconsistent with human impacts explanation |
| Strength of association | Low $P(e,b)$ | Incidence of the human activity is associated (not necessarily consistently) with a 'particularly large' change in the response variable, a size of change which is otherwise rarely observed | A 'particularly large' change in the response variable is observed | A 'particularly large' change is not observed |
| Consistency of association | High $P(e,hb)$ | The association between the human activity and the change has always or almost always been observed to occur by other investigators at other times and places | The expected effect on the response variable is observed (may be redundant with Strength of association) | The expected effect upon the response variable is not observed |
| Specificity of association | Low $P(e,b)$ ($= \sim 0$) | The data are only ever seen in the presence of the human impact | The data are observed | The data are not observed |
| Temporality | High $P(e,hb)$ | An observed change in the response variable occurs after onset of the human activity | The expected change in the response variable occurs after onset of human activity | A change in the response variable occurs before onset of human activity |

| | | | | |
|---|---|---|---|---|
| Biological or ecological gradient | Low $P(e,b)$ and High $P(e,hb)$ | There is a strong relation between dose and effect in well-designed studies of the human activity | A dose–response relationship is observed (if gradient design used) | There is no or a poor relationship between dose and effect |
| Biological or ecological plausibility | High $P(e,hb)$ | There is a plausible or known mechanism of cause-and-effect of the human activity | The study to hand meets any requirements for the hypothesized mechanism to apply | The study to hand does not meet the requirements for a plausible or known mechanism of human impact |
| Experimental evidence | High $P(e,hb)$ or Low $P(e,b)$, depending on design | Well-designed experiments show strong evidence of causality between the human activity and the observed change | The predicted effects from the experiments are observed to occur in the human impact study | The predicted effects from the experiments are not observed to occur in the human impact study |
| Analogy | High $P(e,hb)$ | There is a predicted effect known from other stressors that are argued to be the same as the one(s) resulting from the human activity under study | The predicted effect is observed | The predicted effect is not observed |

[a] Low $P(e,b)$ means we are making an argument that there is a low probability of obtaining the evidence in the *absence* of human impact (i.e. with background variability only). High $P(e,hb)$ means we are making an argument that there is a high probability of observing the evidence when human impact occurs.

that are also otherwise rarely seen in nature (e.g. Janssens de Bisthoven *et al.* 1998; Nimmo & McEwen 1994). High numbers of such deformities can provide very strong inference of human impact. Note that monitoring studies of the effects of toxicants using this causal criterion may also require that stressors should be found within exposed organisms for this criterion to be considered fulfilled (Beyers 1998; Suter 1993b).

### Temporality

If a human activity has caused some change, then the change must follow the onset of human activity. If we can show that changes occurred before the onset of human activity, then we have excluded the latter as an explanation. Temporality is thus a particularly useful criterion, because it has the potential to discard explanations – either the human impacts explanation or alternative ones. In studies where we can monitor before the onset of human activity, this will be an important criterion. In others where any human impacts have happened in the past, it may be possible to reconstruct past environments, so that we can still test whether environmental changes occurred before or after the start of human activities (e.g. Davis & Finlayson 1999; Korhola & Blom 1997; Reid *et al.* 1995; Walker 1993).

### Biological or ecological gradient

If we can observe a distinct increase in the magnitude of effect with increasing intensity or frequency of human impact, then we have further evidence of causality. The relation need not be linear but should show an increase in magnitude of effect with increasing intensity of exposure over some of the latter's range. Such gradients of exposure can be generated in laboratory experiments (especially with toxins) but also from field studies. For example, Janssens de Bisthoven *et al.* (1998) showed that increasing proportions of chironomids had deformed mandibles along a gradient of increasing concentrations of metals (copper and lead) in river sediments. Field studies of this kind are relatively rare but can provide fairly compelling evidence (Ellis & Schneider 1997).

### Biological or ecological plausibility

Sometimes we may have a good understanding of how human impacts can cause change, and there may even be evidence that the mechanism is correct in particular studies. In other cases, we may lack this sort of understanding – we may have evidence that certain changes are associated with a human impact, but no data that explain how or why they occur. If we do not have any data that can explain the mechanism behind causality, should we then say the study has failed to meet this criterion? This is a difficult area, because if the rest of the criteria provide strong signals, we would be loath to then use the lack of data on a mechanism as a reason to dismiss causality – absence of evidence, after all, is not evidence of absence. We suggest that researchers may wish to begin with a definition that requires the mechanism be at least plausible (admittedly a subjective measure), even if no current data are available.

*Experimental evidence*

Experiments where the human impact has been manipulated in a controlled fashion can provide extremely strong evidence, especially if done in the field and at spatial and temporal scales that relate well to those of the human impact. Such evidence is probably still relatively rare but may become more common in the future. The liming experiment described in section 8.4 is a good example. Mesocosm experiments – conducted in small enclosures – have been roundly criticized as 'unrealistic' because they are often conducted over scales far smaller than that of human activity or may lack the natural complexity (e.g. diversity) of the ecosystem (e.g. Carpenter 1996). They can, however, sometimes provide detailed pictures of possible mechanisms otherwise not discernible in large field surveys (e.g. Barmuta *et al.* 1990). Experimental evidence can include instances where human activities have ceased and effects monitored (e.g. removal of excessive nutrient loads from sewage outlets).

*Analogy*

We expect that arguing by analogy will be difficult for many potential stressors. It may be most useful for toxins from particular chemical groups that are known to all have the same mechanism of action within particular taxa. For other sorts of stressors, we suspect that variability between ecosystems, regions and hemispheres will tend to make arguing by analogy difficult.

### 9.3.2 Building a 'levels-of-evidence' case for changes associated with the human impact

Armed with our criteria, the next step is to conduct a systematic review of all literature pertaining to the putative human impact at hand and to extract information required by the causal criteria. Our approach to the literature review needs to be systematic – that is, we need to be clear about what sorts of studies we seek and the specific information we will extract from each. We are looking for information on changes wrought by the human impact and we wish to collect specific effect sizes, instances of association between the supposed impact and the change as well as non-association, and so forth. The review will provide us with a list of changes in response variables – effects – that are putatively caused by the human impact. For each effect, we can evaluate the evidence of association against any of the criteria for which we have sufficient information. This should allow us to order the effects from those with the strongest to those with the weakest evidence of association. We can do this by proceeding through a set of six steps. In the following we assume that a review is being carried out to inform a monitoring program for a specific development.

### 1. Set down the characteristics of the human activity

Our first step is to make sure we thoroughly understand the exact nature of the potential impact – its timing, size, spatial area, and so forth – so that we understand the types and likely magnitudes of any impacts. There is little point including studies where impacts are likely to be qualitatively different to the one at hand. For example, a consideration for assessing the effects of water release from dams is that reasonably large dams have a number of impacts of different kinds, depending upon how they were constructed. Large, older dams typically produce cold and anoxic conditions immediately downstream because stored water is released through bottom-release valves, which remove water from the cold and anoxic layer that typically develops at the base of stratified water bodies. Smaller dams, where stratification in the reservoir behind the dam is not as marked, or modern dams able to release water from a variety of positions within the water column, do not produce such impacts (McMahon & Finlayson 1995). Hence if we were collating the effects of small or modern dams, there would be little point in using studies of large or bottom-release dams. Making distinctions between the effects of different sorts of dams might be fairly straightforward, because their physicochemical effects seem relatively well studied. We suspect that it may be much more difficult for less well-studied human activities. In the latter case, it is probably better to not make such distinctions and to include all studies initially, with the option of ruling some out when it becomes apparent that they are not relevant to the situation at hand.

In the uranium mine example (Box 4.1), a main focus is impacts caused by waste waters from the mine. Researchers would need to be clear about the likely volume, timing and frequency of any wastewater discharges into creeks, as well as having information about the nature, concentrations and form of chemicals present, which include uranium, magnesium, radionuclides, suspended solids, hydrocarbons and process chemicals such as manganese and sulphate (Humphrey *et al.* 1995).

*2. Set down the characteristics of the impact location*

As we have done for the impact itself, we need also to decide whether its location will mean we will restrict our literature review only to particular sorts of rivers. Should we restrict our sampling of the literature to papers discussing rivers in similar climatic regimes or having a similar geomorphology etc.? As discussed in chapter 8 (section 8.1.3), there are no simple answers to such questions. In some cases of human impact, the processes are probably sufficiently understood to know the sorts of rivers that will be relevant for the review. In others, our knowledge may be too poor to allow anything but rough guesses about major differences among river types. In the latter case, it is better to include all rivers and to record sufficient information from each study so that a distinction between different sorts of rivers can be made later, if necessary. It makes the review process more tedious, but it reduces the probability that we end up with too small a sample of studies and have to return to the literature to repeat the whole exercise again.

For example, in our mining study, the mine is located in a wetlands area in tropical, northern Australia, which has a six-month wet season when most rain falls and the wetlands flood, and a 6-month dry season, when a lot of creeks dry down to stagnant pools in some locations. The concentration and speciation of pollutants can be greatly affected by such large changes in water column depth, especially as other variables, like oxygen concentration or pH, also change. Hence, mining-affected rivers having comparable seasonal changes in discharge ought to be particularly relevant, but we would probably still be interested in ones where discharge is far more consistent throughout the year.

*3. Clarify the question(s)*

As we described above (section 9.3.1), our questions may be very different in different studies. In some situations we may know something about the kinds of impact expected and be trying to answer specific questions.

For example, there may be interest in knowing whether the impact might cause, or has caused, a loss of macrophyte beds downstream or reduced the size of fish populations. In other situations, we may be interested in any change in any part of the ecological system, with different effects to be sorted secondarily in terms of their ecological and/or social significance (see chapter 11). Clearly, in the latter case we will be cataloguing all changes recorded within studies, whereas in the former case we will be collecting quite specific information about particular taxa.

In the mining example, the directive was to ensure 'no observable effects' on the natural ecosystems. Additionally, an environmental requirement for mining in Australia was to 'maintain biodiversity and ecological systems'. Consequently, there was interest in monitoring all sorts of organisms that would provide information about the state of the ecosystem. However, gill-breathing aquatic organisms and soft-bodied species were thought to be most at risk from water-borne pollutants, so benthic macroinvertebrates and fish communities, or representatives thereof, were selected as the most practical taxa to monitor. Both short-term effects caused by exposure to high concentrations of pollutants and chronic effects caused by small releases over longer time periods were considered important (Humphrey *et al.* 1995). Consequently, a literature review would be targeted toward macroinvertebrate and fish species and both short- and long-term effects collated.

*4. Decide how an effect will be considered to have been 'detected'*

We want to produce a list of effects associated with the impact and an estimate of the proportion of studies that detected each effect. However, we will need to think about effect sizes and the maximum values of $\alpha$ and $\beta$ we will accept for statistical tests. The traditional value of $\alpha$ is 0.05, and 0.20 is becoming an unfortunate standard for $\beta$ (or a statistical power of 80% or 0.80; see section 4.7). However these values are simply arbitrary conventions – we may wish to accept a higher probability of Type I errors than 0.05 for example. Moreover, the traditional preoccupation of researchers with Type I errors has meant that Type II errors and the estimation of them gets little attention. Few studies calculate $\beta$ or consider the possibility that a lack of statistical significance was caused by low power rather than a real lack of effect. This means that, during our review, we cannot simply produce a list of changes that were 'detected' because studies may have had low power to detect a given change (caused by low sample size, for example) or conversely (on very rare occasions!) had high statistical power and detected changes that we might, ultimately, consider to be biologically or ecologically trivial. Moreover, there is a bias in the literature anyway because non-significant results are often accorded little importance and hence sometimes do not get published (Loehle 1987). An unknown proportion of these are real outcomes (i.e. are not caused by low statistical power).

We need to decide then upon maximum values for both $\alpha$ and $\beta$ that we will use to assess different studies. However, we are comparing among effects. As long as we keep conventions of $\alpha$ and $\beta$ the same for every study, then we are generating comparable *relative* error rates for each effect. Arguably, real changes are still likely to stand out as producing consistent effects, and at lower values of $\alpha$ and $\beta$, than effects that are more weakly associated with the human impact. Standards for $\alpha$ are relatively high at 0.05 for most studies, but as we suggested above this convention need not be followed blindly and a higher value for $\alpha$ could be accepted (we discuss implications of fixing Type I error rates in chapter 12). For studies that record a non-significant test, we need to decide upon a value for $\beta$ at which we would rule out a non-significant result as simply being too unreliable. $\beta$ should be less than 0.5 (i.e. there should be a higher probability of correct than false outcomes), so a maximum could be 0.4 (power of 60%) and, clearly, a lower value would be better. Regardless, we certainly should record effect size, probability of test statistics, $\alpha$ and an estimate of $\beta$ for every test.

*5. Decide upon the qualities of studies to be included in the review*

Quality of studies includes a variety of attributes. First, it means including only those that allow some assessment of whether any patterns detected were indeed caused by the human activity in question. Such studies would include information from not only the impact river but also comparable rivers (ideally, controls that have been matched to the impact location) and/or data that were collected both before and after start of human activities. We can place studies on a gradient of inferential uncertainty, as discussed above. Field experiments carried out at realistic scales of space and time, or Beyond-BACI or MBACI-designed surveys will provide information of least uncertainty. Surveys missing one or more BACI elements or experiments carried out in the laboratory or field mesocosms provide less certain information – the surveys for the reasons discussed above and the experiments because they may lack realism. Studies where no attempt has been made to assess the degree to which changes were caused by the human impact are not likely to provide useful information for all the reasons we have emphasized previously (see chapter 5, and the introduction to this chapter). The exception is where the changes are quite specific to the human impact, such as morphological deformities caused by release of heavy metals or other sorts of pollutants. In these cases, we may feel confident that the number of individuals with deformities is caused primarily by the pollutant because background levels of such deformities are often low.

A second quality of studies is the type and amount of information that is reported. We are interested in cataloguing the specific ecological changes detected and each effect size. Effect sizes will be of two kinds. Where the study design contains discrete comparisons (e.g. control vs. impact and/or before vs. after), then effect sizes will generally be the percentage difference between means. If the study has examined the effects of impacts along some gradient (e.g. distance from impact vs. effect, or magnitude of impact vs. effect; i.e. the impact is measured as a continuous variable), then effect sizes will be measures of association, such as correlation coefficients. In either case, we require associated probabilities of test statistics and the information needed to estimate $\beta$, if it is not provided. Unfortunately, some studies do not provide all of the information (e.g. variance, sample sizes) necessary to estimate $\beta$, and some do not provide the values of test statistics or even actual values of means (Box 9.2). Although we could rule out all such studies, we may exclude a large proportion of the available information. It may be best to categorize studies into those with good quality of information and those with poorer quality of information. However, we should choose a minimum level of data required for inclusion – studies where no quantitative data were collected should probably be ruled out. There are also difficulties in calculating the value of $\beta$ in some circumstances, such as when we need to examine interaction terms in a complex model (see worked example in chapter 13) or if papers are using multivariate tests (section 4.10).

### 6. Carry out a broad-ranging review, extracting relevant data

We now have the criteria we need to conduct our review. Our criteria must be set down clearly so that they allow us to make definitive decisions about whether each study will qualify for inclusion in our sample and, if it qualifies, where it falls along our gradient of inferential uncertainty, and whether we regard its data as reliable or having only moderate or poor reliability. Having an unambiguous approach like this is how we ensure we conduct our review in a rigorous, objective manner,

**Box 9.2** A request to editors and authors publishing human impact studies

A review where we hope to generate new data by systematically collating results across studies can only be as good as the quality of reports available in the literature. Unfortunately, current reporting standards are extraordinarily heterogeneous among both individual studies and journals. Very few studies supply all of the data that reviewers require. Many studies do not report even basic levels of information needed to calculate effect sizes and $\beta$. It would be very handy indeed if most journals had editorial policies requiring authors to report at least minimum levels of biological, geographical and statistical data. For example, biological data should include the higher taxonomic groups to which individual taxa belong–currently some studies report species names without ever clarifying what kinds of organisms are under discussion. Geographical information could include a variety of data about locations: latitude and longitude, climatic data, altitude, geology. Information about the impact location should also include estimates of spatial and temporal extents of impact, magnitude of effects etc. Statistical reporting is often very poor; it should include, for each variable, the means together with estimates of variance of samples (either standard errors or standard deviations) and sample sizes. The value of $\alpha$ used for tests should be stated explicitly. Effect sizes (either the absolute difference between means or the difference divided by a common standard deviation–see Cohen 1988) should also be reported. Most importantly, an estimate of the power of each test (or its inverse, $\beta$) should be provided. Not only will this allow reviewers to catalogue the probability of Type II errors, it should encourage authors themselves to consider Type II errors when they fail to reject null hypotheses. It is also important that authors and editors do not fail to report instances where a null hypothesis was not rejected with relatively high statistical power; these studies are equally valuable to those that detect a significant association, and should be published.

not a vague and subjective one. Our sampling of the literature needs to be as rigorous as any other sampling protocol (chapter 4).

Almost certainly we will use electronically based indices to search the literature – these require experience to avoid narrowing the range of studies found. A very wide range of journals should be consulted, and we need to be confident that we will find human impact studies, if any, that report a *lack* of change as well as those that report changes (Box 9.2). Once we have included and categorized a study for its reliability, we record the effect sizes of each ecological change and associated measures of statistical significance of the change (e.g. $\alpha$, $\beta$). We record also whether the effect is thought to be specific to the human impact and whether it contributes information to any of the other causal criteria (i.e. temporality, gradient, experimental evidence etc.). We thus should end up with a list of ecological effects and the levels of evidence associated with each, an imaginary example of which is given in Table 9.5.

A lot of patience and attention to detail is required to do this properly, because there is enormous variability in standards of reporting in different studies (Box 9.2). The same ecological change may be reported using different jargon in different places. Authors sometimes do not provide enough information about the taxonomic identity of species, important aspects of biology that explain why species responded as they did, the exact nature of the impact or of the impact location. Results can be reported in ways that appear ambiguous. The first studies examined may have to be re-examined in light of information uncovered later, and initial progress may be very slow. Nevertheless, if the preliminary groundwork is done properly then the review will gradually become much more routine, and we will have confidence in the reliability of the data we are collating.

### 9.3.3 Collating common sources of natural variance in the response variables

In section 9.3.2 above, we have identified a series of effects – changes in response variables – that we might predict to occur, with more or less confidence, in the instance of the human impact of interest. Our next step is to look at the possible alternative explanations for such changes. We want to find out if the changes *are commonly observed in situations other than the human impact at hand*. We do this systematically by taking each of the changes identified in the step above, and conducting a second review of the literature, this time seeking sources of the same changes in the *absence* of the human impact we are examining. We repeat the same steps above of collating specific information about putative causes of changes in each variable and assessing each of them against the causal criteria. We aim to end up with a list of influences on each variable ordered in terms of the evidence evaluating their strength of inference of causality. These influences will provide us with a list of plausible, possible alternative explanations for differences we might observe in the monitoring study. Because we use studies of comparable ecosystems not influenced by human activities, we should be accessing a different set of studies than the ones observed above (although there may be some overlap if we use control or reference sites from human impact studies). Because our predictions were drawn from two separate reviews, we can argue that they constitute independent lines of evidence for making inferences.

To continue our example, take the effect of decreased egg output by snails. We would seek any studies examining changes in egg outputs of snails not exposed to uranium waste waters. We collate the evidence of the causes of these changes, along with the levels of evidence that establishes them, in the same way we documented above.

### 9.3.4 Cataloguing effects

Our next step is to return to our first review and to distinguish effects that have a high strength of association with the human activity from those where evidence of association is poor. An effect associated with a human impact that fulfils many of the criteria will inspire more confidence than one that fails on many or most criteria. Collectively then, we can put together, for each effect, a summary of the criteria that were fulfilled and examine how many criteria were met. Our second review helps by showing whether any effects are otherwise rarely seen in nature, i.e., do have low $P(e,b)$. An imaginary example of the kind of information we can gather is provided in Table 9.5.

Effects on our list will be of different value. Of particular interest are effects where we can argue both high $P(e,hb)$ and low $P(e,b)$ – that is, effects rarely seen in circumstances other than when human impact has occurred but which also have a high probability of occurring if the human impact is present. These are the most valuable because the changes, if observed, provide strong inference of human impact.

We may find that other effects are established with low $P(e,b)$ criteria alone. These are effects that are specific to the human impact but are not necessarily commonly observed – an example would be unusual deformities that result from a toxin, but with such deformities only

Table 9.5. *An imaginary 'levels-of-evidence' catalogue of biological effects associated with releases of uranium-contaminated wastewater from mining operations*

| Biological effect | Strength | Consistency | Specificity | Temporality | Gradient | Plausibility | Experimental |
|---|---|---|---|---|---|---|---|
| Decrease in number of eggs laid by snail species | 100%–200% relative decrease below mine | 10 of 15 studies show relative decrease downstream of mines | No | 3 studies all show decrease after mine constructed | 5 of 8 studies show increase with distance downstream | Mechanism of action of uranium on gonads of molluscs known | 5 of 8 toxicity tests have molluscs with lower egg output |
| Increased mortality of larval fish | 30%–80% relative increase in mortality rates | 15 of 25 studies show relative increase in mortality below mines | No | 5 of 8 studies show drop occurred after mine start-up | No information | Mechanism of action of uranium on physiology of fish known | No information |
| Increase in mortality of juvenile snails | 10%–100% relative increase in mortality rates | 7 of 12 studies show relative increases below mines | No | No information | 1 study shows evidence of gradual decline | Mechanism of action of uranium on physiology of molluscs known | 3 of 8 laboratory-based toxicity tests show increased mollusc mortality |
| Increase in average dissimilarity among macroinvertebrates | 10%–40% relative increase | 5 of 8 studies show relative increases | No | No information | No information | 'Sensitive' taxa are killed downstream of mine | No information |

| Increase in concentration of chemicals in tissues of freshwater mussels | 10%–25% relative increase | 2 of 5 studies show relative increases | Yes | No information | No information | Filter-feeders are exposed through bioaccumulation | 10 of 20 laboratory-based studies show increases |

*Note:* We have drawn these biological effects from those examined in the Alligator Rivers Region mining example (Box 4.1), so the effects are real. However, the reported numbers of studies and effect sizes are completely imaginary and *not* based upon an actual literature search; we provide numbers simply to illustrate the type of information sought during the review phase. Given these data, uranium effects on egg output of snails is better established (six of seven criteria fulfilled, assuming effect sizes of 100%–200% decreases are not otherwise seen) than effects upon macroinvertebrate dissimilarities (two, possibly three criteria fulfilled). Additionally, if the second literature review reveals that decreases of 100%–200% in egg output are almost never seen under natural circumstances, then this effect has both high $P(e,hb)$ and low $P(e,b)$ evidence, making it a valuable prediction compared to ones for which we can only marshal some high $P(e,hb)$ evidence.

noted from a very few instances of exposure. Such effects can be useful – they can provide evidence of impact – but they are less valuable because no inference can be drawn from observing a lack of the effect. That result does not rule out the human impact.

Other effects may be established only with high $P(e,hb)$ criteria alone. These are effects that are associated with the impact but could also be explained by other factors. The degree to which the latter is the case should have been revealed by our second review, where we will have determined whether changes of the same magnitude and direction are common. If they are common and we observe these effects during monitoring, we may end up with ambiguous results (i.e. data consistent with both the human impacts hypothesis and one or more alternative explanations). Use of effects established only with high $P(e,hb)$ criteria is also problematic if our second review simply provides no information on other factors that can cause those changes because they have rarely been studied. Another difficulty occurs if the effect is not very strongly associated with the human activity, because failure to observe the effect cannot rule out the human impacts hypothesis. Hence, effects established with high $P(e,hb)$ criteria alone are also less valuable than those established with both sorts of argument.

Finally, we may have some effects for which the evidence of association with the human activity is too weak for us to have much confidence that they form useful predictions. Additionally, where evidence comes from studies with low powers of inference and relatively high values of $\alpha$ or $\beta$, then we will have less confidence than those where inferential certainty and/or reliability of data were high.

Should we distinguish more formally between well-designed and poorly designed studies, and should we try to weight criteria? For ecological risk assessments, Menzie et al. (1996) suggested 11 different sorts of attributes of 'measurement endpoints'. Measurement endpoints are outcomes that might carry a significant risk to the environment, and there are multiples of these for any particular human impact. For example, a measurement endpoint could be the concentration of a chemical of concern in sediments relative to levels reported to be harmful. Each measurement endpoint gets a score between 1 and 5 for each of the 11 attributes, according to tabled descriptions of how such scores might be gained. Each attribute is assigned a scaled weight reflecting its importance – from a weight of 1.0 for degree of biological linkage demonstrated (this would integrate across several of our causal criteria) to 0.2 for attributes like whether standard methods exist to evaluate the line of evidence (this will reflect some measure of the usefulness of the evidence). The overall evidence can then be assessed by multiplying scores by scaling values and summing for each line of evidence, which then produces a weighted average for each. A process such as this allows some formal method of assessing the overall quality and applicability of the evidence behind any putative effect. However, we suspect that a fixed scoring system for evaluating effects across different criteria, as well as weights for evaluating the quality or usefulness of the evidence, might be very hard to do across all sorts of stressors. The approach seems to have been developed mainly with toxicants in mind, and that literature has some standardized methodologies. For other sorts of stressors, we think the evidence in the literature might be heterogeneous in quality and difficult to group into common standards, beyond what we have suggested above.

### 9.3.5 Predictions and ways of ruling out alternative explanations

We have now a series of predictions we can make concerning the effects of the human impact. However, as we described above, some of these predictions may not be unique to the human impacts explanation – our second review may have revealed major natural influences on monitoring variables that produce changes of the same magnitude and direction. How can we rule out the latter as explanations if these predictions are found to be correct? Ruling out alternative explanations is important for all designs, but is particularly critical when we lack one or more of the logical elements (e.g. controls, before data, replication), because we cannot make a prima facie case that alternative explanations are less likely.

For the latter effects, there are ways we can potentially disprove any alternative explanations. First, an examination of our control versus impact locations (or our before vs. after periods) may show that these alternative factors do not vary between them and hence cannot be a potential explanation for differences. For example, suppose our human impact causes increased sediment movement into streams, and that our review of natural causes revealed that the geology of catchments is the most important natural influence on sediment supply into streams. If the geology of control and impact catchments is all the same, then geology *per se* cannot be an explanation for differences between them. We have ruled out a major, alternative explanation.

Second, it may be possible to structure alternative factors into the monitoring design to reduce their influence. Suppose the monitoring design is one where no before data are possible but there are potentially multiple control locations. Recall in the previous chapter (section 8.1.2), we discussed ways of developing criteria for the selection of controls in which we seek to maximize their comparability to each other and to the impact location. The process we have just been through in 9.3.4 above is a formalized, rigorous way of developing control criteria, as we have identified the major natural influences on the variables of interest. If we are able to match control and impact locations, so that these natural factors do not vary among them, it will increase our confidence that high $P(e,hb)$ effects are due primarily to human impacts, not to those alternative factors. To continue the example above, we may be able to match control and impact locations for catchment geology, again removing geology as an explanation for variability in sediment supply. We can pursue an analogous procedure if we have before vs. after comparison but no controls (i.e. if we can choose our before and after periods, we may be able to rule out other alternative factors that vary in time). Suppose we have identified floods as a major natural influence on some of our high $P(e,hb)$ effects, so that the instance of floods in one period and not the other can be an explanation for differences. If we have a choice of sampling times, we might be able to choose those having similar probabilities of floods in both the before and after periods. Ultimately we have no control over whether and when floods occur – we may be unlucky – but we will have increased the chance that floods are less likely to be an explanation for before vs. after differences.

However, we may have the situation where an important natural influence on monitoring variables is exactly confounded with our control vs. impact (or before vs. after) comparison and we cannot use either of the above strategies to remove it. That is, we have no choice about control locations or sampling times. In this case, we can use another strategy to strengthen the argument that human impacts cause any changes rather than alternative factors. For a particular effect we can choose specific taxa that are predicted to respond to the human impact as well as a suite that is predicted collectively to *not* respond. If some taxa are known to be sensitive to the impact and other, related species are known to be insensitive, then we strengthen our argument by predicting, ahead of time, which taxa ought to show the effect and which should not. The key to this strategy is that taxa ought to be sufficiently similar to each other that we can argue that any alternative explanation is unlikely to explain any systematic differences between our sensitive and insensitive species. For example, in the previous chapter, we discussed the effects of liming streams to reduce acidity (section 8.4). Because researchers had preliminary data on the acid-sensitivity of individual taxa, they were able to make predictions regarding *specific* taxa that ought to show increases under liming as well as those that ought to show no systematic differences among limed and unlimed streams. All taxa were otherwise fairly similar in use of habitat, taxonomy, and so forth. Unless acid-sensitivity is exactly correlated with the effects of some alternative factor at the species level, this approach strengthens the argument that it is the liming, not some other explanation, that causes shifts in densities of some macroinvertebrate taxa. We are making use of the argument that it is unlikely that something as specific as acid-sensitivity is correlated exactly with sensitivity to some unrelated, natural influence. Our confidence in this finding increases with the number of taxa that we predict, successfully, to show the expected change and those that should not. Examples of this approach are relatively rare because few researchers have perhaps realized that systematically collecting data that will help negate alternative explanations is as important as marshalling evidence in support of the human impacts explanation. However to provide at least one non-pollution example, Keough and colleagues (Keough *et al.* 1993; Keough & Quinn 2000) adopted this approach to look at the effects of human harvesting on intertidal mollusc populations. No before data were available, but mollusc densities and body sizes could be compared between several beaches accessible to the public and a section of shoreline protected for decades by a rifle range. Initial studies of human collecting behaviour (and use of studies from the literature) showed large molluscs were more likely to be collected than small, and certain mollusc species more likely to be collected than others. Keough *et al.* (1993) showed that average body sizes of collected species were much smaller on accessible beaches than on protected shores. Species that were not collected – either because they were too small overall or were cryptic and hence harder to find – showed no systematic differences in body sizes between accessible and protected shores. This lack of shift provides further evidence of human impacts because it is unlikely that all uncollected species are systematically different from all collected species in ways important to molluscan growth rates. This result provides a second line of indirect evidence that body size shifts were due to human collection over the last century and not due to some historical or current physical difference between impact and control locations.

Finally, we can also look for other predictions that the alternative hypothesis makes that are different from those made under the human impacts explanation. Suppose that for the alternative hypothesis to be the correct explanation, several other prediction(s) (i.e. other than the one(s) overlapping with the human impacts explanation) must be found to be correct as well. If these latter predictions are incorrect, then overall corroboration of the alternative hypothesis is low. In essence, we would apply the whole procedure described above in section 9.3.2 to the alternative explanation – that is, rather than simply looking at one set of predictions it makes (those that coincide with the human impacts explanation), we would catalogue whole suites of other predictions as well. If these latter predictions are then not supported by the data that we collect in the monitoring program, then the alternative hypothesis has not been well corroborated. This may strengthen an argument that a human impact has occurred, providing that predictions of the latter explanation are better fulfilled.

### After impact with data from the impact location only

To this point, we have discussed situations where some comparable data are available – either from control locations or from the impact location prior to the start of human activities. The most difficult (and unfortunately, common) situation is where no such data exist. These are often large ecosystems with a long history of possible human impacts of different kinds. Arguably there are either no controls or possible controls are likely to be similarly affected by human activities. Impacts, if any, were started decades (if not centuries) before, and no data from before that are available. Hence we are left with collecting data after impacts have occurred from the impact location alone. Nevertheless, given that these systems have certainly been affected by human activities, we need to address questions such as: To what degree has this ecosystem been shifted, from what would otherwise be its 'normal' level of variation, by human activities? We will probably have to accept that some of these questions are unanswerable, and that any case we can build will certainly be a weak one because of the poorer degree of inference we can draw, but there are some strategies we can pursue.

We should ensure that we *have* been through a structured process for identifying any potential controls (or reference) locations, such as described in section 8.2.3. We should distinguish between questions about impacts that do affect the entire ecosystem versus those that affect only parts of it. In the latter case (e.g. localized pollution), it is possible that control locations for those parts of the system can be found elsewhere within the same ecosystem. In the former case (e.g. reduced flows caused by dams, acid rain), the whole ecosystem is likely to be affected and we do need to search outside it for controls, where possible. In many cases, the result may seem a foregone conclusion, but it is important *not* to make hasty and subjective decisions. Given the improvement in inferential strength when we have any comparable data, it is worth investing time and effort into the possibility of finding comparable locations. Additionally, going through a structured process forces individuals to reveal their biases. There is some evidence (e.g. Table 8.2) that many ecologists implicitly believe that large ecosystems are, by definition, idiosyncratic and hence not comparable to each other. However, in most instances an objective and structured review of the literature (as described above) has never been attempted, hence such beliefs are not always based upon reliable information. Assuming, however, that we lack control or reference locations, we are faced with making absolute predictions about the state of the system, rather than relative ones based on comparative situations (see Table 9.1).

Our first step will still be to carry out the reviews described in sections 9.4.2 and 9.4.3 and to ask whether we can construct a strong case that particular impacts do cause particular changes. The most difficult step here will be step 2: deciding whether and how other rivers are comparable to the system at hand. For large ecosystems, we may be forced to use a lot of information from ecosystems in other continents and other hemispheres. The question is whether markedly different regional histories (e.g. glaciation events, climatic changes, continental drift) create such variability as to swamp any possible similarities among ecosystems on different continents. That question is probably yet to be answered for most systems. We should certainly expect that any similarities work only at taxonomic levels above that of species or for taxa playing well-understood ecological roles. In any case, reviewers will need to pay strict attention to peculiarities of both ecosystems and the taxa within them, because the need to pick up on idiosyncrasies is particularly important. This will mean recording a lot of information about both ecosystems and taxa during the review.

When we have finished the reviews, one possibility is that we may have established some changes with enough evidence and a sufficiently strong case that we might argue, from these data alone, that the effects are very likely to have occurred in the system at hand. It may even be possible to make absolute predictions (mentioned above), in which case data can be collected to test their validity. For example, some entire taxonomic groups are predicted to be absent under some situations because of intrinsic biological characters. Mayflies and stoneflies are thought to be usually absent from systems where sediment deposition is high, for example, because they possess delicate, external gills that are easily smothered. However, many such beliefs are not necessarily based upon well-structured, systematic reviews of the literature (see section 10.1.2) or the literature used may not be relevant to the ecosystem under consideration. We will have more confidence in such absolute predictions if we uncover any we can make with some confidence during our own review. Nevertheless, it is likely that we will end the review phase with many effects only weakly established because of the difficulty of getting reliable data on large ecosystems or because of persistent differences among systems.

There are at least three sources of inference that we can draw upon. First, any historical data on our ecosystem is extraordinarily valuable. There are some taxa where we can get information about past abundances or diversity (through fossils, tree rings, pollen, diatom sub-fossils etc.). Any ability we have to test predictions resulting from our review with historical data are valuable, simply because those tests help establish any validity of the review's predictions. If we have a couple of predictions that can be tested with historical data and are warranted, then it increases confidence in other predictions.

A second way we can proceed is to use any gradients in the strength or magnitude of impacts within our ecosystem to examine gradients of ecological response. We argued previously that weakly affected versus strongly affected comparisons can provide much information, and analyses of gradients of impacts are useful ways of building a compelling case (Ellis & Schneider 1997). Certainly extrapolation beyond measured levels – here to zero, or the no-impact situation – is uncertain. However, if our review has suggested that a particular effect is likely to occur and we are able to show that we get the expected ecological response along a gradient of impact, then we have strengthened our case that the impact has had an effect within the ecosystem. The key will be whether we can find ways of disentangling the effects of one impact from another, as human impacts often co-occur. Mesocosm-style experiments may be particularly useful in this situation as well.

We can use the argument above (section 9.3.5) about predicting taxa that should not change as well as those that should to make predictions about the relative performance of species within our ecosystem. Such reasoning underpins, of course, the use of indicator species and ratios of taxonomic groups to each other as measures of human impact (see sections 6.1 and 6.3.3). Our use of these variables can be somewhat different from previous studies in at least two ways. First, we will have established the worth of predictions in a systematic review, rather than relying on predictions that may have been based on weak evidence or river systems not relevant to the one at hand (section 10.1.2). Second, we need not obtain estimates of level of impact by dividing abundances of one taxon or set of taxa by another to create an index or ratio, which has had various attendant problems (section 6.3.3). Instead, we could set down hypotheses where we express and test predictions about the relative abundances of a taxon (or set of taxa) to another (or others). We can also, where possible, examine rivers nearby that suffer the same impact. Possibly we would argue that these rivers are not strictly comparable to our impact location, but they may provide information about effects upon specific taxa present in that larger ecosystem. If they also show the same relative differences between sensitive and insensitive taxa, then we have strengthened our case.

### 9.3.6 Assessing the predictions

At this point, we will need to make a decision about how the evidence will be used collectively to infer that changes were or were not caused by human impact. In the best possible scenario, we would have a series of predictions associated with the human impacts explanation that are likely to be seen and that are improbable outcomes under alternative explanations, i.e. have both low $P(e,b)$ and high $P(e,hb)$. This seems most likely when we have a good knowledge base and/or if we are able to design our study so that we reduce the chance that major natural sources of variability are confounded with the human impacts explanation. Nevertheless, we will still need to decide which of our predictions ought be fulfilled and how many. We might rank each of our hypotheses by the strength of evidence associating a change with the human activity at hand. Those hypotheses concerning variables that fulfilled the most causal criteria, especially where both low $P(e,b)$ and high $P(e,hb)$ criteria are included, might be considered more important to fulfil as predictions than those where the evidence of association was weaker. Other than this however, we prefer not to make any specific suggestions about how many or what kinds of predictions ought to be fulfilled. There are simply not enough examples of a levels-of-evidence approach to make such advice useful at this point.

The most difficult situations are where we have only very weak predictions or many of our predictions are consistent with both the human impacts explanation and plausible, alternative explanations, and we can discover no way of separating those explanations. If we have few predictions that can provide evidence that will either corroborate the hypothesis of human impacts or provide us with the evidence that would lead us to reject that as an explanation, then we should probably return to the literature to see if anything has been missed or to search for other predictions of the alternative explanation that we could test.

## 9.4 SOME FINAL COMMENTS ON THE PROCESS

A levels-of-evidence process must be carried out in a rigorous fashion. If there is a lack of structure in approaching the reviews, then we have the possibility that results can be unconsciously manipulated to generate some desired result. This may be a genuine problem, in particular, if the review is not done prior to monitoring and analysis, as there then may be a real incentive to generate specific findings from the literature. Stating the predictions a priori and then testing them is preferable. The outcome is then less likely to be tainted by bias. One way to achieve this is to have different people conduct each of the reviews and, considering the problem with 'confirmation' or 'wish' bias, to get competent reviewers who have published little in the specific literature under consideration. If the reviewing process is done separately from data collection and analysis, this will also increase the independence of these two parts of the process.

Second, we have made some suggestions concerning how to measure the criteria and how to generate a levels-of-evidence case. These are only suggestions. What is needed are multiple attempts at this whole process to expose the problems with it, especially for human impacts that are little studied. Methods for quantifying criteria and what constitutes adequate evidence to establish causality needs lots of rigorous – and informed – debate.

Finally, we do need to keep in mind that we are still dealing with fairly weak evidence. An unreplicated design where there are no controls and/or no before data will have higher levels of uncertainty associated with it than a replicated BACI-type design. The causal criteria cannot (usually) definitively exclude particular explanations; it is important to keep in mind that many of them constitute weak evidence. It should also be apparent that conducting a levels-of-evidence approach is not simpler (nor necessarily less expensive, in some cases) than doing a replicated BACI-type design given the exhaustive, systematic examination of a potentially very large literature that will be required (see Box 9.3 for an example). We imagine that where MBACI or Beyond-BACI designs are possible, these will provide the primary basis for inference, with other lines of evidence providing further corroboration.

## 9.5 IMPORTANT ISSUES

- Many monitoring programs have designs in which there is inferential uncertainty (i.e. there are alternative hypotheses, besides that of the human impacts hypothesis, that can potentially explain our

**Box 9.3** An example from the ecotoxicology literature

An example of the way in which evidence from published studies can be reviewed is provided by Calabrese & Baldwin's (1997) review of the evidence for hormesis (see section 11.2). After clarifying the definition of what constituted hormesis (described in section 11.2 and Figure 11.3), they surveyed nearly 4000 potentially relevant articles of dose–response relationships, of which approximately 350 showed some qualitative evidence of the effect. They then examined these articles and scored each study against a number of quantitative criteria that were established a priori. These criteria included aspects critical to the experimental design (in this case the number of concentrations measured below the No Observable Effect Level (NOEL), and how well the NOEL was estimated), and aspects of the strength of the effect (e.g. statistical significance of the test, magnitude of the response relative to the control and reproducibility of the data by other studies).

Details of the scoring system and how the scores were combined are given in Calabrese & Baldwin (1997); the important point for our discussion here, though, is that this exemplifies how published evidence from a disparate literature can be combined and scored in an objective fashion to evaluate the evidence for a particular response or phenomenon. In this particular example, the authors were looking for evidence of a generalized phenomenon rather than say the response of a single variable (e.g. snail fecundity) to a particular disturbance (e.g. concentration of radionuclides), so the basis for their literature survey was very much larger than we would expect for many examples of disturbances in rivers. Nevertheless this example does demonstrate that data can be evaluated systematically in situations where formal meta-analyses are impossible. It also demonstrates that systematic, quantitative reviews of published evidence can be very time-consuming and require careful definition and quantification of the criteria used to make the evaluation.

data equally well). This issue is particularly the case where we cannot carry out replicated and complete BACI designs.

• A levels-of-evidence approach, using causal criteria (effectively, a set of circumstantial arguments), which have been developed particularly well in the field of epidemiology, can collectively improve inferential strength. The causal criteria are strength of association, consistency of association, specificity of association, temporality, biological or ecological gradient, biological or ecological plausibility, experimental evidence and analogy.

- Causal criteria form two types of arguments: those where there is a low probability of observing some data in the absence of human impact, and those where the evidence is something that follows from impact. Strongest inference is gained when we can marshal evidence that uses both forms of arguments.

- Evidence is gained by reviewing the literature with an objective and carefully designed sampling process to gather information about changes associated with the putative human impact. For each change, we collate evidence under each of the causal criteria.

- A second, and independent, review of the literature will then reveal whether changes associated with the human activity are commonly observed in other situations and, if so, under what circumstances.

- From our reviews, we can construct a list of changes associated with the human impact, which can be posed as predictive hypotheses. These hypotheses can be ranked from those that will provide the strongest inference (e.g. should allow definite exclusion of the human impacts hypothesis if no impact occurs), to those that provide some inference but which cannot necessarily definitely exclude one explanation over another (e.g. where changes may be equally explained by both the human impacts hypothesis and an alternative explanation). For some of the effects on our initial list, we may ultimately decide that the evidence of association with the human activity is too weak to provide useful predictions.

- Where we have identified factors (other than the human activity) that may affect our response variables, we can improve our monitoring design by factoring in these factors so that their effects will not be confounded with any changes resulting from the human activity of interest.

- Overall, we should decide, before collecting any data, how many and which predictions must be fulfilled for us to draw an inference that a human impact has occurred.

- It is extremely important that every effort is made to avoid the potential for bias, by having reviewers who are independent of each other and by having the review process separate from data collection and analysis.

# 10

# Variables that are used for monitoring in flowing waters

From the previous chapter, it should be apparent that choice of variables to use in a monitoring program is a critical decision and not one to make arbitrarily or hastily. However, many biomonitoring texts (e.g. Davis & Simon 1995; Karr & Chu 1999; Rosenberg & Resh 1993) focus upon only particular taxonomic groups or particular kinds of variables. Specific biomonitoring studies are spread over many different journals, from those specializing in particular taxa to ones on whole communities, and from those publishing mainly in areas of basic ecology to those directed specifically at applied ecology or environmental management. It is difficult for any individual to keep track of all of this literature, and researchers can often be unaware of developments outside their immediate field of expertise. 

Here, we summarize briefly the characteristics of useful variables. Although much of this material is discussed in other chapters (chapters 6, 9 and 11), we summarize it here simply to emphasize that useful monitoring variables are not necessarily associated with particular taxonomic groups and that researchers are advised to look widely across the taxonomic and ecological range. From chapter 9, it should be clear that dogmatic advice about which taxa should be monitored should be avoided. Although this may disappoint those who prefer simple recipes, choice of response variables requires measured consideration of a number of attributes. We provide tables that summarize the sorts of taxa and variable types that have been used in flowing water studies to illustrate the sheer variety available for consideration (see Tables 10.1 and 10.2). There is an enormous literature on biological and ecological responses of different taxa (Tables 10.1 and 10.2) to different sorts of stressors. Such a wealth of potentially useful information underscores the need to approach the choice of response variables with an open mind.

Table 10.1. *An overview[a] of some of the general advantages and disadvantages of using each of the main taxonomic groups in human impacts assessments in flowing waters*

*Bacteria and fungi*
Overviews: Finlay *et al.* (1997), Leff & Lemke (1998)
Example papers: Issa & Ismail (1995), Lemly (1998)

| Potential advantages | Potential disadvantages |
| --- | --- |
| Can provide direct measures of very important ecological processes (e.g. decomposition rates) Known importance in detrital food web loops May provide good signal of nutrient status of water No ethical problems associated with experimental exposure to stressors Extracellular enzymes are detectable as indicators of bacterial activity | Usually need high level of expertise Highly technical and sometimes complicated laboratory work required to collect information Serious lack of knowledge about the activities, responses and significance of most taxa Negative social image might hinder their use |

*Protozoans*
Overview: Jack & Gilbert (1997)
Example paper: McCormick *et al.* (1997)

| Potential advantages | Potential disadvantages |
| --- | --- |
| Small size means mesocosm-style experiments are very feasible No ethical problems associated with experimental exposure to stressors | Usually need high level of expertise Highly technical and sometimes complicated laboratory work sometimes required to collect information Very little information available about their biology, distribution, or responses to stressors |

Table 10.1. (*cont.*)

*Plankton*

Overviews: Havens & Hanazato (1993), Boon *et al.* (1994), Gehrke & Harris (1994), Basu & Pick (1996), Kobayashi *et al.* (1996), Reynolds & Descy (1996)
Example papers: Locke *et al.* (1993), Havens (1994), Anderson-Carnahan *et al.* (1995), Bonecker *et al.* (1996), Ruse & Hutchings (1996), Thorp *et al.* (1996), Bass *et al.* (1997), Vranovsky (1997)

| Potential advantages | Potential disadvantages |
| --- | --- |
| Common component of food web of larger, lowland rivers | May not occur much in upland streams or middle-order rivers |
| Large amount of information available on some species' responses to toxins and other stressors | Considerable laboratory work required for identification and enumeration |
| Small size means realistic mesocosm-type and/or field-enclosure experiments are possible | Relatively high level of expertise required |
| No ethical problems associated with experimental exposure to stressors | High between-sample variability requires compositing or sub-sampling techniques to be developed |

*Benthic microalgae*

Overviews: Hart *et al.* (1990), Lewis (1995), Reid *et al.* (1995), Whitton & Kelly (1995), Lowe *et al.* (1996), Pan *et al.* (1996), Stevenson (1997)
Example papers: Kutka & Richards (1996), Lobo *et al.* (1996), Deegan *et al.* (1997), Korhola & Blom (1997), Reavie & Smol (1997), Medley & Clements (1998)

| Potential advantages | Potential disadvantages |
| --- | --- |
| Common and diverse | Not observable in field – requires laboratory work to process samples |
| Many cosmopolitan genera and species | Often very high variability among samples over small scales – requires compositing or sub-sampling techniques to be developed |
| Can provide univariate or multivariate data | |
| Some species (e.g. diatoms) occur as fossils or subfossils so some information about past environments may be retrievable | Often high temporal variability |
| | Laboratory work to identify and count species can be reasonably complicated |
| Sessile, so can provide information about localized impacts | Usually high level of technical expertise required |
| Some well-established relationships with certain stressors e.g. diatom density or diversity, and measures of water quality (e.g. nutrients, salinity) | Possibly somewhat negative social image associated (often wrongly) with algal blooms |

Table 10.1. (*cont.*)

| *Benthic microalgae (cont.)* | |
| --- | --- |
| Potential advantages | Potential disadvantages |
| Reasonably well-established taxonomy | |
| Reasonably good information about the use of artificial substrata to collect samples simply | |
| No ethical problems associated with experimental exposure to stressors | |

*Aquatic macroinvertebrates*
Overviews: Hart *et al.* (1990), Rosenberg & Resh (1993, 1996), Resh *et al.* (1995)
Example papers: Mersch & Pihan (1993), Chessman (1995), Growns *et al.* (1995), Plénet (1995), Rundle *et al.* (1995), Wright (1995), Coimbra *et al.* (1996), Barton & Farmer (1997), Bervoets *et al.* (1997)

| Potential advantages | Potential disadvantages |
| --- | --- |
| Often numerous and diverse | Sampling equipment works over |
| Can provide both univariate and | small spatial scales and can produce |
| multivariate data | large between-sample variance |
| Widespread – occur in most rivers | requiring compositing with |
| and streams | sub-sampling |
| Collectively form a very significant | Taxonomy may be quite poorly |
| part of many lotic food webs | developed in some regions or for |
| Some established relationships with | some taxa |
| certain impacts (e.g. organic | May require relatively highly trained |
| pollutants) | staff to process samples (depending |
| Currently expanding knowledge base | on taxonomic level required) |
| regarding relations between habitat | Can be high rate of errors in |
| types and presence/absence of species | identification and enumeration for |
| Many species relatively immobile and | species-level identifications |
| show impacts occurring over small | If population data required, |
| spatial scales | relatively little known of the |
| Many species respond rapidly to | requirements of the (mostly) |
| impacts | terrestrial adults |
| Equipment to collect larvae relatively | Most species require significant |
| inexpensive and simple to use | amount of laboratory work to |
| Identification to taxonomic levels | generate reliable numbers from |
| above species may be sufficient to | samples |
| detect effects | |

Table 10.1. (*cont.*)

========================================

*Aquatic macroinvertebraes (cont.)*

| Potential advantages | Potential disadvantages |
| --- | --- |
| Depending on identification levels, concerned public groups could be involved in data collection<br>At present, no ethical problems associated with experimental exposure to stressors, except for some crustaceans under some jurisdictions | |

*Macroalgae and bryophytes*
Overviews: Lewis (1995), Whitton & Kelly (1995), Lowe & Pan (1996), Carr *et al.* (1997)
Example papers: Mersch & Johansson (1993), Engleman & McDiffett (1996), Klein *et al.* (1997)

| Potential advantages | Potential disadvantages |
| --- | --- |
| Commonly occurring with many cosmopolitan species<br>Ability to provide both univariate and multivariate data<br>Some understanding of the relations between species abundances or occurrences and water quality (e.g. nutrients)<br>Sessile on hard surfaces, therefore provide reliable signal for small-scale impacts<br>Can be enumerated (as % cover) and often identified in field, therefore no or little laboratory work required and non-destructive or repeated sampling possible<br>Equipment to collect samples usually inexpensive and simple to use<br>Useful in toxicity tests<br>No ethical problems associated with experimental exposure to stressors | Abundance may be naturally very low in some systems, hence power to detect changes may potentially be low<br>Some species (e.g. filamentous green algae) may not be readily identifiable in field<br>Initial identifications may take considerable taxonomic expertise<br>Some species (e.g. bryophytes) may be very slow growing and hence slow to respond to perturbations<br>If population data required, relatively little may be known of other stages of life cycle (e.g. sporelings, spores) |

Table 10.1. (*cont.*)

*Macrophytes*
Overviews: Hart *et al.* (1990), Carbiener *et al.* (1995), Lewis (1995), Whitton & Kelly (1995), Carr *et al.* (1997)
Example papers: James & Hart (1993), Van Den Brink & Van Der Velde (1993), Schmieder (1995), Aguiar *et al.* (1996), Klein *et al.* (1997), Malthus & George (1997)

| Potential advantages | Potential disadvantages |
|---|---|
| Can be enumerated (as % cover) and often identified in field, therefore no or little laboratory work required and non-destructive or repeated sampling may be possible<br>Some remote sensing of abundances may be possible<br>Sessile, therefore should provide reliable signal for local impacts<br>Some known relations between macrophyte presence or density and environmental conditions<br>Free-floating plants may provide a good signal of nutrient status of water<br>Useful in toxicity tests and other laboratory or mesocosm experiments<br>No ethical problems associated with experimental exposure to stressors | Absent or highly patchy in distribution in some rivers and streams and there may be many reasons for their absence<br>Taxonomy may not be well developed in some areas<br>Effects of specific environmental stressors may be known for only a few species<br>Rooted species may signal sediment characteristics rather than water quality<br>Negative social image (as 'water weeds') in some places<br>Identification, if related to reproduction, may be only possible in some seasons |

*Riparian invertebrates (e.g. Insects, spiders)*

| Potential advantages | Potential disadvantages |
|---|---|
| Organisms like spiders may provide signals about abundances of in-stream invertebrates | Very little relevant work on these groups – many unknowns regarding patchiness, sampling protocols, reliability of information and connection to specific perturbations |

Table 10.1. (*cont.*)

*Riparian vegetation*
Overview: Hart *et al.* (1990)
Example paper: Bren (1992)

| Potential advantages | Potential disadvantages |
| --- | --- |
| Can supply both univariate and multivariate data | May not provide any reliable signals about within-channel conditions or aquatic populations |
| Information on reproductive success (seed set, seedling density) may be relatively easy to obtain | Population densities may show only slow responses |
| Many species relatively long-lived and can integrate signals over long time periods | May not respond to some acute impacts |
| Tree rings provide information about past environments that may be relevant | In the absence of remote sensing, can be very labour-intensive to collect reliable, quantitative data |
| Could provide signals about bank condition | Initially, taxonomic identifications may require considerable expertise |
| Some remote sensing of abundances may be possible | |
| Often well-developed taxonomy and many species relatively easy to identify | |
| Involvement of concerned public groups possible | |

*Fish*
Overviews: Moyle (1993), Boon *et al.* (1994), Gehrke & Harris (1994), Gray (1995), Harris (1995), Harris & Gehrke (1997), Bain *et al.* (1999)
Example papers: Lemly (1993, 1996), Bishop *et al.* (1995), Weatherley *et al.* (1997)

| Potential advantages | Potential disadvantages |
| --- | --- |
| Commonly present and often widespread | Mobility means they may flee from point impacts or travel routinely between impacted and non-impacted areas, i.e. at larger scales than the study |
| Can provide univariate data and, in some areas, multivariate data as well | |
| Some species at top of food chains and hence potentially integrate signals from species at lower trophic levels | Some species occur in low densities or are highly aggregated (e.g. schooling fish) so that distribution is highly patchy |

**Table 10.1.** (*cont.*)

*Fish* (*cont.*)

| Potential advantages | Potential disadvantages |
| --- | --- |
| Mobility means some species can integrate effects of non-point impacts occurring over relatively large areas | May be very labour-intensive to sample in some rivers or for some species |
| High longevity of some species means they integrate impacts over time | Some sampling equipment (e.g. electrofisher) requires expertise, specific training and possibly licencing |
| Many species with high social and economic significance | |
| Taxonomy usually well established and species often relatively easy to identify | If population data required, little may be known of larval stages, which often inhabit different environments to adults |
| Possibility of involving community groups (e.g. angling clubs) in data collection | Possibly some ethical problems in exposing individuals to stressors in experiments |
| Some sampling methods (e.g. non-fatal electrofishing, direct observation) mean areas can be sampled repeatedly | Not very diverse in some water bodies |
| Relatively good information about the relations between environmental variability and species' abundances or occurrences in many regions | |
| Relatively good information about responses to a variety of perturbations | |

*Aquatic or riparian vertebrates (e.g. amphibians, reptiles, mammals)*
Overviews: Croonquist & Brooks (1991), Scott & Grant (1997)
Example papers: Sinitsyn (1992), Lemly (1993), Thurmond & Miller (1994), Welsh & Ollivier (1998)

| Potential advantages | Potential disadvantages |
| --- | --- |
| Mobility of many species means they can provide information that integrates effects of non-point impacts occurring over relatively large areas | Those that are highly mobile may flee from areas of impacts or travel routinely between impacted and non-impacted areas (i.e. at larger scale than the study) |

Table 10.1. (*cont.*)

*Aquatic or riparian vertebrates (cont.)*

| Potential advantages | Potential disadvantages |
| --- | --- |
| Many species are at the top of food chains and hence signal perturbations like accumulating toxins | Usually very labour intensive to obtain reliable, quantitative data |
| Usually of high social and/or economic significance | Many species have unknown responses to perturbations[b] |
| Many species are relatively easy to identify, with well-established taxonomy | Territorial species may have large ranges and low densities and hence offer potentially low power to detect effects |
| May be possible to involve local community groups (e.g. bird watching groups, Frog Watch groups) in data collection | Relatively long life cycles for some species may mean populations respond only slowly to impacts |
| | Population responses of riparian species may be only weakly related to condition of stream or river channels *per se* |
| | Ethical issues with handling and experimentation |

*Waterbirds*

Overviews: Weller (1995), Kingsford (1999)

Example papers: Baxter (1994), Kingsford & Johnson (1998), Weller & Weller (2000)

| Potential advantages | Potential disadvantages |
| --- | --- |
| Mobility of many species means they can provide information that integrates effects of non-point impacts occurring over relatively large areas | Those that are highly mobile may flee from areas of impacts or travel routinely between impacted and non-impacted areas (i.e. at larger scale than the study) |
| Many species are at the top of food chains and hence signal perturbations like accumulating toxins | May respond to variation outside of scientists' experience (e.g. water availability in semi-arid regions causes absence closer to coast and no birds to assess) |
| Usually of high social and/or economic significance | |
| Many species are relatively easy to identify, with well-established taxonomy | Many species have unknown responses to perturbations[b] |
| May be possible to involve local birdwatching groups in studies | Relatively long life cycles for some species may mean populations respond only slowly to impacts |

Table 10.1. (cont.)

| Waterbirds (cont.) | |
| --- | --- |
| Potential advantages | Potential disadvantages |
| Reproductive (i.e. sub-lethal) responses to water quantity and quality may occur each year | Requirements for feeding and breeding may be separated geographically and so hard to integrate in a site-based assessment Ethical issues with handling and experimentation |

[a] The table is adapted from a similar summary for marine taxa given in Keough & Mapstone (1995). Also given for each taxonomic group are references that provide an overview of the ecology of those species re monitoring uses as well as papers that provide examples of recent applications of that group to examine responses to a specific impact. The disadvantages and advantages are meant as fairly general indications, and what is a disadvantage for some monitoring programs can be an advantage for others, depending on the exact question (e.g. slow responses to a stressor would be disadvantageous for detecting short-term, pulse disturbances but would be an advantage for long-term, press disturbances). Some taxa have only a few advantages or disadvantages listed; these have been relatively little studied, so it is unknown whether the advantages or disadvantages listed for other taxa apply or not.
[b] Because of expense and ethical considerations, most freshwater vertebrates (with the exception of fish) are used only rarely in laboratory experiments (e.g. toxicology experiments) that examine direct responses to environmental stressors. Coupled with the difficulty of gaining field data on these groups, this results in far fewer data, relative to other taxa, that provide direct evidence of responses to specific environmental perturbations.

## 10.1 CONSIDERATIONS FOR CHOOSING VARIABLES

### 10.1.1 Questions addressed by the monitoring program

Choice of variables hinges upon the questions posed, and such questions will vary widely depending upon the reason monitoring is to be carried out (see chapter 3). For example, a program concerned with monitoring a population of an endangered species may pose questions regarding the latter's habitat use and food resources in natural and modified environments. In contrast, a surveillance program designed to detect changes in an ecosystem might be more concerned with overall shifts in diversity or

Table 10.2. *Some examples of the types of ecological variables used in human impacts assessments in flowing waters*

| Examples of types of variable[a] |
| --- |
| *Single species attributes* |
| Mating success/fecundity (e.g. egg output) |
| Larval recruitment rates |
| Mortality rates |
| Sub-lethal effects (e.g. biomarkers, deformities, fluctuating asymmetry) |
| Measures of abundance |
| |
| *Attributes of multiple species* |
| Taxa richness |
| Diversity indices |
| Ratios of one or more taxa to other taxa |
| Biotic indices using scores related to stressor tolerance |
| Functional group ratios, e.g. feeding types |
| Matrices of (dis)similarity indices |
| |
| *Ecosystem processes* |
| Detrital processing |
| Oxygen production |
| Nutrient fluxes |
| Respiration or photosynthetic rates |
| |
| *Assessment of habitat features* |
| Scores for substrate characteristics |
| Scores for sediment deposition or transportation |
| Scores for condition of banks |
| Scores for type and density of riparian vegetation |
| Specific requirements for focal species (e.g. snags for fishes) |

[a] See chapter 6 for more information about past use of variables and their associated problems and some examples.

biotic composition than with the abundances of single species. Clearly, the ultimate aims of the monitoring program impinge directly on the type of variables that are appropriate.

In either case, the language of questions should be as clear and precise as possible, and the environmental stressor(s) should be specifically named. Nebulously worded questions make variable selection unclear and increase the probability that, ultimately, the information gathered will not address the question. For example, a relatively clear question might be: How will stressor *A* change the density of species *X* in locations 1 and 2 over the next $n$ years? Here, the variable of choice will be the density of species $X$, or any variable that can act as a good surrogate of its density (as discussed below). An example of a poorly defined question might be: What is the effect of impact $A$ upon river health? 'Ecosystem health' is an example of a term that does not have a universally accepted definition (see Box 3.1). Those definitions that have been suggested often use characteristics of ecosystems that are not quantifiable. Consequently, use of the term 'health' (or similar sorts of terms) in a monitoring question often results in considerable confusion about what should actually be measured. Note that it is possible to be more rigorous: an example of variable selection made against set criteria for indicating aspects of catchment health is Cranston et al. (1996).

### 10.1.2 Causality, mechanisms, inference

As we stressed in the last chapter, we seek variables that are related causally and/or are strongly associated with the impact to hand. If our variables do not have this characteristic, then two interpretations are possible if we detect only small changes. The first possibility is that, given a monitoring program of adequate statistical power to detect change, we can conclude that the stressor had little effect on the ecosystem. The second and undesirable possibility is that change occurred but went undetected because the chosen variables did not respond to the stressor. For example, researchers often use species richness as a variable because there is a common belief that human impact invariably causes a decline in richness. This need not necessarily be true. Profound changes in faunal or floral composition can occur without any concurrent alteration to overall numbers of species, and these changes should be ecologically important (Keough & Quinn 1991). Zampella & Bunnell (1998) found similar numbers of fish species in reference and disturbed rivers but the main difference was the incidence of non-native species in rivers subject to human modification, and they considered this to be ecologically significant. Coimbra et al. (1996) found that species richness of macroinvertebrates was actually highest in polluted stream sites in some seasons. Examples like these show that we need to assess the evidence upon which a purportedly effective response variable is based, and we have described a method for doing this prior to monitoring in section 9.3.2. It may be that a variable provides a good signal only for certain types of perturbations, for some taxa, or in some sorts of rivers. It is also possible that belief in particular response variables has developed because the evidence supporting that contention has not been properly critiqued. For example, diatoms are believed to be reliable signals of water quality, but many relations between diatoms and specific environmental stressors are untested or based on weak evidence because of relatively poor survey designs (as pointed out by Medley & Clements 1998). Similar problems have dogged applications of the 'indicator species' concept (section 6.3.3). When we have only poor evidence that a variable is causally related to or associated with a stressor, then we do not have the grounds to interpret small changes as 'unimportant' (chapter 11); in fact, we do not know what it indicates. It is possible that important changes occurred but that we picked insensitive variables to monitor.

### 10.1.3 Ecological and socioeconomic significance of change

We will want to choose variables that can signal changes that are 'important', a topic we discuss in the next chapter. Because importance includes societal wishes, choice of variables is not driven solely by ecological knowledge and so choices should not be made just by scientists.

### 10.1.4 Efficiency

Efficiency of variables will be dependent upon several aspects: their sampling variance and how easy or difficult they are to measure. Sampling variance is the variation seen typically between replicate observations taken at the same time and location (see Box 4.2 and section 4.4). The greater the variance, the more replicates are required to gain a particular level of precision, and therefore the more effort is needed to gain reliable information. Variables differ greatly in sampling variance. Large sampling variance is common in estimates of organism density, for example (section 6.3.1). The causes of high sampling variance are various: for example, large small-scale variability in space or time relative to sampling unit sizes (Schneider 1994); equipment that samples the environment with rather low precision (such as the kick sample for gathering invertebrate species); or where a high degree of technical expertise is required, which may result in a higher rate of operator errors (see Turner & Trexler 1997 for examples of the different performances of aquatic marsh samplers). Additionally, variables requiring considerable laboratory processing costs (e.g. measurement of nutrient concentrations, counting and identification of invertebrates) have more steps where errors of measurement can occur. This is especially true where a high level of technical ability of staff is required but not available. Mistakes in measurement, recording or identifications can be considerable (see Norris & Georges 1993 for an example). Also, we can reduce mistakes in identification by coarsening the required taxonomic resolution (e.g. Wright *et al.* 1995).

Some variables are much more difficult to deal with than others. For example, some are difficult to quantify simply because of the nature of the environment. Extreme water depths or velocities, and hazards like snags, typically make large lowland rivers much more difficult to sample than shallow, upland streams. The densities of some organisms are inherently difficult to quantify because the equipment to do so efficiently has not yet been invented or is labour-intensive to use. Examples are organisms, like some waterbirds, that routinely travel large distances over a few days and make getting accurate counts very difficult. Animals or plants that are highly cryptic, secretive or that inhabit underwater burrows or other hidden environments are also difficult to sample precisely. In each case, the use of many field hands to gather data or the use of advanced equipment like radio telemetry (where animals are tracked with individual radio transmitters) can help, but these requirements will also drive up costs. Rare species pose at least two other problems as well. First, repeated sampling of locations might itself reduce population densities and, in the worse situation, cause local extinction. Second, such organisms may generate many zeros in the data set. Skewed distributions caused by large numbers of zeros pose difficulties for analysis, because the data are unlikely to meet distributional requirements for analysis and may compromise statistical power.

Finally, some variables can act as surrogates for others or as 'piggybacks'. Useful surrogate variables are those that are highly and reliably correlated with a variable of interest, but are much easier to measure, which means that almost the same quality of information can be gathered for far less effort. For example, if members of a taxonomic group all respond very similarly to the stressor, an easily collected, abundant species could be examined rather than a rarer species that may be of more interest but labour-intensive to collect. It is also possible that the extent of key habitats can be used as a surrogate of biodiversity. Piggyback variables are those that can be collected simultaneously with others for very little extra cost or effort. For example, many water-quality meters can routinely measure several aspects of water quality. Once in the field to gather information on one variable, virtually no extra cost is incurred for measuring all the other variables as well. Piggyback variables may also be particularly pertinent where automated sampling is possible (e.g. data loggers can often sample several aspects of water quality without much increase in costs). The same may be true of satellite imagery.

In some cases the easiest variables to measure may not involve direct observations of organisms. For example, ecology often uses variables based on indirect indications of presence or activity of an organism (e.g. burrows, nests, tracks, scats, extracellular enzymes of bacteria). These feature in many indices of relative abundance (see Krebs 1999). Indications of activity *per se* of organisms (rather than their presence alone) are particularly important where the focus is on the dynamism and function of ecological processes (see Fairweather 1999a,b). The expected impact being monitored for will often determine the choices to be made.

A useful pragmatic choice in dealing with multi-species assemblages is to use a coarser level of taxonomy than species for identification with savings being put into processing more samples or sampling more locations. Studies in both marine soft-sediment (see Warwick 1993 for review) and stream (e.g. Wright *et al.* 1995) habitats have shown little loss of power to detect impacts, at least in cases of extreme pollution, even when resolved to merely family or order.

Mixed strategies of measuring a suite of easy variables routinely but supplementing these data with occasional measurement of variables that are more difficult may be particularly cost-effective.

## 10.2 RELATIVE WEIGHTING OF ATTRIBUTES

The attributes described above (summarized in Table 10.3) are all useful, but each should not be considered in piecemeal fashion. It is important to weigh up attributes collectively. Overall, two general issues should influence decisions about choosing variables: the strength of inference the information will supply versus the total effort (hence cost) involved in collecting the information. As we have reinforced throughout this book, monitoring programs fall along a continuum of inference from strong to weak, rather than being simply at either end. The inferential quality of a monitoring program is improved if variable selection is weighted toward those where there is good evidence of association with the stressor. Costs are low if variables provide a good signal-to-noise ratio while not demanding a lot of labour or equipment to collect information. Ideally, all variables should do this, but it is more likely that many variables will require us to trade off attributes. A variable that costs more per datum than another may be warranted if it provides much

Table 10.3. *General attributes that make response variables useful in studies of human impacts*

| Desirable attributes of variables |
| --- |

*Causality, mechanisms and strength of inference*
Responses to stressors are well-understood and predictable because they have been established with a broad array of evidence (chapter 9)

*Significance of changes* (chapter 11)
Of socioeconomic concern
Ecologically important
Processes or functions where biological importance is well established

*Efficiency*
Organisms that do not move much or disperse from impact area, are not cryptic, secretive or difficult to sample
Organisms that are reasonably numerous
Organisms that are not rare or at least will not be wiped out by repeated sampling
Piggyback variables that can be collected at the same time as others with little extra effort
Subsets of variables that survive scrutiny for redundancy or a lack of hypothesized impact from the stressor under examination
Technology for measurement is readily available, methods are reliable and the expertise needed to collect data is available
Measurement is particularly cheap or cost-effective
Can be surrogates for other variables, especially where the latter are difficult or costly to measure or collect
Automated measurements and remote sensing are possible and cost-effective

stronger inference of impact. That sort of conclusion, however, requires a balanced consideration of overall attributes, a topic we consider specifically in chapter 13.

Difficulties occur where social or economic requirements clash with choices of variables that are more desirable for other reasons. For example, in Denmark, the Saprobiensystem of metrics, biochemical oxygen demand and biological indicators were written into law as measures for water quality, thus obviating some choice among response variables (Cairns & Pratt 1993). In other cases, there may be considerable political or social pressure to monitor particular species. The current intense fascination with biodiversity or with large mammals and birds in many sectors of the community is a good example. Mammals and birds, for example, may be otherwise very poor choices for some monitoring programs. Conventions also occur within the scientific community. In freshwater biomonitoring, there is often an expectation that monitoring studies should use macroinvertebrates to detect human impacts, even though the efficacy of this taxonomic group has rarely been compared with others. There is also some pressure (e.g. Fore *et al.* 1996) to keep decision-making processes (and therefore choice of variables) simple so that they are interpretable by lay persons or those without ecological or statistical training. Unfortunately, variables that are simple to explain or are appealing to a lay audience may not possess other useful attributes. The use of socially acceptable but otherwise poor variables, or variables that are used purely out of convention, is to be avoided where possible. If such variables are used, then the reasons why they were selected should certainly be made explicit in any reporting of the monitoring program.

### 10.3 IMPORTANT ISSUES

- Variables chosen for monitoring should be efficacious: relevant to the questions asked; strongly associated with the putative impact; ecologically and/or socially significant; and efficient to measure.
- There is a tremendous diversity of *potential* response variables, and decisions about which are to be used should be based on a careful examination of each variable's efficacy. Variables should not be chosen purely because of convention, habit or social pressure.

# 11

## Defining important changes

306 In previous chapters we have referred to the 'effect size' of an impact. This refers to the size of the change in a variable that constitutes the greatest level of acceptable change. We therefore need to define what sort of change is important in the context of the question being asked. We avoid the term 'significant change' in this context to prevent confusion with 'statistical significance'. In formal terms this is the amount of departure of the data from the null hypothesis (i.e. that the potential impact has resulted in no important change in the variable) that we need to observe before we favour the alternative hypothesis (i.e. that the potential impact has resulted in an important or unacceptable change in the variable).

To those who are new to environmental monitoring and assessment, speaking of 'important' or 'acceptable' changes in a variable seems like a form of newspeak: a euphemism covering a failure of will on behalf of the scientists or managers who should really be insisting on 'no change' in the variable in question. In the next section we justify why effect sizes need to be specified at all. This then leads us, in subsequent sections, to discuss the kinds of change we may need to detect, the risks and consequences of setting effect sizes, and then some practical guidelines on how we can set 'effect sizes' or 'important changes' in rivers and streams.

### 11.1 WHY DO WE NEED TO DEFINE CHANGES IN TERMS OF 'EFFECT SIZES'?

At first glance it seems reasonable and environmentally cautious to insist on a criterion of 'no change' in the variable being measured relative to control conditions. For example, to determine whether the effluent from a pulp mill has been diluted sufficiently, it seems simple to stipulate that hepatic function of fish (as measured by assays of EROD (7-ethoxyresorufin 0-deethylase) activity) be the same in both control and potentially impacted locations. Unfortunately, this criterion cannot be implemented because it requires us to prove the null hypothesis – something that is impossible in the framework of formal hypothesis-testing (see discussion in sections 14.2 and 14.3.1). Remember also, that without consideration of the effect size, the methods discussed in this book are meaningless: we can always reject a null hypothesis, even for the most trivial difference, with a large enough sample size.

Instead we specify some level of change (the critical effect size; see section 4.7) in a variable within which it is not important to reject the null hypothesis of 'no change'. In our hypothetical example, we may have information about how much change in hepatic function can be tolerated by the fish before its foraging behaviour or fecundity is affected. This requires us to be explicit about what level of change in the variable is regarded as harmless or acceptable. In formal terms this process involves specifying a critical effect size (Cohen 1988; and see section 4.7). However, specifying an effect size is not a simple procedure. Although science is involved in quantifying relationships, the strength of impacts and the variables being measured, there are social and economic values that need to be made explicit when trying to decide what constitutes harmless or acceptable change.

Sometimes the criterion of 'no change' is specified in terms of there being no *detectable* change in the variable. For example, the objective of the aquatic biological monitoring associated with the Ranger uranium mine in the Northern Territory of Australia (see Box 4.1) is that there be 'no observable effects upon (the selected indicator) organisms in a comprehensive and sensitive biological monitoring program' (Humphrey & Dostine 1994). This approach may seem reasonable given analogous criteria in toxicology (see Box 11.1). Unfortunately, it is no more rigorous than simply specifying 'no change' because any change, no matter how minute, can technically be detected provided the sampling effort is large enough (Cohen 1988). Conversely, it is easy to not detect change by deliberately using a design with low statistical power (i.e. a high probability of erroneously accepting the null hypothesis; Fairweather 1991b; Keough & Mapstone 1997).

Setting a critical effect size, therefore, requires us to identify a value of our chosen variable which constitutes what we and the stakeholders in the decision regard as the limit of acceptable or unimportant change from control conditions. The response of any biological variable

**Box 11.1** Toxicological attempts to define 'no effect'

The release of potentially harmful substances to the environment
has prompted toxicologists to try to stipulate a concentration of
the substances at which no harmful effect will occur. One
approach that has been used is that of No Observable Effect
Concentration (NOEC), which is sometimes also called No Observed
Adverse Effect Level (NOAEL) or No Observed Effect Level (NOEL;
Rand 1995). Conceptually it is defined as the highest concentration
at which there is no observable effect in the response variable.
Trying to make this concept operational leads to definitions such
as this: 'The highest concentration of a material in a toxicity test
that has no statistically significant adverse effect on the exposed
population of test organisms compared with the controls' (Rand
1995, p. 944). Again, we are trying to 'prove' a null hypothesis,
which is logically impossible. Moreover, we can easily and trivially
retain a null hypothesis if we have a small sample size and low
power in our test. Conversely, massive sample sizes may find
biologically unimportant differences between exposed and control
populations which are, nevertheless, statistically significant.

   Several methods are used to try to derive a NOAEL for a given
substance, all of which are controversial. One sophisticated route
is to derive a concentration–response curve and use mathematical
models to extrapolate the curve to low concentrations and link the
estimated values with formal risk analyses. Even this approach is
fraught with difficulties because some reviewers hold that
concentration–response curves are usually formulated on high
concentrations of substances and thus are poor measures of
responses at low concentrations (Calabrese & Baldwin 1999). As a
result, hormesis may occur at low concentrations (see section 11.2
and Fig. 11.3) with implications for how assays and tests are
designed, and how this information is translated into formal risk
assessments and regulatory criteria (Sielken & Stevenson 1998).

to an impact is likely to be a continuous function of the strength of that
impact, whereas the decision about whether an impact has occurred is a
point on that continuum (Mapstone 1995; see discussion concerning Fig.
11.1 below). Deciding at what point along a continuum that an impact is
deemed to have occurred depends on the questions being asked (Keough
& Mapstone 1995, 1997; Mapstone 1995).

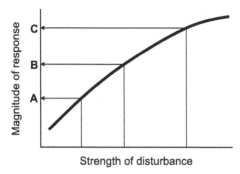

Fig. 11.1 Hypothetical relationship between the response of an indicator variable (a measure of hepatic function) and the strength of a disturbance (concentration of an effluent). The points A, B and C refer to different potential decision points on the continuum. See text for explanation.

For example, an approach based on early detection of impact will have a different emphasis from one geared towards assessing the ecological importance of an impact that has already occurred. Early detection requires a decision to be taken *before* the level of change becomes harmful, otherwise the change may be irreversible, especially if the pollutant is persistent. By contrast, assessing the importance of, say, an accidental spillage involves deciding whether the level of acceptable change has been exceeded and by how much. In this situation the decision rule is at the point of harmful change rather than some smaller, ostensibly harmless value.

If we return to the hypothetical example of hepatic function in fish and concentration of pulp mill effluent, Fig. 11.1 shows that, as the concentration of pulp mill effluent increases, so does hepatic activity. Suppose laboratory observations and experiments have shown the following: that the fish start to be more sedentary and spend less time foraging when hepatic activity reaches point A; the fecundity of females has been observed to decrease by 15% when hepatic activity reaches point B; and when the activity reaches point C the fish are usually in very poor condition with high mortality rates. Whether we choose A, B or C as our criterion of 'important change' depends on the aim of the assessment, which in turn depends on the social and economic contexts of the decisions that need to be made. If the goal is to minimize the impact of the pulp mill on the fish because they are an endangered species then the criterion of 'important change' needs to be set at least at point A in Fig. 11.1. However, even this value may be too high if there is a long lead time between detecting this change in hepatic activity and the engineers running the mill being able to alter the concentration of the effluent. Thus for early detection, the criterion of 'important change' may need to be set at much lower levels of hepatic activity than point A. If the circumstances of the decision are different (e.g. the fish species is not at all endangered, the public does not perceive the river as being of high environmental value and the pulp mill is seen as crucial to the economy by all stakeholders) then a relatively modest decrease in fecundity may be tolerable and the decision criterion set at point B in Fig. 11.1. Finally, setting the effect size at point C would probably only be defensible in a punitive context, such as defining a threshold for compulsory fines after an accidental spillage.

### 11.2  KINDS OF CHANGE, RISKS AND CONSEQUENCES

There are two components of an effect size: its magnitude and its form (Cohen 1988; Mapstone 1995). The previous section introduced some of the problems with defining the magnitude of an effect size. The form of an effect involves decisions about the type of parameters (e.g. means, variances) that are expected to differ between control and impact areas, and the pattern of differences or trends that it is necessary to detect (Green 1989; Stewart-Oaten et al. 1986; Underwood 1991a,b).

The pattern of differences that might be expected depends on whether the disturbance is likely to be a pulse, press or ramp (see chapter 3). Each of these types of disturbance requires fine-tuning of the survey design (e.g. Walters et al. 1988; and see chapter 7). Another aspect of the pattern of change is the likely relationship between the values of the variable we are measuring and the strength of impact. In toxicology, this relationship is often called a concentration–response or dose–response relationship, and conventionally it is thought to follow one of the patterns in Fig. 11.2. The simplest is some linear or simple monotonic relationship between the variable we are interested in and the dose or concentration of the substance under consideration (Fig. 11.2a). A qualitatively different relationship is some form of threshold or step function (Fig. 11.2c), where below a certain concentration or dose there is no relationship between the variable and the concentration of the substance. A more common model used in toxicology is a sigmoid response (Fig. 11.2b), where a fairly rapid change in the response variable takes place over a relatively small range of concentrations.

Researchers in public health have discussed (reviewed by Calabrese & Baldwin 1999) the implications of choosing 'threshold' models (Fig. 11.2c) over linear or monotonic dose–response relationships (Fig. 11.2a). If such a threshold exists, then the best management option would be to ensure that the dose or concentration of the substance

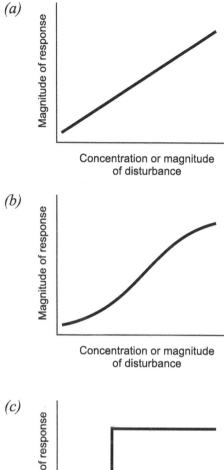

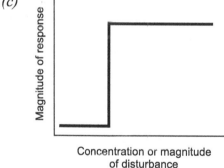

**Fig. 11.2** Three generalized concentration–response or dose–response curves: (a) is a linear, (b) a sigmoid and (c) a step function. Some concentration–response curves are modelled as monotonic functions, many of which can be transformed to linear scales. The sigmoid function is often handled statistically using logistic or logit models or probit analysis.

always remained below the threshold. However, if the dose–response relationship is continuous, no such threshold exists – the risk of a deleterious change rises with increases in dose or concentration, and there is no 'safe' level to manage for. Thus we need to decide where on the continuum to set our criterion of important change. This requires us to quantify risks and make judgements about the timing of detecting a change, the likelihood of its reversibility, and the consquences of the change for ensuing management actions.

Before leaving the topic of the pattern of changes, it is worth emphasizing that not all concentration–response relationships follow the patterns in Fig. 11.2. Calabrese & Baldwin (1999) have reviewed such relationships extensively and found that another pattern is quite common (Fig. 11.3). They term this non-monotonic relationship 'hormesis' (from the Greek 'to excite') because a reasonably common pattern is that the value of the response variable sometimes 'improves' at low concentrations of a stressor, and then deteriorates as the dose or concentration increases. For metabolic or physiological variables, a potential mechanism might be that low concentrations of a harmful substance can stimulate immune or detoxification responses resulting in a modest improvement in response. As the concentration increases, however, the defence mechanisms become overwhelmed and the change becomes deleterious. Hormesis is not confined to physiological or metabolic variables, and has been documented for population characteristics of freshwater organisms (Fig. 11.3a; Bodar et al. 1988). Whether or not the stimulatory effects of mild increases in dose or concentration are to be regarded as beneficial is, of course, a value judgement and, again, will depend on the context of the decision that will be made as a result of detecting the change as well as the interests of a given stakeholder in that decision.

Clearly, science has a role in describing the pattern of changes in the variables we are interested in and in quantifying the risks that are related to these changes. Nevertheless, what constitutes an important change is a societal concern as well (Fairweather 1999b). It involves value judgements, the context of any decisions and the consequences of the decisions made. We are not arguing that decisions on critical effect sizes should be passed on to mysterious covens of power-dressed executives operating secretively in smoke-filled rooms with a few anointed scientists. We do note a tendency in the literature to gloss over the issue of who decides what constitutes an important change; recommendations range from 'the agency biologist must pick the critical value' (McDonald & Erickson 1994, p. 194) to suggesting a conservative 'default' percentile

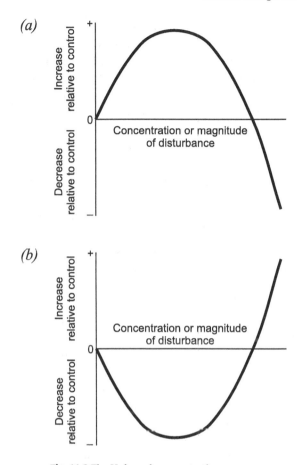

**Fig. 11.3** The U-shaped concentration–response curve shows the change in the value of a response variable relative to control conditions as a function of increasing dose or concentration of a substance (Calabrese & Baldwin 1999). In their review, Calabrese & Baldwin (1999) suggest that the concave downward (upside-down U) curve (a) characterizes phenomena such as reproductive processes and their consequences. Two freshwater examples they cite are the fecundity of *Daphnia magna* in increasing concentrations of cadmium (Bodar *et al.* 1988) and changes in the biomass of algae with increasing concentrations of effluent from a fertilizer factory (Joy 1990). A concave upward (U) curve (b), they suggest, characterizes such things as the incidence of cancer, where, if there were already high rates of cancer under control conditions, mild increases of the harmful agent can actually decrease the incidence of the cancer.

in the range of values exhibited by the variable and revising that value in the light of local knowledge and experience (Hart *et al.* 1999). We would prefer that the interests and concerns of all stakeholders in a decision be made explicit as part of the process of negotiating what constitutes the minimum effect size that needs to be detected in order to allow timely intervention by management (Mapstone 1995). The really important aspect of this process is that *the magnitude and form of an important change is defined up front before the assessment program is implemented* (McDonald & Erickson 1994; Mapstone 1995).

To make this discussion concrete, contrast the different contexts for the release of water from the aforementioned (in Box 4.1) tailings dam containing radionuclides in Kakadu National Park in Australia (Humphrey *et al.* 1990) and the liming trials in Wales aimed at ameliorating acidification (section 8.4). In the case of the release of water from the tailings dam, any potential impact needs to be detected early because the consequences of releasing too high a concentration of radionuclides is irreversible. Because Kakadu is listed as a World Heritage Area and because of the public concern about having a uranium mine in a national park, the 'trigger' for management action needs to be very sensitive. This will allow sufficient time for the mine operators to adjust the release of tailings water well before the effects have any chance of becoming unacceptable. For the liming trials, the aim of the management intervention is to reduce the acidity of impacted streams so that their community structure resembles that of reference streams. The consequences of the failure of liming to produce the desired result are less dire than in Kakadu. Although it would be a waste of resources if liming wrought no important changes in the streams, the management intervention does not preclude other strategies being pursued. However, the pattern of change will strongly influence our perception of the success of this intervention: what if the time-scale of the liming is too short, a mere 'pulse', when what we really should be doing is managing for a very long-term ramp or press?

### 11.3 PRACTICAL STEPS, AND SOME DIFFICULTIES, IN SETTING AN EFFECT SIZE

#### 11.3.1 A caricature of how this seems to work for drinking water

Setting an effect size sounds deceptively easy, especially for physicochemical variables when they are being used to assess river water for a specific end-use, such as drinking water. Guidelines, standards or criteria are used in many countries to define acceptable levels of sub-stances in drinking water, but how are these acceptable levels set? Usually a variety of data from epidemiological, physiological and toxi-cological studies are compiled. The levels at which (presumably statisti-cally) significant effects occur are noted, and then a 'safety factor' is applied so that the acceptable level is set at a value well below that at which deleterious effects have been noted (Rand 1995). The aim here is to define an effect size smaller than that deemed to have an important effect on human health so that management action will be taken before effects become deleterious to humans.

Although this process sounds 'scientific', there are a number of imponderables. First, humans will differ in their susceptibility to the substances in question so that the effect size specified may not be adequate for every consumer. Second, decisions about what constitutes 'important effects' on human health often rely on incomplete or con-flicting evidence. Some studies may have employed small sample sizes resulting in tests that were later demonstrated to be too weak to detect deleterious effects on human health (see Box 11.1). Therefore, in prac-tice, the definition of nearly all effect sizes will involve some element of judgement and negotiation among stakeholders. More recent ap-proaches to setting water-quality guidelines are now seeking to include assessments of values and risks as part of identifying appropriate 'trig-ger values' for management action (see review by Hart *et al.* 1999).

So, the process of setting a critical effect size can be divided into two phases. The first involves collecting scientific information that tries to link changes in the measured variable to the strength of the human disturbance. The second involves judgements about the social, economic and scientific values of the consequences of decisions and management actions that might be pursued. Although we will now describe each of these phases in turn, we are not advocating that they be sequential, separate steps quarantined from each other. Often the process would be iterative and, importantly, we feel strongly that many potential indi-cator variables should be explored simultaneously because trade-offs will inevitably need to be made between scientific tractability of the variables measured, the costs of the program and public perceptions about their credibility and utility (Mapstone 1995; and see chapters 12 and 13).

### 11.3.2  Quantifying the relationship between the response variable and the potential impact

Several strategies could be pursued to describe these relationships. This is not an exhaustive list, and other strategies may arise as researchers and managers get more experience in planning programs with these procedures.

Perhaps the most straightforward cases of setting critical effect sizes for biological variables are for single species that are being managed to prevent their extinction, in either absolute or commercial terms. If there is sufficient information about their population dynamics, then critical densities below which the population should not drop (e.g. to ensure survival of a species or sustainable use of a commercial species) could be established. Species that are nuisances (e.g. some cyanobacteria; Hart et al. 1999) can be dealt with in a similar, but converse fashion; in these cases we are interested in setting densities that the populations should not exceed. For both situations, setting an effect size that is sensitive enough to allow timely intervention will, of course, involve some subjectivity, although the social value judgements that need to be made about effect sizes are very closely tied to the science describing the population dynamics.

Most of the other variables used in environmental assessment can be regarded as 'surrogate' or 'umbrella' measures, where changes in the variable are held to be responses to changes in the environment. Quantifying this relationship in some transparent way is, therefore, crucial. If the relationship is linear, then decisions about critical effect sizes may be more arbitrary than if a threshold can be identified. How, then, to get this information?

We should be able to learn from past experience. Existing impacts elsewhere can provide information about the relationship between the variable and the size of the potential impact, especially if the existing impacts can be arranged on a gradient from mild to extreme. To take a previously mentioned example, Janssens de Bisthoven et al. (1998) regressed various indices of deformities in chironomids against concentrations of various heavy metals found at a number of sites in lowland rivers in Belgium and The Netherlands (section 9.3.1). Although some responses were linear, there were two non-linear responses; two severely polluted sites were strongly influential in these regressions and may be mostly responsible for these patterns. A more common form for such data to take is examination of variables at fewer sites, often at varying distances downstream of a large impact (e.g. Gibbons et al. 1998), although some also include multiple controls in separate rivers (Gagnon *et al.* 1994). Such studies are akin to assembling concentration–response relationships, although the number of studies that can discriminate rigorously between the different models depicted in Figs. 11.2 and 11.3 is surprisingly small. Calabrese & Baldwin (1999) attribute this paucity to an overemphasis on characterizing responses at high concentrations of toxins and pollutants.

Formal meta-analysis, where lots of separate studies are assembled and analysed, is a further source of information that could be used to derive relationships between indicator variables and the strength of human disturbances. Unfortunately, different studies use different protocols and are carried out over different time spans, which hampers our ability to make valid comparisons. However, with increasing ingenuity and sophistication in defining ecologically meaningful measures of critical effect sizes (Gurevitch & Hedges 1999; Osenberg *et al.* 1999), the future of this approach looks promising.

Where time permits, pilot data are invaluable. It is easy to visualize that a pilot manipulation would be an excellent tool for setting effect sizes in rehabilitation projects such as the liming study in Wales, but would we dig a pilot mine in a national park just to provide information to assist in deciding a critical effect size? Obviously not, but field experiments using mine effluents were used in designing assessment programs for further mining proposals in the Kakadu region (Faith *et al.* 1991). Similarly, outdoor mesocosm experiments (reviewed by Cooper & Barmuta 1993) offer opportunities to characterize how variables respond to different intensities of the anticipated impact. Mesocosms share some problems with conventional laboratory-based toxicological studies as well. Extrapolation to large spatial and temporal scales is always controversial (Cooper & Barmuta 1993; Carpenter 1996), and the complex interactions in real-world effluents can dramatically alter the conclusions reached from bioassays of their consitutents (Rand 1995). However, it is pleasing to note that some recent mesocosm experiments have used both complex 'real effluents' and documented patterns in concomitant field surveys to bolster the inferences made from those experiments (e.g. Dubé & Culp 1996; Dubé *et al.* 1997).

The obvious way to circumvent problems with extrapolation is to manipulate entire systems, as has been done conspicuously for whole lakes (e.g. Schindler 1990), and for smaller stream catchments (e.g. Hall *et al.* 1978; Likens 1984; Rundle *et al.* 1995). Although problems of spatial scale are solved, generalizing the results to other systems is problematic (Cooper & Barmuta 1993). Although it is unlikely that whole-system

manipulations will become a common tool for preliminary investigations into effect size, we would like to point to some opportunities. First,
some large-scale manipulations happen anyway (e.g. timber harvesting,
cropping) and these can (and often are) exploited by biologists to identify
the responses of indicator variables (e.g. Hall *et al.* 1978). With some
careful planning, many of these data-gathering exercises could be improved to provide better-focused information that could allow people to
make some informed choices about effect sizes. Second, large-scale experiments can provide insights into processes, and the results can be
used to synthesize and validate predictive models of system behaviour.
For example, large-scale nutrient enrichment of Canadian lakes was
important in constructing and validating process models that ultimately have been useful in defining effect sizes for nutrients and light
climate in many other, often dissimilar, aquatic ecosystems (e.g. Hart *et
al.* 1999). Even where such models are found to be inadequate, they
nevertheless provide the stimulus for more sophisticated, locally applicable models to be developed (Hart *et al.* 1999).

Finally, some human-induced disturbances will happen anyway,
and these can be viewed as opportunities to conduct large-scale experiments within a framework of adaptive management (Walters & Green
1997). Some prominent recent examples include the liming experiments
in Wales that we have used as an example in this book (Rundle *et al.* 1995;
see section 8.4), environmental releases of water in the Grand Canyon to
restore riparian vegetation (Collier *et al.* 1997), and various wetland
restoration options in the Florida Everglades (Gunderson *et al.* 1995).
Although 'management as experimentation' can provide information to
decide critical effect sizes and can yield scientific insights into underlying mechanisms, not all stakeholders will be convinced of the worth of
the additional risks and costs (Walters & Green 1997). We now turn our
attention to negotiations about values, risks and consequences.

### 11.3.3 Negotiating about values, risks and consequences

Once we have some idea of how the candidate variables respond to the
potential impacts, the issue of defining acceptable change needs to be
negotiated, preferably with input from interested stakeholders. It is not
always thus, and almost daily the popular press reflects public disquiet
over a wide variety of developments affecting rivers. Moreover, there are
numerous examples of supposedly scientific monitoring or assessment
systems that are ill matched to management systems (Rogers & Biggs

1999). How, then, do we operationalize stakeholders' expectations with rigorous statements of effect sizes?

Where this has been attempted, a number of ingredients seem essential. Managers need to be able to articulate clear goals in lay terms; a full range of plausible management options needs to be identified; a variety of potential indicator variables from which to choose should be available; the candidate variables need to be clearly linked to the changes or thresholds that are inherent in the management goals; meaningful public consultation needs to occur; an explicit and transparent process deciding what constitutes important change needs to be identified; and a clear vision of what management actions should ensue if an important change is detected needs to be articulated (Fairweather 1999a,b; Hart *et al.* 1999; Maguire 1995; Maguire & Sondak 1996; Rogers & Biggs 1999).

This list is not exhaustive, but two recent published case studies (Grayson *et al.* 1994; Rogers & Biggs 1999) describe some apparently successful attempts at trying to incorporate public perceptions of important change into assessment programs in rivers. The difficulty for many scientists has been in bridging the gap between publicly articulated goals such as 'maintaining ecosystem health' and rigorous, scientific procedures that are meaningful (Boulton 1999; Calow 1992; Rogers & Biggs 1999) – although this is not always the fault of the 'non-scientists' (Fairweather 1993, 1999a,b; Rogers & Biggs 1999).

The formal tools of risk assessment (e.g. Bartell *et al.* 1992) and decision analysis (Clemen 1996) can be combined with structured techniques to elicit and codify values (e.g. Keeney 1992) to provide a framework to identify the goals pursued by different stakeholders, define a suite of variables that might be used to measure how well the goals are being met, assess the relative costs and benefits of different scenarios, and identify priorities amongst the goals (see Maguire 1995; Maguire & Sondak 1996 for some examples). The formal procedures of Adaptive Environmental Assessment and Management (AEAM) provide further examples of how to clarify these issues where many different stakeholders with often divergent views are involved (Holling 1978; Walters 1986). Such formal procedures can be codified and extended so that the costs of particular monitoring programs and management actions can be identified and contrasted with the benefits and values inherent in different scenarios (Walters & Green 1997). Although these procedures may be technically demanding, they are worth pursuing so that courses of action that may be beneficial in the long term are not dismissed prematurely because of perceived high short-term costs (Walters & Green 1997).

Consultation with stakeholders is, nevertheless, hard work. Despite considerable progress in risk assessment and decision theory, Maguire & Sondak (1996) provide a comprehensive résumé of the hiccups and disasters that can beset a consultative process (see Box 11.2). We would prefer you to view this as a list of mistakes to be learned from rather than a deterrent. Nevertheless it is worth noting that a final assessment program may include variables that would seem to have little to do with public aspirations. In Kakadu, a variety of variables is employed depending on the particular facet of the assessment process (Humphrey & Dostine 1994); measures of the community structure of benthic invertebrates and fecundity of freshwater snails are not as charismatic as counts of the massive seasonal migrations of fish, but each of these variables addresses different aspects of public concern about protecting the aquatic ecosystems of a World Heritage-listed national park. The important thing is that the reasons that particular variables are chosen and what constitutes important changes in those variables are made explicit.

It is worth reiterating that once the level of acceptable change has been negotiated, the degree of change in the variable may need to be set to a smaller value so that management actions can be implemented before harmful and irreversible effects occur. Issues such as the fate and persistence of the pollutant and time lags between a pollution event and a measurable change in the biological variable need to be considered in determining the effect size that needs to be stipulated for the monitoring program. Allied to these issues are both the selection of appropriate variable(s) and assessment of the relative costs of erroneously missing an effect of the stipulated size (Type II error) and erroneously concluding an impact occurred when, in fact, it did not (Type I error). Decisions about the form and size of 'important change' cannot be divorced from these other issues of designing an assessment program. It is prudent, therefore, that many candidate variables be considered during these negotiations so that the inevitable iterations in this process are carried out as efficiently as possible.

## 11.4 IMPORTANT ISSUES

- It is impossible to prescribe universal effect sizes for biological variables for two reasons. First, information about the relationships between stressors and biological variables under field

**Box 11.2** What can go wrong in public consultations

Eliciting information from the public about their aspirations for the environmental qualities of a river can range from rudimentary public meetings or questionnaires through to highly sophisticated social research and dispute resolution techniques. Even at the sophisticated end of the spectrum, things can go wrong. Maguire & Sondak (1996) review a number of case studies from a variety of environmental disputes. The potential problems that need to be addressed in a consultative process are:

- A plethora of stakeholders: there will be many more than just two points of view, and some stakeholders will change their viewpoint as the negotiations unfold. On the other hand, large sections of the public may not be accurately represented by organizations or lobby groups, or spokespersons for a group may be too opinionated to represent their constituency accurately.
- Inherent inequalities among the stakeholders: poor single parents are less able to participate or to form lobby groups than rich, well-educated retirees, for example. This problem extends to organizations that purport to represent the interests of several groups of stakeholders.
- History may be important: long-running antagonisms between an agency and a conservation group are common barriers to effective negotiation. There may be other divisions in the community unrelated to the immediate issue in hand that nevertheless get expressed through these negotiations.
- Facilitators may fail to gain the confidence of one or more of the stakeholder groups.
- Stakeholders may hide or strategically misstate their real goals (e.g. by grossly exaggerating expected outcomes) when trying to identify values and try to play the negotiation process as a zero-sum game rather than pursuing joint gains.
- Bias or cognitive limitations of a stakeholder may lead to misrepresentation of their true views or interests. It is not uncommon, for example, for community groups initially to equate potability of water (for human consumption) with a 'healthy' ecosystem. This may often be true, but there are

examples of streams in wilderness areas being unfit for human
consumption (because of pathogens carried by wildlife or high
mineral content from groundwater), or rivers that carry drinkable
water but whose flora or fauna is far from undisturbed.

conditions is scarce. Second, deciding the degree of change in the
variable that is important depends on the environmental values
that stakeholders are seeking to protect.

- There are two components to setting an effect size. The first is
characterizing how the indicator variables are likely to respond
to changes in the strength of the disturbance. The second
involves incorporating societal values into deciding how large a
change should be before it becomes unacceptable.

- Characterizing the response of the indicator variables is easiest if
preliminary or published data from other sources can be
synthesized into statistical or process-based models that can then
be used to inform the consultative phases of the process.

- Incorporating societal expectations into deciding how much
change is acceptable is difficult, but appears to have been
successfully achieved in a few instances in river systems. There
needs to be sufficient flexibility and choice of candidate variables
within the process to ensure that the outcomes are easily
explicable in lay terms and that any trade-offs in Type I and Type
II errors and costs are transparent and explicit.

# 12

## Decisions and trade-offs

Statistical decision theory has a long history and can basically be viewed
in two ways. Classical statistical hypothesis-testing in the Neyman–
Pearson form (Neyman & Pearson 1928), which we described in chapter
4, emphasizes decision errors from the test of a null hypothesis, and
these errors have a frequentist interpretation. In contrast, what is
termed modern statistical decision theory has a strong Bayesian influence (Berger 1985; Hamburg 1985; Pratt *et al.* 1996) and has emphasized
monetary costs and benefits from decisions in an economic and management context. Nonetheless, all statistical decision problems have certain
characteristics in common (Box 12.1; Hamburg 1985; Neter *et al.* 1993). In
this chapter, we will focus on errors associated with the components of
the decision-making process and how the choice of criteria for making
decisions interacts with the design of the monitoring program.

### 12.1 MAKING STATISTICAL DECISIONS

We need to examine the errors possible from a statistical decision-
making process in an environmental monitoring context. In chapter 4
(see Table 4.4) we defined two possible types of error. These errors arise
because we are making decisions about the truth or otherwise of hypotheses about unknown population parameters from imperfect
samples. If we could record an entire population, such as all the possible
locations upstream and downstream of the mine, then we could make
decisions about the truth of hypotheses about those parameters without
sampling error. Errors of inference may still arise, due to measurement
error and confounding.

The first decision error is a Type I error where we reject the $H_0$
when it is actually true. For example, we reject the $H_0$ and conclude that

**Box 12.1** Components of a decision-making process

Consider the simple monitoring design described in chapter 4,
based on Faith *et al.* (1995). They wished to assess the effects of the
proposed gold–platinum–palladium mine, at Coronation Hill
within the Kakadu National Park in northern Australia, on aquatic
biota in the South Alligator River. They used eight locations on the
river: two upstream of Coronation Hill and six downstream. If this
monitoring design was used after mine operations commenced,
our prediction would be that some aspect of the biota (say, species
richness of macroinvertebrates or algal biomass) would differ
between locations upstream and those downstream. The converse
is the null hypothesis that there is no difference in biota between
locations upstream and those downstream.

The decision-making process has specific components:

- Outcome states or events, which represent the 'states of
  nature' outside the control of the decision-maker. In classical
  analyses, these outcome states are binary because the null
  hypothesis is either true or it is false (e.g. algal biomass is
  different downstream of the mine or it is not). One of the
  arguments in favour of the Bayesian framework is that we
  are not restricted to binary outcomes – we might have
  numerous alternative states; for example, a range of actual
  differences (0%, 20%, 50%, 100%) in algal biomass between
  upstream and downstream of the mine.
- Alternative actions, between which the decision-maker
  chooses based on specified criteria. These actions may be to
  reject or not reject the null hypothesis, or in a more
  sophisticated Bayesian context, a choice between a range of
  possible actions based on which difference in algal biomass
  we consider more likely.
- Consequences that occur under the combination of a specific
  action and a specific outcome state. In Bayesian decision
  theory, these consequences can be in terms of costs, usually
  quantified as a loss function, or pay-offs, usually quantified
  as a utility function. In classical hypothesis testing, these
  consequences are long-run frequencies and costs of two
  types of error: rejecting the $H_0$ when it is actually true or not
  rejecting $H_0$ when it is false (chapter 4). Conversely, there

will be benefits from correct decisions, particularly rejecting the $H_0$ when it is false. For example, the consequence of a Type I error may be a serious cost to the mine operator because a false impact has been detected and might initiate modification of operating procedures or even unnecessary remediation. The consequence of a Type II error is more serious to the environment because an impact exists but is not detected, resulting in ongoing environmental degradation.

- Decision criterion, which is the basis, usually in terms of probabilities, by which the decision will be made. For example, the convention in some scientific disciplines is that if the probability of obtaining our sample data or data more extreme when the null hypothesis ($H_0$) is true (the P-value) is less than 0.05, we reject the $H_0$ (chapter 4). Other cut-offs besides 0.05 can be used, especially if devised using our recommended approach of scalable decision criteria where the costs of the two types of error are explicitly considered. Bayesian criteria are more complex, often involving maximizing pay-offs or minimizing losses under a worst-case scenario. With clearly defined alternative hypotheses, Bayes factors are ratios of the posterior odds of one hypothesis over another and guidelines for interpretation of these Bayes factors are available (Ellison 1996).

The debate over the relevance of classical statistical analyses, P-values and Type I and Type II errors to statistical decision-making is ongoing. Many statisticians would argue that only the Bayesian approach provides the flexibility of incorporating prior information into an adaptive decision-making framework. Stewart-Oaten (1996a) argued that statistical hypothesis testing based on P-values is more suited to drawing conclusions from data rather than making decisions. Nonetheless, the classical approach can provide a flexible mechanism for making simple decisions about effects of human activities once the constraints of fixed significance levels are abandoned and careful consideration of effect sizes and consequences of the different errors is included in the design process.

there is a difference in algal biomass between upstream and down
stream locations when there is really no difference (within the level of
accuracy of our measurements). We set the probability of Type I errors
in repeated sampling with our a priori significance level ($\alpha$) and there is
a convention in many disciplines, including ecology, to fix $\alpha$ at 0.05 or
5%. This means that we are willing to accept a long-run probability of
Type I errors of 0.05 when we reject a null hypothesis. As discussed in
chapter 4, this conventional setting of $\alpha$ to such a low value of 0.05 reflects the concern of scientists about incorrectly concluding an effect of
an experimental treatment when there isn't one. In environmental
monitoring, the $H_0$ is usually that there is no effect (on the measured
variable) of a particular human activity. For example, no difference between control and impact locations or that the average difference between control and impact locations is the same before and after the
activity begins. Therefore, a Type I error is the conclusion that there is
an impact when, in fact, there isn't. The other error is a Type II error,
where we do not reject the $H_0$ when it is false. In the environmental
monitoring context, a Type II error is concluding that there is no evidence for an effect of a human activity when, in fact, the activity really
is having an effect. We explained in chapter 4 that the probability of a
Type II error can only be calculated for specific alternative hypotheses
(i.e. effect sizes). For example, what is the probability of not rejecting
the $H_0$ if there really is a 50% change in algal biomass between locations upstream and downstream of a mine? Note that in any decision
process, we can only possibly make one of these errors. The $H_0$ is either
true or it is not; if the former, we can only ever make a Type I error, if
the latter we can only ever make a Type II error.

The issue of a fixed significance level, particularly 0.05, is a controversial one in applied statistics. We acknowledge the arbitrary nature of using $\alpha$ equal to 0.05 and our preferred approach, described
below, is to recognize the two types of error and use a decision framework that balances the costs and consequences of both types of error.
We also note that some decision criterion is necessary whether a classical or Bayesian approach is adopted, proponents of the latter still having to decide between competing posterior probabilities, often based
on Bayes factors.

## 12.2 BALANCING TYPE I AND TYPE II ERRORS

The issue, then, is the balance of these two types of error when testing
the $H_0$ that there is no effect of a human activity. There are two con-

straints on determining this balance. The first we have already discussed: it is relatively easy to set the level of Type I error in advance with our significance level ($\alpha$) but harder to set the level of Type II error because the probability of this error depends on the effect size (chapter 11). The second constraint is that, for given values of all other components of a monitoring program, such as effect size, sample size and variability between sampling units, the probabilities of the two types of error are inversely related (refer back to Fig. 4.3). We can reduce the probability of a Type I error when $H_0$ is true by simply making $\alpha$ smaller than 0.05 (e.g. 0.01 or 0.001). This would push the vertical line in Figure 4.3 to the right, reducing the area under the curve to the right of this line and therefore reducing the probability of a Type I error. However, if the $H_0$ were actually false, we would be increasing the area to the left of this line and thus increasing the probability of a Type II error. So reducing $\alpha$, the acceptable probability of a Type I error if we reject $H_0$ when it is true, must increase $\beta$, the probability of a Type II error if we do not reject $H_0$ when it is false.

Our decision-making framework must provide a reasonable and flexible balance between Type I and Type II errors. In monitoring programs designed to detect environmental impacts, the relative seriousness of making each type of error might be somewhat different from traditional areas of science. For example, failure to detect a real environmental impact (a Type II error) might be considered a more serious error than incorrectly rejecting a null hypothesis of no effect (a Type I error), as discussed by Peterman (1990), Fairweather (1991b), Shrader-Frechette & McCoy (1993), Mapstone (1995) and Keough & Mapstone (1995). The former error may well result in further environmental degradation until the impact is finally detected, with resulting increased costs for remediation. There may also be further costs if, once the impact is realized, compensatory damages need to be paid. In contrast, the latter error is more precautionary in terms of the environment, with the likelihood of no further, or even reduced, impact occurring. In practice, most monitoring programs that set $\alpha$ at 0.05 (5%) will have a high probability of a Type II error if there really is an impact (Mapstone 1995).

The consequences of making each type of error are also different depending on vested interests. Clearly, it is advantageous to individuals and organizations that benefit economically and/or politically from the human activity to minimize the probability of Type I errors. A decision that there is an effect of a human activity may result in a requirement to modify or even cease that activity, with consequent economic and political cost. For example, a Type I error would have serious consequences for the operators of the gold/silver mine (used as an example in chapter 4) because a conclusion that there is an effect of the mine on the biota of the river may require them to modify and even cease operations at an economic cost. In contrast to Type I errors, there is an advantage for those organizations and individuals whose concern is environmental protection to minimize the probability of Type II errors. Not detecting a real impact means that the human activity will continue unmitigated, potentially exacerbating over time the unacceptable effects on the environment. Cynics might even argue, therefore, that proponents of a human activity would wish to minimize the probability of rejection of the $H_0$ even when there really is an effect (i.e. maximize Type II errors) because they would then not need to modify the activity. We will assume for the moment, however, that both sides of the argument wish to minimize the relevant error and be confident of making a correct decision.

Conventional practice, both for science in general and for monitoring in particular, is to fix $\alpha$ at some level, say 0.05, and therefore let $\beta$ (and power) vary depending on effect size. We suspect that most monitoring programs in fresh waters have low power to detect effects that would be considered ecologically important, because sample sizes are usually small and variability between sampling units for most biological variables is relatively large. Therefore, fixing $\alpha$ to a low level implies that Type I errors are more important than Type II errors. This practice favours the interests of the proponents of the human activity rather than the environment (Keough & Mapstone 1995). The burden or onus of proof (Constable 1991) is left to those concerned with protecting the environment because we are unlikely to falsely declare an impact (because we set a low Type I error rate of 0.05) but are often likely to miss impacts because of low power and hence high Type II error rates.

There are three approaches to making this balance of Type I and Type II errors more flexible and sharing the burden or onus of proof more equitably between the proponents of the human activity and those charged with protecting the environment. While we will consider these approaches in the context of monitoring, they are also more generally applicable to scientific practice.

### 12.2.1 Fixed $\alpha$, adjust $n$

The first approach retains a fixed $\alpha$ but designs the monitoring program, especially the sample size ($n$), so that effect sizes of environmental relevance can be detected with reasonable power. For a fixed $\alpha$, $\beta$ is minimized by using an adequate $n$ to detect a predefined effect size (Fig.

12.1a). This idea has been around since Neyman & Pearson first proposed the concepts of null and alternative hypotheses, and Type I and II errors (Neyman & Pearson 1928). The methodology became more accessible after Cohen's (1973, 1988) publications on power analysis and with the recent availability of software to do the calculations. Another arbitrary convention that seems to have developed in ecology over the past decade is to strive for power of 0.80 and hence a Type II error rate of 0.20. This emerging convention may reflect a compromise given the impractical sample sizes required to achieve a power of 0.95, and hence a Type II error rate that matches $\alpha$ of 0.05, given the high level of variability in most ecological systems. With a fixed $\alpha$ of 0.05, a Type II error rate of 0.20 still rates the seriousness of a Type I error at four times that of a Type II error.

The calculations for determining $n$ a priori are usually termed power analysis and are based on a rearrangement of equation 4.5:

$$n \approx \left( \frac{(1 - \beta)s}{ES\alpha} \right)^2 \tag{12.1}$$

where

$(1 - \beta)$ is the required power of the specific hypothesis-test

$\alpha$ is the significance level (the probability of a Type I error we are willing to accept if we reject the $H_0$)

ES is the effect size we decide is environmentally important and we wish to detect if it occurs (chapter 11)

$s$ is our estimate of $\sigma$, the standard deviation between sampling units used in the monitoring program.

Note that equation 12.1 is not used in practice because the definitions of ES and $s$ depend on the nature of the $H_0$ and the statistical test being used.

These calculations require some estimate of the variation between sampling units, usually from a pilot study or from previous work in a similar freshwater system. If our estimate of $\sigma$ is not very good, then our determined $n$ may not achieve the power required because we underestimated the variability, or may be unnecessarily powerful, and wasteful of resources, because we overestimated the variability. The second component of these calculations is the specification of an effect size that is environmentally important enough to detect. We want to avoid using a sample size so large that we can detect environmentally trivial impacts, nor can we justify a monitoring design that cannot detect important impacts. Either case is a waste of scientific resources. The choice of detectable effect sizes has been discussed in chapter 11.

Clearly, we can shift the burden or onus of proof by setting the

*(a)*

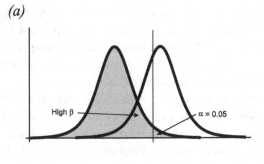

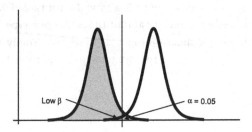

*(b)*

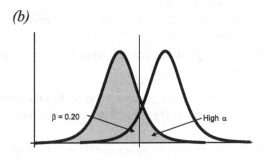

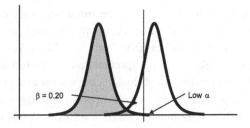

*(c)*

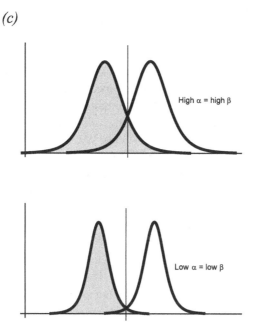

**Fig. 12.1** These graphs represent hypothetical probability density functions for a statistic used for measuring the effects of human activities. The precise shape of these distributions will depend on the test statistic being used (e.g. mean difference in some variable between control and impact sites, $t$ statistic or $F$-ratio statistic). In each case, the shaded distribution on the left is that under the $H_0$ of no impact; and distributions to the right are those for a particular effect size (i.e. $H_0$ false). Vertical lines mark values of the statistic corresponding to the critical decision criterion for deciding whether to reject the $H_0$ or not. (a) Distributions for which the critical decision value is determined by a fixed $\alpha$ (e.g. 0.05), illustrating the trade-off between $\beta$ and sample size and/or variability between sampling units. Top: larger variability and/or smaller sample size result in higher $\beta$. Bottom: smaller variability and/or larger sample size result in lower $\beta$. (b) Distributions for which the critical decision value is determined by a fixed $\beta$ (e.g. 0.20), illustrating the trade-off between $\alpha$ and sample size and/or variability between sampling units. Top: larger variability and/or smaller sample size result in higher $\alpha$. Bottom: smaller variability and/or larger sample size result in lower $\alpha$. (c) Distributions for which the critical decision value is determined by a fixed ratio of errors (e.g. $\alpha = \beta$), illustrating the link between errors and variability between sampling units. Top: larger variability and/or smaller sample size result in higher errors of both types. Bottom: smaller variability and/or larger sample size result in lower errors of both types.

required power to 0.95 and therefore a Type II error rate that matches the fixed Type I error rate of 0.05. However, our experience of monitoring programs in aquatic ecosystems is that the $n$ required to achieve power of 0.95 with a fixed $\alpha$ of 0.05 is usually prohibitive. Nonetheless, we cannot overemphasize the importance of specifying environmentally important effect sizes a priori and then designing the monitoring program, especially the sample sizes, so that such an effect size will have a high probability of being detected if it occurs. Without this aspect, the logical justification for traditional statistical hypothesis-testing with its fixed $\alpha$ is much weaker and, not surprisingly, has been criticized for entrenching arbitrary conventions. Our recommended design strategies outlined below (sections 12.2.2 and 12.2.3) include specification of effect sizes and sample-size determination as crucial components.

### 12.2.2 Fixed $\beta$, adjust $n$ and $\alpha$

The second approach employs not a fixed $\alpha$ but a fixed $\beta$, together with the nominated effect size of environmental relevance. These fixed values ensure a monitoring program design in which the effect size of environmental relevance can be detected with reasonable power. The remaining aspects of design then trade off the costs of nominated $\alpha$ and nominated sample size ($n$) values. An appealing property of this strategy is that this trade-off often may be left to the proponent of the human activity. Figure 12.1b illustrates the trade-off between a higher $\alpha$ (which implies greater Type I error, and so possible economic costs of unnecessary stoppage of activity or remediation of the environment) and a larger sample size (which implies possible economic costs to the proponent, as when time before commencement is a basis for sample size in BACIP designs).

### 12.2.3 Scalable decision criteria

The third approach differs from those previously described primarily by not fixing a priori $\alpha$ or $\beta$ to some conventional level (e.g. 0.05). The levels of $\alpha$ and $\beta$ are both negotiated in advance depending on the relative costs of making either of the two types of decision error. This idea has been championed by Mapstone (1995, 1996) for environmental monitoring, although the general principle of setting $\alpha$ and $\beta$ depending on costs dates back to Neyman & Pearson (1928) who also argued that the balance between probabilities of Type I and Type II errors should be set by the decision maker. Mapstone (1995) called this approach 'scalable decision criteria' and his protocols are based on negotiation among the parties involved in the monitoring program. For the example of monitoring algal biomass upstream and downstream of a mine, these parties might include mine operators, groups concerned with the state of the river and its floodplain (e.g. recreational users, indigenous land owners, government environment departments, fishers) with mediation provided by a relevant regulatory body (Keough & Mapstone 1995).

There are two major components of scalable decision criteria that influence the design of the monitoring program. First, the effect size we wish to detect if it occurs needs to be set (chapter 11). This effect size should be negotiated by the parties involved in the monitoring and may well incorporate political and social information, as well as ecological advice. The crucial aspect is that this choice of effect size should not be influenced by the cost of monitoring nor by the decisions made about $\alpha$ and $\beta$. Once stipulated, the effect size should not be varied at any stage of the implementation or analysis of the monitoring program.

The second component is the designation of the critical levels of $\alpha$ and $\beta$ without either being fixed by convention. The level of $\alpha$ chosen may or may not be the level of $\alpha$ we use to make a final decision about whether we reject the $H_0$ of no impact (see below). These chosen levels of $\alpha$ and $\beta$, termed $\alpha^*$ and $\beta^*$, will then be used to design the monitoring program, especially for determining sample sizes.

The first step is to determine the relative costs of making a Type I and a Type II error, which Mapstone (1995, 1996) termed $C_\alpha$ and $C_\beta$, and then the ratio of these costs, $k = C_\beta / C_\alpha$. The values for $\alpha^*$ and $\beta^*$ used in the design process should be set in relation to the costs of making each type of error. Keough & Mapstone (1995) suggested a number of issues that might help estimate these costs. These include:

- The economic costs to the proponent of a human activity if an impact is inferred (i.e. the costs of a Type I error). This may entail ceasing the activity, modifying the activity or remediation of the environment. For example, reducing the effect of a mine discharge on algal biomass in a lowland river may require more expensive wastewater treatment. The implications of detecting an impact rarely seem to be considered when monitoring programs are designed but should be an essential component of calculating the costs of a Type I error.

- The economic costs to the proponent of a human activity of failing to detect a real impact (i.e. the costs of a Type II error). Again using the mine discharge example, an undetected impact might prove very costly later, assuming it will be detected, because the environ-mental damage will be severe and the cost of remediation and compensation potentially great.

- The social, political and secondary economic costs of ceasing or modifying the human activity. For example, if mine discharge to a river had to be stopped, then the wastewater would need to be diverted elsewhere, such as on to land, which may be very unpopu-lar with people living nearby. Increasing the level of wastewater treatment may divert government funds from other activities, resulting in reduced living standards etc.

- The social, political and secondary economic costs of serious dam-age to the environment if impacts are not detected. For example, not detecting increased algal biomass resulting from a mine dis-charge may mean continuing declines in stocks of recreationally and commercially important fish, or continuing damage to habi-tat and water quality. This may eventually affect values for ecotourism, aesthetics or other recreational activity.

Mapstone (1995, 1996) acknowledged the difficulty of determining these relative costs and suggested that, in the absence of better information, the costs of the two types of error should at least be considered equal and therefore $k$ set to 1.

The second step is to negotiate the maximum risks of a Type I and a Type II error each party in the monitoring program would be prepared to accept. For example, the proponents of the human activity would wish to set a maximum Type I error rate and the groups concerned with protecting the environment would wish to set a maximum Type II error rate. These maximum acceptable error rates, along with the value of $k$, are then used as the basis for negotiated agreement on the final values of $\alpha^*$ and $\beta^*$ to be used. Note that at least one of the error probabilities will need to be negotiated and the negotiation process is critical in the whole strategy of scalable decision criteria. The advantage of requiring this negotiation is that the entire design and decision process of setting effect sizes and error rates is transparent and open to scrutiny without non-negotiable conventions or hidden criteria. Now the monitoring program can be designed based on the effect sizes and agreed levels of $\alpha^*$ and $\beta^*$. Power analysis is used to ensure that the sample size is adequate to detect the designated effect size given the agreed levels of $\alpha^*$ and $\beta^*$.

As emphasized earlier, fixing $\alpha$ at 0.05 nearly always favours the proponents and the burden of proof rests on managers and those charged with environmental protection. Mapstone (1995, 1996) and

Keough & Mapstone (1995) have pointed out that, in contrast to any of the fixed-$\alpha$ strategies for decision-making, the scalable criteria do not favour either the proponents of the human activity nor the protectors/managers of the environment. For example, proponents will be reluctant to set very low values for $\alpha$ because this will also produce low values of $\beta$ (if $k$ is close to unity) and a powerful monitoring program that might detect even trivial changes.

Finally, we must make the decision about the $H_0$ of no impact once the monitoring program has been completed, or at least once it has reached a stage that hypotheses about impacts can be tested. The original values of $\alpha^*$ and $\beta^*$ may not have been realized by the monitoring program because, for example, the estimate of error variability used in the power calculations may not have been close to the true variability, or the pattern of variability may have changed between the pilot study and the actual monitoring program. If we fix $\alpha$, any such inadequacies in the monitoring program would result in reduced power and increased $\beta$, meaning environmentally important impacts might not be detected. If we fix $\beta$, $\alpha$ is increased, meaning we run the risk of unnecessary intervention with development. Scalable decision criteria require the value of $\alpha$ used in deciding about $H_0$ to also be adjusted if the monitoring program did not realize its desired error rates. This then shares the consequences of an inadequate monitoring program equally between the proponents and managers. It is important to realize that the agreed ratio of $\alpha$ and $\beta$ is not changed, just the absolute values depending on how well the monitoring program reached its objectives (Fig. 12.1c).

Mapstone (1995, 1996) and Keough & Mapstone (1995) proposed a mechanism by which these decision criteria might be set:

1.  First, set $\alpha$ equal to $\alpha^*$ determined a priori
2.  Using the effect size nominated a priori and the actual estimate of the error variability from the monitoring program, calculate the realized $\beta$
3.  Compare the ratio $\alpha / \beta$ with the ratio $\alpha^* / \beta^*$ nominated a priori
4.  If $\alpha / \beta$ is less than $\alpha^* / \beta^*$, increase $\alpha$; if $\alpha / \beta$ is greater than $\alpha^* / \beta^*$, decrease $\alpha$
5.  Then using that new value of $\alpha$, recalculate the realized $\beta$ and compare the ratios again
6.  Repeat these steps until $\alpha / \beta$ equals $\alpha^* / \beta^*$. The value of $\alpha$ that achieves this equality is then used for the relevant hypothesis-test from the monitoring program.

We emphasize again why this process is an improvement over the

traditional approach of fixing $\alpha$. Under the fixed-$\alpha$ scenario, if the realized power of the monitoring program was less than expected, then the
probability of Type II errors must increase for the specified effect sizes.
Using scalable decision criteria, the ratio of Type I to Type II errors is
fixed so that the probabilities of Type I and Type II errors would be
affected similarly by a monitoring program that did not achieve expected power (Fig. 12.1c).

### 12.3 COST–BENEFIT ANALYSIS AND DESIGN

Another component of the design process is the allocation of limited
resources to multiple spatial or temporal scales of sampling. For
example, consider the nested sampling program we discussed in chapter
4 for monitoring the effects of a mine on algal biomass at locations
upstream and downstream of the mine on a lowland river. The main
factor is upstream versus downstream, with randomly chosen locations
upstream and randomly chosen locations downstream, and randomly
chosen snags at each location. We might have additional levels of subsampling, such as subsamples from each snag. The important level of
replication for detecting an impact is locations within upstream or
downstream. Other scales of sampling (e.g. snags) are included so that we
can better estimate the mean algal biomass at an individual location.
The basic sampling unit for algal biomass is a snag but we expect algal
biomass also to vary at the spatial scale of locations. Clearly the number
of levels of the upstream–downstream factor is fixed but we have a
decision to make about allocating our sampling effort to the other levels.
Do we use more locations or more snags within each location for the
same total cost? There are two criteria we use to decide on this relative
allocation: first is the precision of the means for locations and snags
within locations or, conversely, the variance of these means; second is
the costs, in terms of money and/or time, of sampling each snag and each
location.

A number of textbooks (Snedecor & Cochran 1989; Sokal & Rohlf
1995; Underwood 1997) provide equations for relating costs and variances to determine the optimum number of replicates at each level of
sampling to minimize the variance of the mean at a particular sampling
level and we won't repeat those equations here. Keough & Mapstone
(1995) made a number of sensible recommendations for deriving and
using these values for sample size at each level of subsampling. First, the
calculated sample sizes depend on the quality of the pilot data, particularly the variance estimates, and how well the variances in the real

monitoring program will match those from the pilot study. It is important, therefore, that the pilot study is done in similar locations and at a similar time (e.g. season) to the monitoring program. It is also important to check that these variance estimates still hold once the monitoring program has started and to adjust the sample sizes if necessary. It is much easier to reduce sample sizes during an ongoing monitoring program than to increase them, so the initial sample sizes should be generous. Second, the sample size values will usually not be integers so they should be rounded up to the nearest integer. Finally, the calculations may recommend sample sizes of less than one, because the variance at that level is so small or costs are cheap. However, some level of replication is necessary for sensible inference and, remembering that pilot studies may underestimate the true variance, we recommend that more than one replicate at any level should always be used.

We provide a worked example of using scalable decision criteria at the end of chapter 13, which introduces the idea of optimization: the development of a monitoring program that leads to unbiased decisions but with, also, an effective use of available resources.

## 12.4 FURTHER VARIATIONS ON BALANCED DECISION-MAKING

The general framework for balancing errors allows for some interesting variations. We have discussed the necessity for a nominated critical effect size when estimating Type II errors. A low Type II error means it is probable that an impact will be inferred whenever an impact as large as that nominated critical effect size occurs. But a different kind of 'critical' effect size may be recognized by the proponent, in the context of a modified form of Type I error. The proponent may want it to be improbable that impact is inferred whenever the impact is as small or smaller than this new critical effect size. If this smaller effect size is interpreted as effectively defining a modified notion of 'no-impact', that improbability can be thought of as a low Type I error. This modified Type I error could be viewed also as resulting from a null distribution shifted slightly to the right (Fig. 12.2a).

This consideration might dominate the decision-making process on occasions when there is no nominated effect size associated with Type II error. For example, in the Kakadu monitoring program (Faith *et al.* 1991) Type II errors were only explored with a range of hypothetical effect sizes, and no single critical effect size was identified. In such a case, the design might focus only on the need for a low Type I error

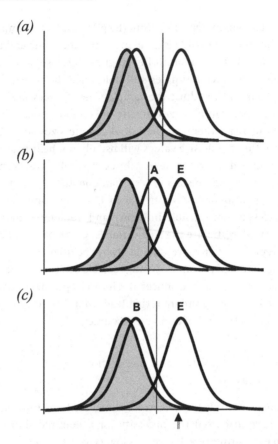

**Fig. 12.2** These graphs represent hypothetical probability density functions as described in Fig. 12.1. (a) The distribution just to the right of that under the $H_0$ is for some small effect size that can be used as an alternative to calculate Type I errors. (b) When impact is inferred because $H_0$ is rejected, the impact might be some effect size A, smaller than the critical effect size E. (c) The arrow shows a hypothetical value of our statistic from monitoring. It is improbable by chance that this observed evidence would be found when there was an impact with an effect size as small as B.

associated with a nominated small effect size. The developer then has some assurance that impact will not be inferred if the impact is as small as that effect size, and so may readily agree to face costs of activity stoppage on any occasions when impact is detected because the effect is larger than this critical effect size.

The consideration of other effect sizes can influence decision-making in other ways as well. To see this, it is useful to consider the corroboration of an impact hypothesis arising when the null hypothesis is rejected. Finding that our evidence for impact is improbable by chance (so rejecting the $H_0$ and corroborating the hypothesis of impact) does not rule out that the impact might have some effect, A, smaller than the critical effect size, E (Fig. 12.2b). Low Type II error may have guaranteed that, if an impact was as large as E, an impact was inferred – but it does not mean that any impact inferred must have been as large as E. What then can be concluded about the size of impact? Distinctions among different effect sizes are possible when we consider the actual observations from monitoring, shown hypothetically by the arrow in Fig. 12.2c. It is improbable by chance that this observed evidence would be found when there was an impact with an effect size as low as B but probable if the effect size was E (Fig. 12.2c).

Suppose that M is the largest effect size that can be excluded in this way. We can interpret this as providing a degree of corroboration of an hypothesis that the impact has an effect size as large as E. This hypothesis is better corroborated to the extent that M is nearly as large as E. We can say that it is improbable to have observed our evidence by chance not only under usual no-impact conditions, but also under impacts as large as M.

Thus, attention to the actual observed outcome opens the door to further inference (inferences compatible with Mayo's (1996) use of 'severe tests', discussed in chapter 4). Such inferences can feed in to decision-making. In the simple cases described above, we had a dichotomous decision based on reject/not-reject. While non-rejection of the null hypothesis may well continue to describe simply 'no action', rejecting the null hypothesis can lead to a range of actions, depending upon the conclusions about the likely magnitude of impact that are warranted given the actual observed monitoring outcome.

## 12.5 IMPORTANT ISSUES

- Components of any decision-making process include outcome states or events, alternative actions we might take, consequences that occur, outcome state probabilities and decision criteria.
- We set the probability of Type I errors in repeated sampling with our a priori significance level ($\alpha$) and there is an arbitrary convention in many disciplines, including ecology, to fix $\alpha$ at 0.05 or 5%.
- The probability of a Type II error can only be calculated for specific alternative hypotheses (i.e. effect sizes).

- Our decision-making framework must provide a reasonable and flexible balance between Type I and Type II errors.
- It is important to specify environmentally important effects sizes a priori and then design the monitoring program. Sample sizes are critical considerations in ensuring that a nominated effect size will have a high probability of being detected if it occurs.
- The advantage of requiring negotiation is that the entire design and decision process of setting effect sizes and error rates is transparent and open to scrutiny without non-negotiable conventions or hidden criteria.
- In contrast to strategies for decision-making that fix $\alpha$ or $\beta$, scalable criteria do not favour either the proponents of the human activity nor the protectors/managers of the environment.
- Further evaluation based upon the actual observed outcome from the monitoring design allows for additional inference that can feed in to decision-making. While non-rejection of the null hypothesis may well continue to describe simply 'no action', rejection of the null hypothesis can lead to a range of actions, depending upon the conclusions that are warranted concerning the likely magnitude of impact, based upon the actual observed monitoring outcome.

# 13

## Optimization

In the previous chapters, we have outlined the logical and statistical bases for environmental monitoring, and discussed some of the practical constraints that may be encountered along the way. While an understanding of these principles is a key ingredient, the ultimate aim is to establish a monitoring program that will lead to reliable, or at least unbiased, decisions with a known risk, and that is an effective use of the available resources. We have termed the processes of translating these principles into a workable sampling program **optimization**.

### 13.1 WHAT WE MEAN BY OPTIMIZATION

Optimization has three important components.

- Developing an idealized sampling program, satisfying a set of target criteria, independent of cost. This idealized program details the number of samples that need to be taken to attain the desired level of confidence in the decision-making process
- Comparing this program with the financial and logistic constraints of the particular monitoring situation
- Trading off attributes of the monitoring program to produce the best 'constrained' solution.

It is useful to portray these components as a flow chart, as shown in Fig. 13.1, to emphasize the sequence of events, and the iterative nature of the process.

### 13.2 BY NOW YOU SHOULD HAVE . . .

There are several steps you must have completed before you can move to optimizing your monitoring program. You should have:

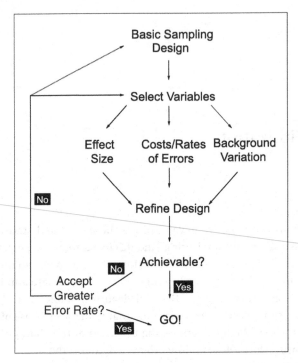

**Fig. 13.1** Flow chart for optimizing a sampling program (from Keough & Mapstone 1997). The diagram shows three key stages – identifying the parameters of the monitoring program and refining the design, in which the desirable number of samples is calculated; determining whether this program is achievable; and making trade-offs.

- Considered the best possible design for your monitoring situation, using the principles outlined in chapter 5 and taking into account any immutable constraints (e.g. no Before data)
- Specified the statistical model that you intend to fit to the data (chapter 7)
- Identified a set of criteria for control locations, with a notional set of potential locations satisfying these criteria (chapter 8)
- Identified variables that are likely to provide a clear indication of any impact (chapters 9 and 10)
- Thought about what an important change would be for those decision variables, and negotiated the kind of change that will be a trigger for management action (chapter 11)
- Considered the consequences of Type I and Type II statistical errors, and negotiated the relative rates of these errors (i.e. the $\alpha/\beta$ ratio) (chapter 12)

Table 13.1. *Variances that must be estimated to optimize two common monitoring designs*

| Design type | Variance estimate required |
| --- | --- |
| BACIP | Variation in $d$ (i.e. difference between Control and Impact) values through time |
| MBACI | Space–time variance – variation among control locations through time periods (at scale of Before and After periods). This variance is a measure of the extent to which the group of control locations 'track' each other through time |

- Designated a desirable overall level of risk for the decision-making (i.e. the absolute values of $\alpha^*$ and $\beta^*$) (chapter 12).

### 13.3 YOU WILL NEED AN ESTIMATE OF VARIANCE

Whether you are calculating the number of samples as part of a formal power analysis associated with a test of an hypothesis, or as part of a procedure to obtain a confidence interval of a particular size, the calculations require some idea of how variable the data are likely to be.

Environmental monitoring covers a wide range of designs, which often have complex statistical models. These models often include several kinds of space, time and space–time variances. As part of the optimization procedure, we need to identify the kinds of variation that are to be used in assessing the impact. In hypothesis-testing procedures, especially the ANOVA models used to illustrate chapter 7, this amounts to identifying the variances that will be used as denominators in $F$-ratios. Because these variances change with monitoring designs, so does the variance that we need to estimate. Table 13.1 lists two of the common designs, and the variances that must be estimated. It should be noted that these variances include variance combinations of space and time.

### 13.3.1 Sources of variance estimates

There are two main ways we can obtain variance estimates:

1. The most common way of estimating variances is through the collection of pilot data, in a structured way, from a few places and/or times. Pilot studies are useful also for compiling an initial inventory making predictions in a legally mandated assessment stage, or proofing field methods. We note that pilot studies vary

enormously in quality and the data from them should be scrutinized carefully before use. Hazards associated with the use of pilot data will be lessened the more pilot studies are formally planned as prototypes of the expected monitoring program.

2.    Existing scientific literature also can provide a guide (see section 11.3.2) if there are data from the same biological system and same geographic region – preferably the same catchment/watershed. However, data simply describing variation often are not of sufficient interest to warrant publication in widely circulated journals, and the relevant information often must be obtained by calculation from published figures, or, more often, from the grey literature (see section 9.3.2). Unfortunately, this literature tends to be poorly indexed, and the material not easy to locate. This situation may improve as more such data are published, at little cost, on the World Wide Web.

In getting pilot estimates of variance, an important point is that the scales of pilot sampling should match scales to be used in the monitoring program. If we are asking questions about control and impact locations, and the variation between control locations is used to assess the impact, we must obtain pilot variances at the spatial scale of those locations. Similarly, if we are seeking to detect changes over time-scales of years, then we need to understand variation on that scale.

It can be tempting, when time or resources are limited, to use estimates of variances from other temporal or spatial scales. In particular, obtaining an estimate of spatial variance is often far easier than estimating a temporal variance, and we may try to use this estimate. It is important to note that there is no scientific reason why such disparate variance estimates should bear any relationship to each other – it is not difficult to think of species that are consistently common over a wide area, but which show strong temporal fluctuations, and species that show substantial and relatively synchronous temporal changes (e.g. annual or seasonal species), but with considerable spatial patchiness. Some of this information is captured in scope diagrams (see section 2.3).

Matching the scales is often difficult – with pressure to commence monitoring, it can be difficult to argue for an extended period just to obtain pilot variances. In a similar way, sampling at the larger spatial scales at which we might expect control locations to occur may well be expensive, and we may struggle to convince those paying for the monitoring to commit enough resources to collect data that are not directly used in decision-making. In these cases, we need to be sure that

Table 13.2. *Outcome of power analysis for two common designs*

| Design | Power analysis to get: |
| --- | --- |
| BACIP | $t_B$ and $t_A$; numbers of times Before and After. In practice, $t_A$ is likely to be fixed, and $t_B$ is the target. |
| MBACI | $I$; number of Control locations (or $l_C$ and $l_I$, numbers of Control and Impact locations). In practice $l_I$ is fixed, as we are aware of no cases in which the number of impact locations was increased to raise the power! |

those overseeing the monitoring program understand clearly that no power analysis or sample-size calculations are possible without variance estimates.

### 13.4 DEVELOPING AN IDEALIZED SAMPLING SCHEME...

With the information described in sections 13.2 and 13.3, we can use power analysis to calculate the desired amount of replication. This replication varies with the design in question, as shown in Table 13.2 for BACIP and MBACI designs.

Doing a power analysis, either at the planning/optimization stage or *post hoc*, requires a precise specification of the underlying statistical model, and the link between fitting a statistical model and the decision-making procedure. This step is not necessarily simple because, as we mentioned earlier (section 4.7), several common monitoring designs have very complex underlying statistical models. Calculation of the optimal level of replication is not possible without this step, however, and, in general, power analysis is not possible without a precise statistical model. This link is often forgotten when notions of power are introduced into broad policies.

Being forced to specify the statistical model also focuses attention on the design, units of replication etc., and we find this a useful aspect, especially for those with limited experience in designing such programs. Specification of the model also brings to mind the likely assumptions of the procedure.

### 13.4.1 Form of output

In its simplest form the power analysis gives a simple answer: the desirable level of replication (to detect a given effect, with a given level of

*(a)*

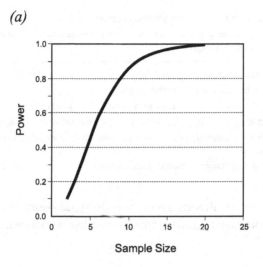

*(b)*

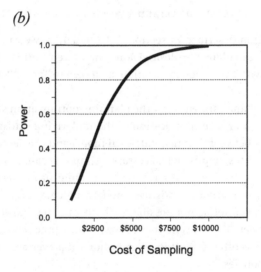

**Fig. 13.2** (a) Sample output from power analysis, relating number of samples to power. (b) The same power curve expressed as a function of the cost of sampling, assuming $500 per sample of invertebrates from Sürber samples.

confidence). However, a single value such as this has limited usefulness, because the next step of the process almost always involves consideration of different intensities of sampling. A particularly useful way to present the results from this stage is as a curve relating power and number of samples (Fig. 13.2a), with an indication of the target level of power. For communicating with less technical participants in the monitoring program, it is worth emphasizing that power is effectively confidence in the monitoring program. The curve shows quickly how changes in sampling will influence the outcome.

In many cases, it is relatively simple to translate the number of samples into a cost of monitoring (Fig. 13.2b).

## 13.5 TRADING OFF

In determining whether the ideal program is feasible, several constraints come into play. As an initial step, the number of samples will be translated into a cost of sampling, and, in the majority of cases with which we are familiar, the costs of this idealized program will far exceed the available resources. In many cases, there will be severe time constraints, and, in particular, there is likely to be little flexibility in length of sampling before the particular activity commences. There is often additional pressure to complete the sampling and come to some decision about the future of the particular activity. These combined pressures frequently result in compromises having to be made, and the process of optimization examines the way in which compromises are to be made. Broadly, we must decide whether to spend more and maintain the sampling, reduce the overall sampling (and raise the risk of inappropriate decisions), or to redesign part of the sampling program to use the resources more effectively – thus making trade-offs. Changing the fundamentals of the design generally is considered to be undesirable, given that the logical basis of the design (and analytical models) should have been agreed to before getting to this point.

In making trade-offs, our emphasis is on minimizing cost of sampling whilst ensuring the selection of appropriate variables (see chapter 10) and sampling regimes that are likely to provide clear-cut results, with minimum ambiguity or uncertainty in tests. The principal trade-off, therefore, is between cost (of monitoring) and probability of errors, bearing in mind that there are costs associated with any actions precipitated by results of analyses (of data). To understand and discuss these trade-offs, the actions that will follow from each statistical inference must be fairly well prescribed so that players can project the likely consequences for them of those actions. Identification of actions from rejection/non-rejection of the null hypothesis in favour of each nominated alternative should also clarify for all players the benefits (for them) of better (or worse!) monitoring.

Assuming (a) there is a desire to minimize risks of erroneous decisions and (b) some proponents will consider that the cost of sampling for the desired $\beta$ is too great, we have several options, which we can often demonstrate to non-technical people involved in monitoring using the kinds of curves shown above – cost ( $\sim$ sample size) versus $\beta$ and cost versus $\alpha$ curves for main variables.

### 13.5.1 Spend more

In some cases, the power curves can show that for a reasonable increase in resources there could be a sharp increase in confidence. In such cases it may be possible to increase the resources to the sampling program and get close to the desired level of power. For instance, in Figure 13.2, increasing costs by 25% (e.g. from four to five replicates) would increase power by 33% (from 0.3 to 0.4).

### 13.5.2 Live with increased risk

The most direct option is to keep the costs fixed, conduct the sampling program, and accept the increased risks of an erroneous intervention or an undetected environmental impact. Again, the power curves are useful ways to explore the increase in these risks – if the resources are close to the desirable level, the increased risk may be minor and acceptable to everyone. However, if the desirable program far exceeds the resources, the increase in risks may be substantial. Figure 13.3 shows several power curves. For curve A, with a target power of 0.80, seven or eight samples per group (e.g. control locations) would be required, while for data sets B and C, the numbers are 11 and 19, respectively. In the event that only seven samples are possible, the power for data set A drops to 0.6–0.7, which might still be acceptable, whereas for data sets B and C, the power values drop to 0.5 and 0.3, respectively. The latter two values would cause substantial problems.

In this latter, probably more common situation, we recommend very strongly against the conventional hypothesis-testing approach, in which the value of $\alpha$ is held fixed. When the monitoring is scaled back, the direct consequence is a rise in $\beta$, so the relative risks of the two incorrect decisions are altered. The scalable approach outlined earlier has the advantage of maintaining the relativity of the two risks (section 12.2.3). If the relative weighting of errors is to be reconsidered, we suggest that the agreed level of $\beta$ should be protected more strongly than the level of $\alpha$. In most instances, the economic benefits of reduced costs of monitoring will accrue to the proponent (who is paying for monitor-

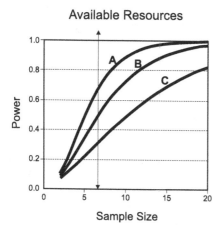

**Fig. 13.3** Power curves for three different data sets (labelled A, B and C), yielding different recommended numbers of samples. The vertical line indicates the number of samples that the monitoring program's resources can support.

ing). Accordingly, where the risks of errors are to be changed, it seems most appropriate to allow the proponent to evaluate the relative benefits of reducing costs of monitoring against the consequences of increased risk most relevant to them – a Type I error. Such an approach may mean that the level of $\alpha$ is increased in order to ensure a constant level of $\beta$ as the costs of monitoring are reduced (see sections 12.2.2 and 12.4). We have found that the graphical illustration shown in Fig. 13.4 conveys these arguments to a wide range of audiences.

### 13.5.3 Maintaining the risk, reducing the cost

Perhaps the most useful approach is to modify the mix of variables and sampling regime to reduce overall costs so that the desired levels of uncertainty can be realized. In particular, reducing the processing costs of each sample can be very helpful.

*Eliminating variables*

Quite often, many observations on a wide range of variables are made from each sample. The total cost of processing each sample may depend on the number of such variables. For example, if we take samples to estimate abundances of benthic macrofauna and to estimate levels of toxicants, the cost per sample may depend strongly on the number of

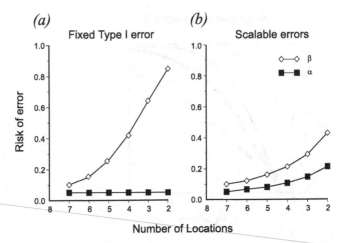

**Fig. 13.4** Traditional versus scalable decision rules versus extent of sampling. The curves show error rates ($\alpha$ and $\beta$) as a function of the number of samples, showing the effects of reducing the number of samples using (a) a traditional rule of $\alpha = 0.05$, and (b) a scalable rule of $\alpha = 0.5 \times \beta$.

different toxicants to be assayed. It may be possible to take a critical look at the list of toxicants, and identify some of lesser importance, and to consider dropping them from the program. Similarly, when processing biological samples in the laboratory, some taxonomic groups may require much more effort at the identification stage. Given limited resources, and acknowledging that our variables represent only a small subset of the possible measurements, we might decide to ignore some taxa. At a broader scale, most sampling programs include a wide range of sampling methods, targeting different components of the environment. Omitting one component might provide sufficient overall cost savings as to raise the power for the remaining components. In cases such as this, we must make professional judgements about the ratio of information gained to sampling effort, and the scientific and political desirability of each component of the monitoring program, and focus on variables that make the decision most clear-cut.

A fallback position to consider is the use of surrogates that are cheaper to measure than the variables originally preferred, but with some loss of certainty.

*Sampling more cheaply*

There are several other ways of reducing per-sample costs, particularly of biological variables.

In processing species-rich samples, a large amount of time is often spent identifying material to finer scales of taxonomic resolution. The time required to identify material is often a function of the taxonomic level desired, and, at least in some habitats, there is a sharp saving in time by working at the next coarsest taxonomic level (e.g. moving from genus to family or species to genus). Consequently, identifying material to coarser levels can reduce costs dramatically. The trade-off in this case is between coarser biological resolution, with high power, and fine resolution, with low power. Whether this trade-off is beneficial may depend on individual biological systems – if finer taxonomic groupings (or functional groups) respond relatively uniformly to a particular disturbance, the signal from these data will be strong, and there will be a clear benefit. However, if individual taxa respond in contrasting ways, pooling them into larger groups may result in variables that have more desirable statistical properties, especially lower variances, but which do not respond to the particular disturbance, and hence are inappropriate decision variables (see section 10.1.2).

In most sampling programs, we have in mind a spatial scale that is used for making a decision. For example, in an MBACI design, in its simplest form, impacts are assessed using variation at the scale of *locations*. We would normally take a series of subsamples of each location for two reasons: (1) in order to ensure adequate characterization of each location; and (2) in order to remove the potential of confounding variation between locations with variation within locations. For example, we might take replicate samples from riffles along one of our control streams. The data from this lower spatial scale (and any other lower-level subsampling) are not directly used in the statistical analysis, and, when fitting a statistical model, these data are averaged or summed. Bearing this in mind, it may not be necessary to process all of these individual samples, and an alternative is to pool the samples, mix them thoroughly – a procedure labelled compositing – and then take a subsample from this pooled sample (discussed in section 8.2.4). The aim of averaging the smaller-scale variation would be met, and there can be considerable savings. For example, compositing may be done on site, reducing time to curate each individual sample. The biggest savings, however, are likely to come in the laboratory, when fewer samples need to be processed. The exact cost savings will depend on how easy it is to mix and subsample the composites, and on the proportion of the composite samples that must be examined in order to get an accurate estimate of their composition.

We suggest that as a routine part of optimization, the sampling design be scrutinized carefully, the units of replication used to assess the impact be identified clearly, and then consideration be given to compositing material below the level of those units. Note, however, that this procedure requires some confidence that smaller-scale effects are of little interest.

### 13.5.4 Accepting larger effect sizes

For key variables that are also very variable (and too expensive to monitor with the desired precision), we need to consider whether to drop them from the program or live with the likelihood that our program will be sensitive only to an increased effect size. We may decide to revisit the earlier discussions we have had about setting an effect size (section 12.2.2). To do this, plots of critical effect size against sample size or cost can be particularly helpful. We emphasize that changing the critical effect size should not be considered lightly (and only ever during the planning stage) since doing so necessitates a reconsideration of what the public, proponents and managers have agreed to be important for this activity (see chapters 11 and 12).

### 13.6 UNCERTAINTY IN OPTIMIZATION

It is important to realize that the optimization procedure is an approximate one, using the best available information, which has considerable uncertainty associated with it.

### 13.6.1 Origins of uncertainty

The main source of uncertainty is in estimation of (error) variance. Unfortunately, our initial estimates of variance are generally based on pilot data, which, by their nature, are quite limited. They are also likely to be from one short time period, and not reflect environmental fluctuations at larger time-scales. We can improve our estimates by increasing the size of the pilot estimates, and by using as many different estimates of variance as possible – rather than using pilot data or published information, use both, and combine them (see chapter 9).

It is also possible to quantify the uncertainty in our estimates, by calculating confidence intervals about the variance estimates. It would then be possible to calculate the required number of samples for the upper and lower estimates of the variance. In deciding the level of sampling, we would then need to weight these values against those obtained from our original (and best) estimate. This is rarely done, and we suspect that most people rely on their estimates being unbiased, and, therefore most likely to represent the true variance.

If we are to have an unbiased estimate, we must assume that the impact does not change the relevant variances. If for example, a particular activity caused a series of impact locations to follow widely divergent paths through time, a consequence would be an increase in the space–time variance used to assess the impact. In an analysis of an MBACI design, for example, using analysis of variance, the relevant denominator is the Locations(Control–Impact) × Before–After interaction, which is an average of the L(C) and L(I) × B–A components. The optimization would have been done using the L(C) component only, which would in this case be an underestimate of the overall variance. Such a problem could be dealt with in the simplest case by adjusting the $\alpha$ and $\beta$ values *post hoc*, as described in section 13.7.

### 13.6.2 Incorporating capacity for readjusting the sampling program

Given uncertainties in the predicted performance of a sampling program (and associated statistical tests), we recommend starting with the expectation that the design may need to change as more information is gathered during the Before period. Accordingly, we should err on the side of caution (e.g. start with more rather than fewer sites etc.) and acknowledge that it may be acceptable to drop sites if variances look better than those from pilot data. Adding in sites may be more difficult than dropping sites, in lots of ways. For example, many of the statistical models rely on sampling the same physical locations through time, and it is difficult to add new locations part way through the sampling program. It is more practical to include 'spare' locations, rather than trying to add new ones later; this procedure may also provide for much easier economic planning by the proponent.

### 13.7 POST-MONITORING 'OPTIMIZATION': IMPLICATIONS FOR DECISION CRITERIA

At the conclusion of the monitoring program, we may find that the actual levels of variation are quite different from those used at the planning stage. There may also have been unforeseen events, such as the loss of some control locations. We know of cases in which control locations disappeared as a result of El Niño events, and even of planes crashing into the water on or near control locations. In this case, our data set at the end of the sampling program may have quite different power characteristics.

As a result, the realized levels of $\alpha$ and $\beta$ may be quite different from those that formed part of the optimization procedure. We therefore recommend that the power calculations be revisited, and also recommend that the scalable decision criteria provide a substantial advantage over traditional fixed-$\alpha$ approaches.

If the data set proved to be less variable than anticipated, and the decision is made using the critical $\alpha$ designated at the planning stage, a direct consequence will be that the value of $\beta$ is reduced. In contrast, if the number of locations is reduced, or if the level of variance is greater than expected, $\beta$ will rise.

To maintain the relativities of these two errors, we would use the revised levels of replication and the improved variance estimate to recalculate the critical $\alpha$ value, as per section 12.2.3. In doing this, it is important to use the original effect size – a common and serious mistake is to use the newly observed effect to calculate power a posteriori. This changes all calculations fundamentally, is circular, and constrains the values that power can take before significance so that power, in this context, becomes meaningless.

### 13.8 A WORKED EXAMPLE – LIMING TO DECREASE ACIDITY OF STREAMS

We will carry on with the example introduced in section 8.4 of the liming of some Welsh streams to increase their pH, continuing to view the liming as a potential 'impact' rather than as a restorative measure. Let us suppose some local fishing businesses wish to add lime to several naturally acidic streams to improve the survival rates of trout in these streams. Trout fishing is locally popular and the proponents argue that increasing the number of streams suitable for trout fishing will greatly increase the number of outside visitors to the area. This will bring a variety of economic benefits to many local businesses and other stakeholders. The acidic streams are naturally low in pH, and consequently the proposed liming, which will increase the pH to approximately circumneutral, may have some sort of detrimental effects upon these ecosystems. The acidic streams have lower species diversity than circumneutral streams, but let us suppose that the species within them occur only in acidic waters. Hence, there is some concern that boosting pH will see invasion of these waters by many other taxa. The protection of low-diversity systems may seem odd, but this is because we typically equate high numbers of species with 'good' and low numbers of species with 'bad'. Nevertheless, if systems naturally have low numbers of species then we should not view the invasion of these systems by more species (such as weedy taxa that occur in many places) with equanimity. It is reasonable to protect the naturally low diversity of such systems.

### 13.8.1 Nomination of an important effect size

Our first step is to decide upon the size of change we will consider to be important. To keep this simple, we will focus on only a single variable – the numbers of individuals of all acid-sensitive macroinvertebrate taxa (given as data in Table 8.9) – even though discussions about important changes would range over a wide variety of other possibilities. In a real situation, we would almost certainly be concerned about decreases in numbers of taxa adapted to living in acidic waters as well as increases in numbers of potentially unwelcome taxa.

All relevant parties with an interest in the liming (the proponents, environmental groups etc.) need to decide collectively what size of increase in densities of species that are sensitive to pH (and hence occur in low densities in the acidic streams) in limed streams would be considered an 'important' change. First, we need to consider how changes in our variable are linked to the magnitude of the human disturbance (see section 11.3.2). pH affects the concentration of filterable aluminium in the water column, which is toxic to biota at high concentrations and thought to be one of the main causes of mortality. There is a negative relationship between pH and aluminium concentration that flattens out at about pH 5.5 and 10–15 $\mu$ equivalents per litre (Fig. 13.5a). Mortality of trout within streams that have high concentrations of aluminium is 100%, falls gradually with decreasing concentrations, and then drops off rapidly to around 20–40% mortality with aluminium concentrations of about 15–20 $\mu$ equivalents per litre (Fig. 13.5b). The responses of acid-sensitive macroinvertebrates to filterable aluminium concentrations show evidence of a threshold effect. Numbers are extremely low at high aluminium concentrations down to around 15–20 $\mu$ equivalents per litre, when they increase 10-fold (Fig. 13.5c). Note that these graphs are from field measures of streams differing naturally in aluminium – we would also want to consider other sorts of data (such as experimental manipulations of aluminium) to make sure we have as much information about the relationships between our variables and the magnitude of human impact as possible (see also chapter 9).

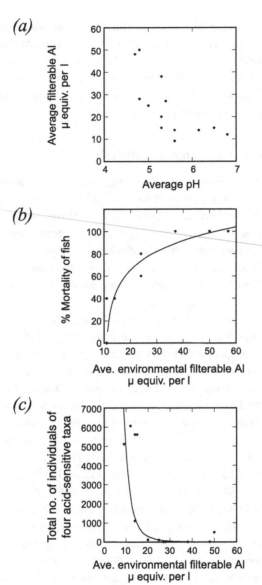

**Fig. 13.5** (a) The relationship between average pH and average concentration of filterable aluminium in different streams. (b) The percentage mortality of caged trout, *Salmo trutta*, exposed in the field for 17 days to nine streams differing in the concentration of filterable aluminium. Each point is one stream. A log curve has been fitted to the data. (c) The total numbers of individuals of four acid-sensitive taxa (*Baetis rhodani*, *Brachyptera risi*, *Leuctra inermis* and *Isoperla grammatica*) found in streams differing in concentration of filterable aluminium. A power curve has been fitted to

However, given these relationships, what sort of effect size might be reasonable? To improve fish habitat, the proponents wish to increase the pH of acid streams to around 5.5–6.0. This would mean decreasing aluminium concentrations of acidic streams to around 10–15 $\mu$ equivalents per litre. From Fig. 13.5c, this would suggest that acid-sensitive macroinvertebrate taxa could increase from a base of $\sim 200$ individuals to anywhere from 1000–5000 individuals – that is, increase at least 1000% in numbers. However, we wish to detect increases in acid-sensitive taxa before they reach such high numbers, so that management can take early action and intervene before densities of these taxa become a serious problem. The steepness of the curve relating densities of acid-sensitive taxa to pH means we want to detect changes that are at the very start of this curve. Consequently, it is decided that an increase of 400% in numbers of acid-sensitive macroinvertebrate taxa – or a four-fold increase – would constitute an important change.

### 13.8.2 Deciding the relative costs of Type I and II errors

The next step is that all interested parties need to debate the relative costs of making Type I errors, $C_\alpha$, and Type II errors, $C_\beta$, and what the ratio of these costs, $k = C_\beta/C_\alpha$, should be (section 12.2.2). Those in favour of liming will be most concerned about the costs associated with Type I errors, which would mean liming would be stopped unnecessarily because of an incorrect detection of an increase in numbers of acid-sensitive taxa. Those concerned more with protecting the natural environment will be most concerned with the costs of committing a Type II error, which would see a real change go undetected.

After considered discussion, the parties agree that the ratio of costs, $k$, should be 0.5 – that is, that the costs associated with a Type I error are felt to be twice as high as those associated with a Type II error. The reasons for weighting the costs and hence the probability of the errors unequally were that:

---

**Fig. 13.5** (*cont.*)

the data. All data are from Stoner *et al.* (1984). Macroinvertebrate numbers were pooled over three 1-minute kick samples taken on each of two dates. Actual numbers of macroinvertebrates were not supplied, as only abundance classes were given – numbers were roughly estimated by taking the middle value of each class for each stream. Note that filterable aluminium concentrations are a better correlate of biotic changes than pH (Weatherley & Ormerod 1991).

- The proponents showed that the economic costs to themselves and other businesses in the area would be high if liming had to be stopped or the effects of liming had to be reversed. They showed that initial costs to undertake liming of multiple streams would be considerable and that it will take time to build up sufficient angling stocks (given that trout take more than a year to grow to sizes suitable for fishing). It also takes time to build the sort of reputation for consistently good fishing that would see sustained increases in the number of visitors to the area.
- There are no implications for human health, as none of the rivers in the area is used for drinking water.
- Other activities on the rivers (e.g. sightseeing, canoeing) would be unaffected by changes in pH.
- The streams do not occur within a state or national conservation area.
- The fauna of the acidic streams is not unique, as there are other acidic streams in the area.

Note that we have constructed the example this way simply to illustrate that there may be cases where it is reasonable to weight the costs of one type of error – in this case, Type I errors – more heavily than the other. We are not suggesting that this should be some sort of standard outcome. Each potential impact must be considered individually. For example, it is easy to imagine alternative situations where the costs of Type II errors would be deemed to be much higher than that of Type I errors. If the development were to take place in a unique natural area of very high social or environmental significance, then it would be reasonable to argue that the costs to the community of degrading that environment are very high, especially if any ecological restoration had to be attempted. Arguably, the example of mining activities in Kakadu National Park (Box 4.1) would fall into this category, given that Kakadu National Park is in a World Heritage area, is considered unique, has high environmental and cultural attributes, and that environmental degradation caused by mining radioactive substances could be virtually irreparable. In other cases, the relative costs of Type I and Type II errors may be quite unclear. We often cannot put an objective and sensible monetary value on many environmental attributes that are still valued very highly by the community. In these situations it may be best to agree to weight the errors equally (Mapstone 1995).

### 13.8.3 Deciding the actual probability of errors

The next step is to negotiate the maximum risks of a Type I and a Type II error each party in the monitoring program would be prepared to accept. The ratio of these errors is set by the relationship $k = \alpha/\beta$, where $k$ is determined by the negotiations in the preceding step (section 13.8.2). The relationship $k = \alpha/\beta = C_\beta/C_\alpha$, sets the costs of committing a particular error in exact inverse relationship with the probability we set for its occurrence. This ensures that as the relative cost of committing one particular error increases relative to the other, its maximum probability of occurrence should decline. Thus, with $k = 0.5$, the maximum probability of committing a Type II error, $\beta$, will be set at twice that of committing a Type I error, $\alpha$, as the parties have already agreed that the relative costs of Type II errors are half those of Type I errors.

In our liming example, the proponents argue that they cannot realistically undertake this venture if $\alpha$ is set relatively high, and wish to proceed with a risk of being halted unnecessarily at no greater than 0.02. This means that the probability of an unacceptable change going undetected must be set to 0.04. After some discussion, the parties agree to these probabilities of errors, so $\alpha^*$ is set at 0.02 and $\beta^*$ at 0.04.

### 13.8.4 Use of pilot data and power analysis to examine the number of locations needed in the monitoring program

The next step is to use pilot data to calculate, for the agreed effect size of a four-fold increase in numbers of acid-sensitive taxa, the number of locations in the monitoring program that would be needed to realize the agreed upon values of $\alpha$ and $\beta$. This step depends greatly upon the analytical model that will be used, because the estimates of variance that are required to calculate power vary between models (Table 13.1, and see chapter 7). Note that power analysis is not simple for complex designs like MBACI, as the calculation of power requires the specification of an alternative hypothesis and the solution of somewhat complex equations whose specific forms depend on the exact analytical model under consideration (see Keough & Mapstone 1995). We have not provided the specific equations to carry out the calculations below because we believe that in most cases a statistician will be needed to carry out the appropriate calculations, but see Keough & Mapstone (1995) for a worked example.

For our liming example, the parties agree to use an MBACI design with replicate Impact locations, and so need an estimate of the variation among Control locations over periods of time equivalent to the expected temporal scales of the monitoring program, where samples will be taken once a year for multiple years before and after start-up. In this example, some data are available from three circumneutral streams (occurring in the same area as the streams proposed for liming), which have been monitored for six years. An estimate of the variance among Control versus Impact locations over time is obtained by imagining that the three locations come from a liming experiment. Two of the circumneutral locations are designated as 'Controls' and one as an 'Impact' location, and the first set of three years designated as 'Before' and the second three years as 'After'. These designations allow an MBACI analysis, which then provides an estimate of the relevant error variance (the mean square from the Locations within Impact vs. Control × Before vs. After term – see Table 8.10). The mean square from pilot data from reference streams is then used to calculate the power of future tests of liming, using an effect size of a four-fold increase in numbers of acid-sensitive taxa at Impact locations compared to Control locations. (In reality, monitoring data from circumneutral streams were collected as part of the original monitoring program, to provide a picture of the reference condition to which researchers hoped to return acidic streams. Nevertheless, these data serve as a good example of the kind of estimation of variance that is required for reliable estimates of power – that is, values of the variables proposed for monitoring collected at the appropriate spatial and temporal scales and in the same general area as the proposed Control and Impact locations.)

Data from the power analysis are shown in Table 13.3 and plotted in Fig. 13.6, which is analogous to Fig. 13.2 except that $\beta$ is plotted on the $y$-axis rather than power (recall that power $= 1 - \beta$ – see section 4.7). The value of $\alpha^*$ that was agreed through negotiation (0.02) requires six Impact and six Control locations to obtain an expected value for $\beta$ that meets the agreed value of slightly less than or equal to $2 \times \alpha$, or 0.04. However, $\beta$ drops well below 0.04 to 0.018. This is true for all the other analyses (which use different values for $\alpha$) and occurs because of the stepwise influences of adding two whole locations (one Impact and one Control) for each value of $n$, given that we cannot add 'fractions' of locations. Consequently, for $n = 6$ Control and 6 Impact locations (12 locations altogether) then $\alpha^* / \beta^* = 0.02/0.018 = 1.11$ for $n = 5$ of each, $\alpha^* / \beta^* = 0.02/0.061 = 0.328$. Hence, neither sample size meets our requirement that the ratio $k = \alpha^* / \beta^*$ should be close to 0.5. Note that we have not considered the possibility of reducing the number of one type of location (i.e. of either impacts or controls) while retaining a higher value

Table 13.3. *Expected value of β if sampling different numbers of locations in each of Impact and Control treatments given an a priori preference for a specific value of α and a desire to have β close to a value of ≤ 2 × α. These are the data plotted in Fig. 13.6.*

| α | Locations | β |
|---|---|---|
| 0.01 | 2 | 0.906 |
| | 3 | 0.626 |
| | 4 | 0.317 |
| | 5 | 0.129 |
| | 6 | 0.045 |
| | 7 | 0.014 |
| 0.02 | 2 | 0.821 |
| | 3 | 0.453 |
| | 4 | 0.182 |
| | 5 | 0.061 |
| | 6 | 0.018 |
| 0.03 | 2 | 0.745 |
| | 3 | 0.347 |
| | 4 | 0.120 |
| | 5 | 0.036 |
| 0.04 | 2 | 0.677 |
| | 3 | 0.274 |
| | 4 | 0.085 |
| | 5 | 0.023 |
| 0.05 | 2 | 0.615 |
| | 3 | 0.221 |
| | 4 | 0.063 |
| 0.06 | 2 | 0.559 |
| | 3 | 0.181 |
| | 4 | 0.048 |
| 0.07 | 2 | 0.509 |
| | 3 | 0.151 |
| | 4 | 0.038 |
| 0.08 | 2 | 0.464 |
| | 3 | 0.127 |
| 0.09 | 2 | 0.423 |
| | 3 | 0.108 |
| 0.10 | 2 | 0.386 |
| | 3 | 0.092 |

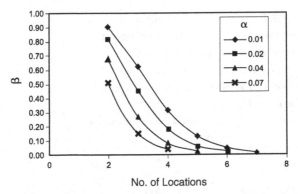

**Fig. 13.6** Values of $\beta$ versus the number of locations (for each of the Control and Impact categories) for different values of $\alpha$, with the target being that sufficient locations should be sampled to ensure $\beta$ is less than or equal to $2 \times \alpha$ ($\beta \leq 2 \times \alpha$). This point is represented by the right-most point on each curve. These data are presented in Table 13.3.

of $n$ for the other (i.e. have unbalanced sample sizes for controls and impacts). Although this strategy could be pursued, it is generally undesirable to have unbalanced sample sizes within designs except under very particular circumstances. To keep our example simple, we will proceed under the assumption that the group desires to keep the number of impact and control locations equivalent.

The next step then is to take the number of locations that are close to achieving the relationship $k = \alpha^* \,/\, \beta^* = 0.5$, and iteratively adjust values of $\alpha$ and $\beta$ until a value for $k$ of 0.5 is achieved. This calculation means, for most nominated starting values of $\alpha$, that the realized value for $\alpha$ ends up less than that nominated a priori (Fig. 13.7). Thus, when beginning with an initially desired value of $\alpha$ of 0.02, $\alpha^*$ sinks below 0.02 and $\beta^*$ ends up below 0.04 when the errors are forced to meet a $k$ of 0.5. Figure 13.7 illustrates very neatly the options available. Either the parties must agree to use $n = 6$ Control and $n = 6$ Impact locations and accept lower values of $\alpha$ and $\beta$ than were agreed upon (see section 13.8.3), or they monitor fewer locations (i.e. reduce $n$) and settle for increased values of both $\alpha$ and $\beta$. For any value of $n$ that the parties choose from Fig. 13.7 however, the ratio of errors that was agreed upon at the start remains the same – 0.5.

### 13.8.5 Trading off costs and risks

At this point, the parties are in a position to calculate the total costs of monitoring a given number of locations for a desired number of years

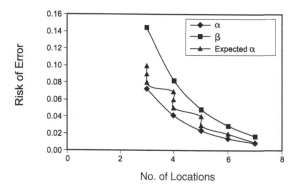

No. of Locations

**Fig. 13.7** Values of $\alpha$ and $\beta$ at which $\beta = 2 \times \alpha$ (i.e. $k = 0.5$) for different numbers of locations sampled in each of the Impact and Control categories. Also shown are the initially desired (or expected) values of $\alpha$ from which the optimized (realized) values derive. As can be seen, the same endpoint in the optimization can result for different desired values of $\alpha$, because of the discrete effects of adding locations in whole pairs.

and to compare this sum to the resources (of money or people) available to meet them. As we indicated above, it is common for there to be insufficient money and other resources (say, of people) to monitor the minimum numbers of locations required to meet the desired levels of risks of errors. Any mismatch requires the parties to discuss whether they are all prepared to live with increased risks and, if not, to examine other options.

In our liming example, the pilot data provide good estimates of the cost of carrying out sampling and processing over three locations, which can easily be used to estimate the costs of monitoring 12 locations once a year for six or more years. When the calculations are done, the amount of money required turns out to be about twice that available to carry out this part of the monitoring program. Consequently, the parties have to explore various options for retaining their agreed levels of risk of errors.

*Spend more money?*

The first option the parties consider is whether sufficiently more money could be made available to allow 12 locations to be monitored at each time. However, the projected cost would virtually double for only a relatively small (although desirable) decrease in probability of errors. Consequently, the parties feel that this sort of expense is not warranted, and that it would prove very difficult to justify such a large increase in cost for a relatively small decrease in risk to those paying for the monitoring program.

### Live with increased risks?

The second option the parties consider is accepting increased risk, given that both the proponents and the opponents face the same relative increase in probabilities of erroneous decisions that respectively concern them. The current budget would allow annual monitoring of six locations (three Impact and three Control) and will produce a value of just over 0.07 for $\alpha^*$ and almost 0.15 for $\beta^*$. However parties on both sides of the issue agree that the risks are higher than they are prepared to accept – the proponents are not willing to accept a 7% chance of the development being halted unnecessarily and those concerned with protecting the environment are not prepared to accept a 15% chance that an unacceptable change would go undetected. Hence, all parties reject this as an option.

### Reduce the cost of sampling?

The next option is to look at the sampling methodology and see if there is any way that costs could be reduced without compromising the quality of the data. The sampling technique at each location, at each time, is to take one 2-minute kick sample from a riffle zone and one 1-minute kick sample from the stream margins. In the sampling protocol used to collect pilot data, riffle and marginal samples were sorted, identified and enumerated separately. Some samples contained many hundreds of individuals of acid-sensitive taxa. Savings could be made if riffle and marginal samples were not enumerated separately and counted completely, but instead combined and a subsampling method developed to limit the total number of invertebrates that need to be counted from each combined sample. It is possible to work out a minimum proportion of sample or numbers of invertebrates that need to be counted per combined sample to provide an estimate of the total numbers of invertebrates with a known level of precision (see Underwood (1981) for examples of the analysis required, and Marchant (1989) for an example of a practical way of subsampling invertebrates). As long as the precision of subsampling is high, there is no great loss of information, because separate values for riffle and marginal samples are not required for the analysis, which uses one value per location per time. Savings are made because the time required to composite samples, subsample them and enumerate each subsample is much less than that required to process all the original samples individually and completely. When the calculations are made, the monitoring costs drop by a third, and it becomes possible to monitor five Impact and five Control locations each year with a slight increase in funds. This produced error rates only slightly above those originally agreed upon – about 0.024 for $\alpha^*$ and 0.048 for $\beta^*$ (Fig. 13.7). Both sets of parties agree that these risks of error are acceptable, and agreement is reached that 10 locations will be sampled yearly using the subsampling technique.

### 13.8.6 Implications from this example

The above example is contrived, but the data we used are real and the outcomes of the power analysis are also real. If the example had also been real, the kind of negotiation over costs and subsampling that we have described would be a likely outcome of looking at the costs and risks of monitoring. However, in our experience a risk of error versus number of locations curve such as that shown in Fig. 13.7, where even only six locations in total produced a value for $\beta$ of under 0.15 (or a power of 75%) is a relatively rare occurrence. It is more typical for power curves to follow the more problematic trajectory depicted by line C in Fig. 13.3, where relatively high numbers of locations are required just to achieve 50% power or $\beta = 0.5$. For many monitoring designs then, negotiations over the relative costs and risks will be difficult and some of the other strategies (such as accepting a large effect size) may have to be explored.

This expected difficulty is why it is important to begin the process with:

* As many variables as possible, with the aim of weeding out those that that are either too expensive or too variable to allow reasonable minimum risks of error to be achieved
* Pilot data that are of sufficient quality and quantity to provide good estimates of error variances, so that we will have confidence in the power analyses and estimates of numbers of locations and/or times required
* Realistic estimates of the costs of collecting data for each variable, including ways of reducing costs (such as compositing or reducing taxonomic resolution, or using cheaper, surrogate variables).

Finally, we should remember that we can use the first two or more times of monitoring our locations to recalculate power curves and determine whether the initial estimates of sample sizes are under or over that required for the agreed levels of risks. The point to remember is that our pilot data and preliminary power calculations do not *guarantee* particular probabilities of Type I and Type II errors – they simply guide us concerning their likely values. We must always keep in mind that there will be uncertainty in the optimization of our monitoring design, and to plan ahead (section 13.6 above) for maximum flexibility for dealing with that uncertainty.

### 13.9 IMPORTANT ISSUES

- Optimizing designs involves explicitly evaluating the trade-offs between what is desirable (ideally) and what is affordable and logistically feasible, given the specific objectives for which monitoring is being done and the risks of error that are acceptable to stakeholders.

- Formal optimization requires good knowledge of the costs of sampling and a priori estimates of key variances – specifically those that represent the background against which future impacts will be inferred. The exact source of variation for which an estimate is needed will be design specific. In MBACI designs, the key variance will be the variation among control locations through time.

- Prior variance estimates can be obtained from dedicated pilot studies, from previous impact monitoring studies for similar developments or from published research on relevant variables. Dedicated pilot studies are usually preferable, but should be well designed to provide robust estimates of variances.

- Compromises in design are inevitable. The main trade-offs involve balancing the level of spending on monitoring against willingness to accept risks of making incorrect inferences about impacts (either that 'unacceptable' impacts have occurred or that there has been no 'serious' impact).

- Optimizations should be done for several variables that are expected to be sensitive to impact. In this way, flexibility in the monitoring design is included from the outset. Variables that require inordinate levels of sampling to meet desired objectives for risk can be excluded in favour of those that are more efficiently measured.

- Evaluating risks requires clarity about what are considered acceptable and unacceptable levels of impact. These decisions should be made up-front. Changes to these levels represent fundamental shifts in objectives for monitoring and should not be made lightly.

- Monitoring designs can be made more economical by reducing the taxonomic resolution with which biota are recorded, thereby reducing the time and cost of sorting multi-species samples. The taxonomic resolution appropriate to each monitoring program must be considered carefully, however, in the context of the main concerns about, and expected nature of, impacts.

- Optimizations of sampling designs are approximations and should be used as guides to best practice rather than prescriptions. Uncertainties inherent in the optimization process mean that monitoring programs should be considered flexible and some changes should be expected as additional information is gathered. Changes in program design, however, should not undermine the basic principles of good design (chapter 5).

# 14

## The special case of monitoring attempts at restoration

Not all assessments involve decisions about detecting an impact. Given that many substantial impacts have already occurred, there will be an increasing number (we hope) of programs geared towards assessing the success of restoration and rehabilitation programs. During the century ahead, we believe that an expanding response to human impacts on riverine ecosystems will be to attempt to restore them to some defined condition, such as a previous, less-impacted status. Such attempts to augment nature have considerable theoretical interest to ecologists as well as obvious practical implications for managers. It has often been said that the ultimate test of our understanding of an ecosystem is to create or repair a habitat and its function.

Ecological restoration is an applied approach to fixing environmental problems once they are diagnosed. It has arisen from a variety of practical starting points (Ehrenfeld 2000) and as such is not, as yet, a coherent and focused sub-discipline within scientific ecology. We see considerable confusion currently about the aims and techniques applicable to restoration monitoring (see also the critique by Chapman & Underwood 2000) but also appreciate the importance of going beyond merely detecting problems to trying to fix them.

We also do not wish to get entangled in subtle distinctions among 'restoration', 'rehabilitation', 'regreening' or 'ecological landscaping' (see Samways 2000) – if you are going to act then the outcome needs to be monitored. Our emphasis here is on being explicit about what you are trying to do regardless of the label you give it. Impact detection, as outlined earlier in the book, should lead to considering possible restorative actions – here we wish to mandate what makes a good design for monitoring the effectiveness of restoration. We also see the assessment phase as important to identify exactly why degradation has occurred, and hence the action needed to fix the problem. Such an ecological restoration was the aim of the liming case study (described in section 8.4) but the general case is also the focus of this chapter.

## 14.1 ISSUES CONCERNING THE STUDY OF ECOLOGICAL RESTORATION

One of the main actions that should follow successful impact monitoring is attempting to fix the problems identified during monitoring, especially by attacking the root cause of the impact. There are also more general concerns with trying to redirect ecosystems to some agreed target condition. Both removing stressors and rebuilding towards a target require monitoring to assess the realized performance of such restorative measures, otherwise the efficacy of restoration can be all too easily misrepresented (as either 'good' or 'poor' by different interests).

In situations where rivers are already suspected or known to have been impacted significantly by humans, the agenda may not be the detection or characterization of those impacts so much as their amelioration. There are programs under way to clean up water quality, remove outdated or obsolete dams, return minimal 'environmental flows' to hydrological regimes in ways that try to mimic the variability of natural flow regimes, and so forth. Restoration relies upon ecosystem resilience (see chapter 3) and how we can harness or direct any successional changes. We may not yet know enough about the function of many riverine ecosystems to apply knowledge about concepts like successional changes to our attempts at restoration.

Nevertheless, we see that restoration studies will become more common and widespread in the new millennium. Economic growth will be based on new industries such as the development of 'clean-up' technologies (e.g. using macrophytes to 'hyperaccumulate' trace metals from sediments; Bargagli 1998; Brooks 1998). We expect that restoration assessment will be the growth sector within environmental monitoring more generally. This creates some challenges because far fewer monitoring programs have so far attempted to evaluate restoration than programs designed just to detect an ongoing impact. Concepts underlying the monitoring are relevant because, in moving from detecting impacts to assessing restoration, we must use different hypotheses; in fact, a *lack* of difference between restored areas and the target is what needs to be detected following restorative action (see below). This means that the objective(s) of any monitoring program must be redefined in an explicit manner.

## 14.2 CAN BACI DESIGNS BE APPLIED TO ECOLOGICAL RESTORATION?

Restoration projects are amenable to analysis by the sorts of BACI designs that we have been discussing, because we are trying to detect a change in some locations over time. Nevertheless, in practice, there is a serious problem in that the fundamental question is, in fact, rather different. A common goal of these programs is to test whether restorative efforts are returning ecosystems back to some prior, defined or more 'natural', 'reference' state (see Box 5.1). Thus our criterion for success is to find *no difference* between impacted locations and these states, rather than to detect differences (see discussion on 'bioequivalence', Box 14.1). This poses a serious logical problem, in that if our hypothesis is one of no difference, then there is no single, logical null hypothesis that we can form. This means that we have to test our hypothesis instead (Underwood 1990). If we then reject our hypothesis of no difference it leads to rejection of our model – here, that restoration is returning rivers back to some prior state. This is different from the usual falsificationist procedure (see chapter 4) where rejection of the null hypothesis provides support for the model. The real difficulty is, however, created when our hypothesis is not rejected. We would like to interpret this as meaning that we have gained support for our model, but unfortunately such a conclusion could be based on a fallacy (see Underwood 1990 for discussion). The same issue occurs for testing whether water quality departs from established guidelines because, again, the goal is to find no difference between water samples and established minimum concentrations of pollutants etc. (see section 11.1 and Box 11.1).

We can proceed if we hypothesize that, during the program, the supposedly restored rivers will differ from their immediate, initial or starting conditions (i.e. as quantified during the time when impacted). This allows us to construct a logical null hypothesis of no difference, which we can test, and ultimately we expect to reject this null if restorative efforts are having some effect. Ultimately, the best solution is for restorative efforts to be restricted to one or more rivers and other comparable (i.e. equally impacted) rivers be left as unrestored 'controls' (i.e. run the restoration program as a controlled experiment, including randomization, as was done for the liming experiment described in section 8.4). This design automatically removes the above problem, as our hypothesis will always be that of difference among the supposedly restored rivers and those left as impacted 'controls'. We use inverted commas here solely because the 'controls' in this context do not have the

**Box 14.1** Bioequivalence

In the context of environmental assessment, bioequivalence refers to the situation where the values of a variable in a location that is being rehabilitated or restored are similar to those in the control location(s). To get around the conundrum of trying to prove a null hypothesis, the undesirable outcome (that the rehabilitated location is still worse than the control(s)) is defined as the 'null' hypothesis, $H_0$. $H_0$ is then rejected if there is sufficient evidence against it in favour of the alternative (that the impacted location is now bioequivalent with the control(s)). Thus a Type I error would result in incorrectly deciding that the locations were bioequivalent when they still differed by an important amount, whereas a Type II error would result in deciding that the locations were not bioequivalent when, in fact, they were similar.

To illustrate how this works, consider a simple example where we are comparing the mean density of fish, $\mu_t$, in a treated location that we are trying to restore, which we hope will result in their density increasing to a level similar to the density, $\mu_c$, in a control location. Let $R$ be the ratio of treated to control means, i.e. $R = \mu_t/\mu_c$. The 'null' hypothesis, $H_0$, is that the treated site is not bioequivalent to the control site; that is, $H_0 : R \leq R_l$, where $R_l$ is the lower limit of an equivalence region and we assume that large positive values of $R$ are desirable (i.e. that the density in the treated location is close to or even larger than the density in the control location). Thus if $R$ becomes sufficiently large, we have to reject $H_0$ in favour of the alternative hypothesis that the locations are bioequivalent, i.e. $H_A : R > R_l$. (Technically, we should avoid calling $H_0$ a null hypothesis, because under $H_0$ the distributions for the two means are not equal.)

So much for the statistical logic. The hard part is deciding on a value for $R_l$ (see sections 11.2 and 11.3). For this example, let's suppose that after some civilized negotiations, all the stakeholders agreed that restoring fish densities in the treated location to 85% of those in the control would constitute recovery or bioequivalence. Thus $R_l = 0.85$. If, after restoration, we measured $R = 0.93$ we could perform an appropriate statistical test to determine whether this was significantly greater than $R_l$; if so, we would reject $H_0$: 'not bioequivalent' in favour of $H_A$: 'bioequivalent'.

Note that this example illustrates a one-sided test for the situation where large values of $R$ are desirable. The situation is reversed in the hypothetical pulp mill example of section 11.1. Here hepatic activity *increases* relative to controls with increasing concentrations of untreated mill effluent, and this increase is undesirable. The treatment of the effluent is designed to reduce hepatic activity to control levels, so we are now interested in setting an *upper* limit to $R$, $R_u$, below which we will have achieved bioequivalence. Thus $H_0: R \geq R_u$ and $H_A: R < R_u$.

Finally there may be circumstances where we require a two-sided test of bioequivalence, in which the values of $R$ should lie between desirable upper and lower limits. Thus $H_0: R \leq R_l$ or $R \geq R_u$ and $H_A: R_l < R < R_u$.

same environmental quality (i.e. good versus poor connotations) as Controls in a BACI design to detect impacts; otherwise they are conceptually like Controls in a BACI design.

There are some ethical issues with planning to restore only a subset of impacted locations but restoring everywhere at once would not leave any controls available. Pragmatically, we see that in most restorative efforts there will be a shortfall of the funds and other resources available compared with what is needed – this shortfall will always allow some control sites. Choosing these randomly from among the available possible locations will strengthen our inferential power and is distinctly preferable to only having Before data to use as the sole reference or target state (cf. a statistical problem due to non-independence of data, see chapter 5).

It is also a better design for the reasons discussed in chapters 7, 8 and 9 – if we have both controls and to-be-restored river locations before and after restoration then we have all the logical elements needed to detect change. It may be difficult to find such 'control' locations (e.g. no places may elude some large-scale restorations). If we have no 'controls', or even no monitoring before restoration, then we encounter inferential problems (section 9.3.5). For example, if we compare the supposedly restored rivers to their starting conditions then we have the difficulty that other events, merely coincident with restoration efforts, may have caused any changes we detect (see Fairweather 1993 for discussion). Such an outcome is impossible to discount entirely, even with a full MBACI design. We therefore have to fall back upon weaker arguments to make our case (discussed throughout chapter 9).

## 14.2.1 The real way that restoration differs

However, we still need to test whether any changes we detect are toward some reference or target state, and this leaves us with the other side of the problem with controls described above. A sailing analogy would be that we could now tell that our boat has left the shore but we are uncertain if it is heading in the right direction. Thus, degraded 'controls' that don't receive restoration efforts are insufficient, by themselves, to test the full restoration model. As convincingly argued by Chapman & Underwood (1997, 2000) and Grayson *et al.* (1999), we also need comparative data from a target that we are aiming for – some reference condition that our restored locations should head toward. Thus it is important to judge the direction of change as well as the degree of change *per se*. For that task, we need three distinct types of 'treatments': (a) those degraded locations to be restored (possibly called 'experimental'); (b) those starting out in a similarly degraded state but that will not be restored (called 'controls'); and (c) those representing the target state (called 'references'). Such 'reference' rivers typically will be relatively unimpacted or at least represent some particular desired state we are aiming for (e.g. not necessarily representative of what was before but instead what we want in the future). This may be a more specific use of the 'reference condition' approach applied in RIVPACS, AusRivAS or BEAST (see Reynoldson *et al.* 1997).

Thus, this need to make two simultaneous comparisons is more complicated than BACI-type impact designs. The unrestored controls are needed to know when you've left the previous degraded state ('the shore'; has anything changed?) and the reference targets are needed to know where you are headed (are we heading in the right direction?). Additionally, given that restoration ecology *per se* is in its infancy, the use of such a strongly rigorous and experimental approach must be highly laudable because comparing restored versus reference versus unrestored 'control' rivers will precisely demonstrate the efficacy (if any) of restoration.

Restoration goals should take the form of explicit targets – essentially minimum performance targets that are set prior to the study commencing. An example might be to return the species composition of benthic infauna to 90% of that found in a riffle before a development took place. Of course, such an explicit goal obviously requires an understanding of what benthos existed before development!

## 14.3 ANALYTICAL TECHNIQUES APPLICABLE TO RESTORATION MONITORING

A traditional approach to monitoring restoration may not be as useful as it might appear, even if the reference sites can be agreed upon (e.g. to give the standard for the 90% recovery in our example in the paragraph above). Tradition would suggest that we are then testing for no difference between the 90% figure and the test statistic (e.g. based on species richness estimates from samples in the restored sites). Unfortunately, poor sampling with limited replication would give us a non-significant result (Fairweather 1991a), almost regardless of the average values if the variances associated with them were very large. Potentially this might be good inference but would lead to a bad outcome in terms of the conclusions drawn (see McDonald & Erickson 1994 for discussion).

Similar problems are faced by medical researchers trying to establish whether alternative treatments are equally effective, and often the issue with testing a new drug is whether it is at least as efficacious as the presently used drug. The commonly applied solution to this problem is to use a test of 'bioequivalence' (e.g. Westlake 1988); hence, we are testing for bioequivalence of effects compared to some reference state (i.e. the current drug's performance). The bioequivalence approach was suggested by McDonald & Erickson (1994) for environmental monitoring (see Box 14.1). In such a scenario, precise sampling would be needed to demonstrate bioequivalence *per se* (see Fig. 14.1). As such, both the inference and the outcome concluded about restoration should be good within a bioequivalence-testing framework (see Box 14.1 and below). It has also been suggested that this would be a more precautionary way to approach assessments because the burden of proof is put upon those that benefit most from restoration succeeding.

The formal hypothesis in bioequivalence testing is expressed as a ratio of values of variables from the restored samples to the target values (Box 14.1, Fig. 14.1). Relatively simple comparisons can be done using confidence intervals (e.g. see McDonald & Erickson 1994) but reasonably sophisticated software (e.g. EquivTest, see http://www.statsolusa.com) also exists to automate the task. Such ratios could conceivably be used in BACI-type assessments although a caveat to this technique is that it is difficult to apply to complex statistical designs (Peterson 1993; USEPA 1989).

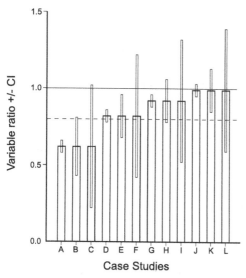

**Fig. 14.1** Assessing recovery using ratios between replicated data from the site undergoing recovery and target sites (i.e. recovery ratio = value at recovering site/target for any ecologically relevant variable). The height of each bar gives the mean value of the ratio for that case and the error bars give the confidence interval (CI). These hypothetical values have been chosen to point out that both central tendency and variation in our estimation of recovery need to be considered. Cases A–C have mean ratio = 0.62, D–F = 0.82, G–I = 0.92 and J–L = 0.99; each triplet varies in terms of the calculated confidence interval around that mean ratio with CI = 0.4, 0.14 and 0.40, respectively. The lower limit line (---) represents a recovery ratio of 0.8 (perhaps a minimum threshold for acceptable recovery) and the upper line ( — ) = 1.0 (i.e. complete recovery). Individual values may exceed 1.0 because it is conceivable that different ecological variables could exceed the value for the target (e.g. numbers of species may be maximal partway through a succession). Thus, in terms of these limits we interpret recovery using both the mean ratio and the uncertainty (confidence interval) around that estimate; we ask 'Does the error bar overlap our target criterion or threshold?' On this basis, then, all cases above, except A, would pass the lower threshold of partial but acceptable recovery, but only C, F and H–L would be considered to be recovered 'completely'. Bioequivalence procedures (see text) provide a way of testing whether such data conform to either of the limits. It remains a challenge for ecologists to suggest acceptable values for thresholds lower than a ratio of 1.0.

### 14.3.1 The logic of specifying an effect size for recovery

Trying to specify an effect size for an assessment program aimed at measuring recovery suffers from the problem first mentioned in section 11.1 and discussed in section 14.2: we are trying to 'prove' the null hypothesis that the mean value in impacted locations equals the mean value in control locations.

Consider this scenario from our hypothetical pulp mill example (discussed in section 11.1). Suppose the pulp mill was established many years ago before any strong environmental legislation. Recent public pressure has been exerted for the mill to treat its effluent so that the effects on fish are minimized. Hepatic activity has been chosen as a variable that best indicates the sub-lethal effects of the effluent on fish. The mill is about to install some very expensive treatment equipment and is required to demonstrate that hepatic activity in the fish will decrease to values similar to fish in an unaffected location on the river before the effluent treatment is deemed to be adequate (not an ideal design – see chapter 5 – but for simplicity let's assume that this was the best design possible under the circumstances). Using the conventional means of framing a null hypothesis, the mill could cheat by implementing a sampling program with low power, and so easily and cheaply demonstrate no statistical difference between control and impact locations. Attempts to get round this difficulty by stipulating large minimum sample sizes are fraught with difficulties (McDonald & Erickson 1994); in fact, insistence on very large sample sizes may detect biologically unimportant or even trivial differences in the hepatic activity.

So, the bioequivalence technique recasts the 'null' hypothesis in terms of the ratio of the value of the variable in the impacted location to that in the control location (see Box 14.1). The two locations are then judged to be bioequivalent if this ratio is outside of the bounds of values that have been agreed as representing recovery. Under this framework the 'null' hypothesis then becomes 'that the impacted location is not bioequivalent to the control location' which is tested against the alternative 'that the locations are bioequivalent'. Note that this technique still requires negotiation of an effect size. In the hypothetical example above, the mill, the community and the management agency would need to agree that hepatic activity in the impacted location needs to be, say, 110% or less of that in the control location before the hypothesis of bioequivalence is accepted. This procedure has been used in toxicology (Erickson & McDonald 1995) and has recently been applied to terrestrial environmental restoration in the USA (McDonald & Erickson 1994). The proponents of this technique claim that its use encourages users to use well replicated, high-power designs; in our hypothetical example it is in the mill's interest to collect sufficient data to reject the 'null' hypothesis of 'no bioequivalence'.

## 14.4 HOW LONG SHOULD WE MONITOR ATTEMPTS AT RESTORATION?

The chanciness of recruitment and recolonization may mean that considerable time elapses before restoration can be seen to be under way. Alternatively, there may be deficiencies in the ecosystem that are still preventing restoration efforts having a positive effect. We need to continue sampling long enough to distinguish these alternatives.

Many monitoring programs that continue into a putative recovery phase are persisted with for too short a time (e.g. Gore et al. 1990; Simenstad & Thom 1996). These programs tend to be stopped as soon as the study location looks right (e.g. the species list might be 'complete' but there is no concern for the age or size distributions within populations), and then recovery declared as officially reached. There is precious little concern then for the subsequent time course of recovery of important aspects of the ecosystem that were being monitored (Grayson et al. 1999). Many ecologists would argue that the age or size distribution has to reflect that of natural locations and, in particular, the populations need to be capable of self-replacement (e.g. through reproduction) before all restoration could be abandoned. In a related debate, Connell & Sousa (1983) suggested that a population had to turn over at least once before it could be termed 'stable'. We suggest that similarly pragmatic definitions of 'sufficient time' need to be argued, taking into account the basic biology of the organisms involved.

To return to the liming case study, we can see in Fig. 14.2 that during 1989 the acid-sensitive benthic invertebrates in the restored ('impact') streams increased in abundance compared with the control streams. But these indicator values had not reached, in any year, the target densities as seen in the reference streams. In terms of our earlier analogy, the boat has apparently left the harbour but has definitely not yet reached the target shore. Thus, we should continue monitoring the recovery that seemed to be under way (at least during 1989) but is nowhere near complete.

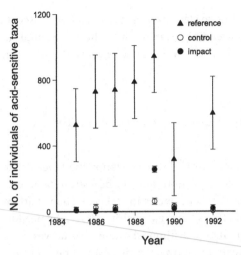

**Fig. 14.2** Data from a long-term field experiment in restoration (originally described in detail in section 8.4) that was designed to test whether liming acidified streams would produce an increase in numbers and diversities of acid-sensitive macroinvertebrate taxa. Plotted are the mean ( ± SE) number of individuals of all acid-sensitive, macroinvertebrate taxa against sampling year. Liming was done in late 1987 and early 1988. 'Impact' streams were limed while 'Control' streams were acidified streams that were not limed – these are the same data plotted in Fig. 8.8b. Reference streams were streams in the area that had natural buffering capacities, so that they had not become acidified. The data demonstrate that, although liming produced a small increase in numbers of acid-sensitive macroinvertebrates in 1989, their numbers did not reach typical background densities, as seen in the reference streams.

## 14.5 THE NEED FOR CLARITY IN DECLARING THE GOALS OF RESTORATION

All programs need to articulate the reference state in quantitative terms to be explicit about the goals. If we do not have the luxury of 'reference' rivers (e.g. there are no unaffected rivers that we can use as targets), then how do we decide what should be the reference conditions to which an impacted river should be returned? As indicated in chapter 3 (see Box 3.1), often we do not have a good enough understanding of the behaviour of natural systems to allow 'absolute' predictions about their states. Commonly, lay people believe that 'naturalness' or what they perceive as the norm equates with constancy, balance and equilibrium, even though ecological science has definitively disproved this idea (Fairweather 1993).

Most ecosystems are naturally very variable in space and time (see chapter 2) but, because we are yet to understand the causes and consequences of this variability, we cannot generally use fluctuations from some norm as a yardstick of 'ill health' or 'impactedness'. Thus, in the situation where a degraded river has no reference rivers to which it can be compared, the public should play an important role in deciding the condition to which it should be returned, for the same reasons that they help decide what changes should be considered 'important' to detect (section 11.3).

Getting the physical appearance right of a restored riverine location is a first and necessary step (Harper *et al.* 1999) but in itself will not represent ecological restoration. The response of different components of the biota in terms of their abundance and activity, and the functional expression of that activity as ecological processes (Fairweather 1999a), need to be assessed to see whether restoration has been effective in an ecological sense.

Habitat creation and repair (*sensu* Gilbert & Anderson 1998) may be an alternative to restoration as we have defined it above. This, mainly European, concept holds that most rivers or streams exist within cultural landscapes that are essentially anthropogenic. As such we should manage disturbed sites so that vegetation flourishes in a semi-natural state (Gilbert & Anderson 1998), and therefore their associated communities could take various forms and functions. People play a part in designing and nurturing the resultant ecosystem in a manner more akin to present agriculture than nature conservation. In such a scenario, we see that the targeted reference state (see above) is simply a conscious choice from many possibilities, all of which are artificial to some degree for any waterway. Such intentional and deliberate management may well become the norm during the twenty-first century.

Finally, we should consider the possibility that ecology may never have the capacity to make exact-state predictions in any case. The makeup of any ecosystem often has a strong element of chance, and historical factors may play a contingent part in determining the exact species composition (e.g. Samuels & Drake 1997). If so, we may need to relinquish any goals of making exact predictions about the species-specific identities of ecosystems. We may be able to predict the existence and importance of functional groups within ecosystems but without knowing the exact mix of species that will carry out these functions. Such considerations will obviously affect the sorts of goals we can realistically set for restoration programs.

Therefore, we argue that assessing restoration must begin with defining its goals and determining what sites are available to use so that we know when we've attained those goals. The specific goals of a program assessing restoration need to be made explicit during the planning phase because they are rarely agreed upon in any specific circumstance (e.g. see Dobson & Cariss 1999 for discussion). Any program that aims to be clear to scientists, managers and the wider public must define its goals and endpoints for the assessment (either in terms of the time it might take or the targets to be reached). In scientific terms, this includes being able to specify the criteria for knowing when the endpoint has been attained. We argue that it is useful to specify the time frame for assessments in terms of the rates of natural change and variability (i.e. some background knowledge is needed). Then, we can use the design principles outlined in this book to detect precisely when those defined criteria are reached.

### 14.6 IMPORTANT ISSUES

- The emerging discipline of ecological restoration requires effective monitoring to ensure that its goals are being reached.
- As well as general design considerations promoted throughout this book (especially the ability to determine if any change has occurred by comparing with unrestored controls), monitoring for restoration requires an additional treatment set. In this we compare putatively restored areas to the ideal that we wish the ecosystem to be restored to, what we refer to as a 'reference' condition. Only this more elaborate monitoring with three sorts of states (restored, control and reference) can allow us to know when our restorative goal is attained.

# 15

## What's next?

Following the arguments and considering the issues presented in this book will allow us to design an effective and flexible monitoring program to detect and evaluate impacts in flowing waters. This includes the negotiations of what effect size is important to detect and what elements can be traded off or even sacrificed (as discussed in chapters 12 and 13). So now we've implemented the monitoring design and presumably detected (or not) some impact with known confidence – but the job is not yet finished. Further negotiations are in order to continue or refine assessment. Truly effective management of impacts requires that some action follows the well-designed studies we have so far advocated in this book. This chapter discusses issues that are central to what needs to be done after the main monitoring task has been completed.

### 15.1 LINKS WITH MANAGEMENT DECISIONS AS POINTS OF NEGOTIATION

We have emphasized the role that input from monitoring data (in terms of results and their interpretation) should have in management decision-making if we are to be engaged in a task that makes any difference. There is an imperative for environmental assessments to become more sophisticated and responsive to societal needs. Monitoring can be reactive (used only once an impact is clearly observed), proactive (seeking to assess impacts before they manifest themselves; see Fairweather 1993; Fairweather & Lincoln Smith 1993) or progress through adaptive learning. The latter means not just trying to benefit from mistakes but also combining elements of learning from both the scientific and management sides. For example, as practitioners of environmental science, we should seek to improve monitoring in terms of what it provides for management purposes. So, what are the minimum steps required to achieve an adaptive monitoring scheme?

In part, the role played by monitoring depends on how early in the assessment process the various likely outcomes from monitoring are specified. Monitoring can be most effective where there are clear feedback loops to managerial activity ('feedback monitoring', *sensu* Gray & Jensen 1993). These need to be agreed upon beforehand so that all parties know what actions may flow from different outcomes of the monitoring and what the specific trigger points are for more intensive studies. We believe that, by explicitly considering the power to detect impacts of given magnitudes against known background variability, we are well placed to designate desirable feedback loops. Fewer 'surprises' should result from such a monitoring regime. A major step forward is to document beforehand what actions will follow which outcomes; that is rarely done now in our experience.

There is an important public dimension to such negotiations. Because the lay public are important stakeholders in the environment generally, and certainly users of most resources stemming from it, negotiated agreements concerning the management of any impacts should have some wider public endorsement. This may include the formal consideration of adaptive environmental assessment and management (AEAM) models, which specifically include input from members of the general public as stakeholders (e.g. Grayson *et al.* 1994).

### 15.1.1 Dangers from losing institutional memory

Comprehensive and regular review of findings is needed to adaptively fine-tune a monitoring program so that it stays cost-effective. Otherwise, we risk suffering by not learning from our mistakes (Fairweather 1989) and we may miss the opportunity to incorporate the latest understanding or techniques. Loss of 'institutional memory' in an organization undertaking environmental assessments occurs where there is continual restructuring of management arrangements, high staff turnover, and scant regard for work that was done before the current regime or fad. In such cases, we routinely see each proposed developmental activity being treated as novel (over and above any peculiarities of its location), and reference being made only to the so-called grey literature (if any scientific references are given at all). Where data from previous assessments of the type under consideration are not made available, we have no opportunity to apply the novel approaches outlined in chapter 9, and our assessments are so made even more tentative.

How do we ensure enough 'institutional memory' is available into the future to understand the past context of present impacts (e.g. present generations may not know what natural features they are missing; see Safina 1998)? This is especially worrying when the curation of paper records is being discarded in favour of purely electronic storage of current data, but few funds are available for updating older forms of storage. Future institutional records are likely to 'begin' at the change-over point with the erosion of experience from before electronic storage as staff turnover.

## 15.2 CHANGING MONITORING OBJECTIVES WITH PROGRESS IN UNDERSTANDING

How long do we continue to monitor once an impact has been established? Certainly examples exist where monitoring has continued too long. By this, we mean cases where action has not been triggered even though previous results of monitoring indicated large or widespread impacts (i.e. the situation was put into the too-hard basket). This undesirable outcome results from having no agreed decision-making process and no triggers set. Time lags do need to be considered, however, because the impacts that are detected early in a BACI-type design may not stay the same during the operation of, say, an industrial outfall. This is because thresholds may be reached only after considerable time, so that alternative ecosystem states are reached once sufficient time has passed. Likewise, sampling has to be regular enough so as to not miss any transitory effects of genuine interest. In contrast, it is also possible that many programs monitoring recovery do not continue for long enough (see chapter 14).

In essence, how long to continue monitoring relates back to the original aim of the monitoring and any a priori decisions (feedback monitoring, *sensu* Gray & Jensen 1993) that were agreed upon, depending on the realized outcome(s). Eventually a shift may be needed to focus upon any intended repair of 'damage' to the environment caused by the human activity, as done through restoration works (see chapter 14).

## 15.3 ROLE OF EXPERIMENTS IN VERIFYING MECHANISTIC UNDERSTANDING OF AN IMPACT

In many instances where impacts are detected by our monitoring, we still do not have a thorough understanding of how they have come about. What stressor has caused the change? What is the specific causal pathway? Are interactions involved with secondary causal factors? Such aspects of causal mechanisms are still not known even for most well-documented impacts – thus, experiments may be needed to demonstrate causality *per se*. In essence, this book has been about shoring up our correlational investigations of putative impacts as much as possible; but the bottom line is that these inferences are still based on mere correlations. In completely planned experiments, we often apply the various treatments in a random fashion to prevent systematic but obscure bias in the estimation of experimental effects. In almost no case of impact monitoring is a stressor from human activities randomly 'applied' and so we cannot be sure that there isn't some other explanation of trends we see and ascribe to the stressor (i.e. as an 'impact'; see section 9.1 and Table 9.1). This is due to the increased risk of confounding our explanation based on the intended experimental design with any other, but unknown, extraneous factors. It may also be true in randomized experiments but their likelihood of such confounding is much reduced because of the experimental control available.

The issue of how we conduct such experiments will need careful and distinct consideration for the type of stress/impact and variables (e.g. indicator organisms) under deliberation. There are some aspects over and above such case-specific needs that are pertinent and useful to mention here. We advocate rigorous experimental designs for testing how impacts affect the target biota (i.e. testing specific hypotheses about the mechanism of the putative impact). The prime ways of producing better mechanistic understanding of impacts are the very stuff of experimental science. The design principles discussed earlier (chapter 5) all apply, plus we would seek to randomize the application of different treatments across experimental units.

Should the experiments be done in the laboratory or field setting? This probably should be determined more by the size of the organisms involved and the time course of the putative impact than the convenience of doing so in either arena. Our preference is for field experiments to maximize the realism of the experimental test (Cooper & Barmuta 1993), but there are also acceptable compromises in the form of mesocosms or artificial streams (e.g. see Lamberti & Steinman 1993, but see Carpenter 1996 for a dissenting view about small-scale tests).

Incorporating more sites or regions into experiments for explicit test should also serve to increase the generality of the conceptual models of impact that we eventually build. These more general models could provide savings by being exportable to larger portions of a country or continent. The issue of whether it is valid to extrapolate at all to other places is a pertinent one for many modern monitoring schemes because of the implicit promise of addressing many problems by funding just a few research programs. Without explicit comparisons across sites, we will never know whether we have achieved that generality. We also see large-scale, experimental tests (often called 'whole-ecosystem experiments') as the key way to trial management actions designed to ameliorate impacts (see chapter 14).

## 15.4 HOW DO WE EVALUATE THE EFFECTIVENESS OF A MONITORING PROGRAM?

Often there is no money allocated for follow-up monitoring (i.e. the 'A' in BACI) to assess how good were our predictions of impact from assessments made before the development decision. Likewise, active restoration (e.g. mine site rehabilitation) often has a very limited time frame so that companies may be able to walk away after only a portion of the projected succession has occurred. Many ecologists suspect that monitoring programs are typically too short to show the true dynamics of any successional change. In such cases we do not really know when restoration has worked.

How do we compare alternative designs or competing impact detection technologies? How can such comparisons be incorporated into pilot programs? For example, where development funds are largely spent on one approach to monitoring (such as RIVPACS, IBI or AusRivAS in Australian riverine monitoring; see section 3.2.1), how can others compete? How do we organize a fair comparison trial (Fairweather 1999b) to see whether the favoured approach is in fact the 'best' way of proceeding? These questions need a lot of attention. For example, we should use prior experience of similar projects for the reasons given in chapters 9 and 11.

We may require different detection schemes to distinguish emerging impacts from longer-standing environmental issues. For example, Fairweather (1999b) extended the ecosystem health analogy to advocate different environmental indicators as done in medical testing, with at least three tiers of different tests used in general practice, follow-up tests and specialist diagnosis. These tests increase in costs and specificity but follow on from one to the other in an attempt to pinpoint what is happening with the patient.

## 15.5 WHAT RESEARCH COULD COMPLEMENT MONITORING PROGRAMS?

As this book makes abundantly clear, the science of monitoring impacts has evolved rapidly in recent years and our search for excellence in design and analysis continues. We are sure that future research will sustain this trend and that managers interested in value for money will always encourage this search. Therefore, we expect that new innovations will be devised to attack the problems discussed herein. Some areas need particular attention.

Both scientists and managers could benefit from more easily applied methods of meta-analysis to demonstrate collectively (i.e. synthesized from several to many studies) any subtle, slow-emerging or diffuse impacts. This could perhaps be applied also to evolving scenarios of wider impacts where the effect sizes (as described in chapter 11) are likely to be, at least initially, quite small.

How can we best utilize what we already know collectively from past monitoring? In this book, we have advocated good use of prior knowledge in effect sizes. Research that does synthesize past results, because it does not look at each impact in isolation, can provide a basis for quantifying informed prior probabilities in a Bayesian approach.

The next wave of monitoring may involve further application of Bayesian approaches, which are not well developed as yet but are worthy of more investigation. We need to know how to best incorporate prior probabilities and to determine how this approach will change the method of inference in complex models of impacts.

How do we develop novel indicators for specific or unforeseen impacts from changing human activities in the future? The choice of variables (see chapter 10) must always be able to evolve to keep pace with our advancing scientific understanding and shifts in societal concern.

## 15.6 REITERATING THE PRINCIPLES OF THIS BOOK

Applying the general principles outlined here (see Box 15.1) is an ever-present challenge to ensure that the consequences of our actions are understood and to be a basis for action in improving our rivers and streams. We suggest that the flow chart in Figure 1.1 shows the minimum required approach. As we bring this book to a close, we wish to emphasize that good monitoring design is evolving all the time and we look forward to the exciting developments ahead of us in the twenty-first century.

**Box 15.1** Twenty important issues in good monitoring design

Effective monitoring programs comprise multiple elements that collectively provide sound use of logic and result in rational decision-making without a profligate use of resources.

| Issue | Primary chapter(s) |
|---|---|
| Current levels of human use and abuse of water resources means we need to implement good monitoring design as an essential – not a luxury – requirement for their further use and management. | 1 |
| Good monitoring design requires us to understand how ecosystems work. In flowing-water ecosystems, structure and function are strongly dependent on the operation of longitudinal and predominantly unidirectional linkages (upstream–downstream), and on lateral linkages (channel–floodplain); these linkages affect how we can apply monitoring designs. | 2 and 8 |
| *Perturbation* of a system consists of two sequential events: the *disturbance* to the system and the *response* of the system to that disturbance. Effective monitoring requires understanding the nature, and temporal and spatial scales, of both the disturbance and the response. | 3 |
| Monitoring may be done for different purposes, and these serve different management needs as well as posing different questions to be answered. | 3 |
| The logical principles of designing a monitoring program to detect the effects of human activities apply irrespective of whether a frequentist or a Bayesian approach to statistics is adopted. Currently, hypothesis testing via frequentist statistics offers the best-developed and most widely used tools for making decisions about impacts but we expect more development of Bayesian approaches. | 4 |

| Issue | Primary chapter(s) |
|---|---|
| The key strategy for inference of impacts is to find some evidence for impact that cannot easily be explained away by various other processes, such as natural variation in the system. Support for an impact hypothesis is only found if the probability of that outcome is small, under normal circumstances, in the *absence* of impact. This pursuit of improbability provides the rationale for specific aspects of monitoring design. | 4 |
| These design aspects include sampling control and impact locations, both before and after putative impact (so-called Before–After–Control–Impact (BACI) designs) together with proper replication of each of these four elements, where possible. Replicated BACI-type designs allow us to separate, with relatively high confidence, human-caused effects from natural processes. | 5 |
| It is important to recognize, from the outset, that deficient monitoring designs usually cannot be rescued regardless of the quantity and sophistication of statistical analysis applied to the data. | 5 |
| We can illustrate the importance for good monitoring design by looking at how river biologists have addressed questions about human impacts in the past. Historically, there has been a number of issues that have prevented river biologists from implementing designs with the strongest possible inferential base. Some of these issues have been within the control of biologists, and some are external constraints imposed either by the geographical peculiarities of the river under study, or by socioeconomic factors. | 6 |
| There are different types of BACI designs, which result in distinctly different analytical models that address different questions. These conceptual | 7 |

| Issue | Primary chapter(s) |
|---|---|
| models are all justifiable approaches to the detection of impacts under particular circumstances, but it is *essential* that anyone implementing one of these designs be aware of the differences between them, and of the important characteristics of each of them. | |
| Different BACI designs lie along a gradient of inferential certainty from relatively strong to relatively weak, rather than providing either perfect or zero inference about human impacts. | 7 and 9 |
| Applying BACI requires tailoring designs to the specifics of the system (its size and uniqueness) and impact to hand (point or non-point impact), and includes developing criteria that help ensure the comparability of control locations to each other and to the impact location(s). There is a dilemma in that, the more narrowly we define characteristics of control locations, the more similar they are likely to be to each other but the fewer will be the number of places likely to meet our criteria. | 8 |
| We can implement a structured 'levels-of-evidence' approach to improve the inferential strength of monitoring designs. This approach uses causal criteria (effectively, a set of circumstantial arguments), which have been developed particularly well in the field of epidemiology. Making a levels-of-evidence case is especially important when elements of BACI-designs are missing. | 9 |
| Variables chosen for monitoring should be efficacious: relevant to the questions asked; strongly associated with the putative impact; ecologically and/or socially significant; efficient to measure. | 10 |
| The magnitude and form of unacceptable environmental changes ('effect sizes') should be negotiated and defined ahead of beginning a monitoring program; it is impossible to prescribe universal effect sizes for biological variables. | 11 |

| Issue | Primary chapter(s) |
|-------|--------------------|
| Negotiations should involve all stakeholders, and defining important changes should include societal wishes as well as scientific input regarding the implications and risks associated with those changes. | 11 |
| Decision-making should consider the risks of making two sorts of errors: detection of an environmental change that is not actually real (statistically speaking, a Type I error); and failure to detect an important change (statistically speaking, a Type II error). The risks of making either sort of decision error should be balanced in the decision-making process in inverse relation to the respective costs of committing that error. | 12 |
| Effective monitoring programs should be optimized, in which the number of samples required are compared to the resources available, and trade-offs made that reduce monetary costs without compromising the inferential strength of the program. | 13 |
| The emerging discipline of ecological restoration requires effective monitoring to ensure that its goals are being reached. As well as general design considerations promoted throughout this book (especially the ability to determine if any change has occurred by comparing with unrestored controls), monitoring for restoration requires an additional treatment set. In this we compare putatively restored areas to the ideal that we wish the ecosystem to be restored to, what we refer to as a 'reference' condition. Only this more elaborate monitoring with three sorts of states (restored, control and reference) can allow us to know when our restorative goal is attained. | 14 |
| Monitoring programs must be linked to management decision-making, such that particular | 15 |

| Issue | Primary chapter(s) |
|---|---|
| triggers (e.g. an effect being detected) will result in some action being taken. No one source of information, including this book, has all the answers, and a degree of flexibility of approach is required as the best way of designing monitoring programs; such best practice will continue to evolve and progress. | |

# References

392   Abel, P. D. (1989) *Water Pollution Biology*. Chichester, UK: Ellis Horwood.

Achieng, A. P. (1990) The impact of the introduction of Nile Perch, *Lates niloticus* (L.) on the fisheries of Lake Victoria. *Journal of Fisheries Biology* **37**, 17–23.

Aguiar, F., Moreira, I. & Ferreira, T. (1996) Perception of aquatic weed problems by managers of water resources. *Revista de Ciencias Agrarias* **19**, 35–56 (in Portugese).

Alderman, D. J. (1993) Crayfish plague in Britain, the first twelve years. *Freshwater Crayfish* **9**, 266–72.

Allan, J. D. (1984) Hypothesis testing in ecological studies of aquatic insects. In *The Ecology of Aquatic Insects*, eds. V. H. Resh & D. M. Rosenberg, pp. 484–507. New York: Praeger Scientific.

Allan, J. D. (1995) *Stream Ecology: Structure and Function of Running Waters*. London, UK: Chapman & Hall.

Allen, K. R. (1959) The distribution of stream bottom fauna. *Proceedings of the New Zealand Ecological Society* **6**, 5–8.

Aloi, J. E. (1990) A critical review of recent freshwater periphyton field methods. *Canadian Journal of Fisheries and Aquatic Sciences* **47**, 656–70.

Anderson, M. J. (2001) A new method for non-parametric multivariate analysis of variance. *Austral Ecology* **26**, 32–46.

Anderson, M. J. & Gribble, N. A. (1998) Partitioning the variation among spatial, temporal and environmental components in a multivariate data set. *Australian Journal of Ecology* **23**, 158–67.

Anderson-Carnahan, L., Foster, S., Thomas, M., Korth, W. & Bowmer, K. H. (1995) Selection of a suitable cladoceran species for toxicity testing in turbid waters. *Australian Journal of Ecology* **20**, 28–33.

Andrew, N. L. & Mapstone, B. D. (1987) Sampling and the description of spatial pattern in marine ecology. *Oceanography and Marine Biology, Annual Review* **25**, 39–90.

ANZECC (Australian and New Zealand Environment and Conservation Council) & ARMCANZ (Agriculture and Resource Management Council of Australia and New Zealand) (2001) *Australian and New Zealand Guidelines for Fresh and Marine Waters*. National Water Quality Management Strategy Paper no. 4. Canberra, ACT: ANZECC & ARMCANZ.

Arnell, N., Bates, B., Lang, H., Magnuson, J. J. & Mulholland, P. (1996) Hydrology and freshwater ecology. In *Impacts, Adaptations, and Mitigation of Climate Change: Scientific–Technical Analyses*, eds. R. T. Watson, M. C. Zinyowera, R. H. Moss & D. J. Dokken, pp. 325–64. Cambridge, UK: Cambridge University Press.

Arner, D. H., Robinette, J. E., Frasier, J. E. & Gray, M. H. (1976) Effects of channelization of the Luxapalila River on fish, aquatic invertebrates, water quality and furbearers. Fish and Wildlife Services, USDI, Washington, DC. FWS/OBS-76/08.

Arthington, A. H., Conrick, D. L., Connell, D. W. & Outridge, P. M. (1982) *The Ecology of a Polluted Urban Creek*. Australian Water Resources Council Technical Paper no. 68. Canberra, ACT: Australian Water Resources Council.

Austen, D. J., Bayley, P. B. & Menzel, B. W. (1994) Importance of the guild concept to fisheries research and management. *Fisheries* **19**, 12–20.

Austin, M. P. (1980) Searching for a model for use in vegetation analysis. *Vegetatio* **42**, 11–21.

Austin, M. P. (1987) Models for the analysis of species' response to environmental gradients. *Vegetatio* **69**, 35–45.

Australian Department of Industry, Science and Tourism (1996) *Managing Australia's Inland Waters: Roles for Science and Technology*. Canberra, ACT: Commonwealth of Australia.

Ayres, M. P. & Thomas, D. L. (1990) Alternative formulations of the mixed-model ANOVA applied to quantitative genetics. *Evolution* **44**, 221–6.

Bain, M. B., Hughes, T. C. & Arend, K. K. (1999) Trends in methods for assessing freshwater habitats. *Fisheries* **24**, 16–21.

Barbour, M. T. & Gerritsen, J. (1996) Subsampling of benthic samples: a defense of the fixed-count method. *Journal of the North American Benthological Society* **15**, 386–91.

Barbour, M. T., Plafkin, J. L., Bradley, B. P., Graves, C. G. & Wisseman, R. W. (1992) Evaluation of EPA's rapid bioassessment benthic metrics: metric redundancy and variability among reference stream sites. *Environmental Toxicology and Chemistry* **11**, 437–49.

Barbour, M. T., Stirling, J. B. & Karr, J. R. (1995) Multimetric approach for establishing biocriteria and measuring biological condition. In *Biological Assessment and Criteria: Tools for Water Resource Planning and Decision Making*, eds. W. S. Davis & T. P. Simon, pp. 63–77. Boca Raton, FL: Lewis Publishers.

Bargagli, R. (1998) *Trace Elements in Terrestrial Plants: An Ecophysiological Approach to Biomonitoring and Biorecovery*. Berlin: Springer-Verlag.

Barmuta, L. A., Cooper, S. D., Hamilton, S. K., Kratz, K. W. & Melack, J. M. (1990) Responses of zooplankton and zoobenthos to experimental acidification in a high-elevation lake (Sierra Nevada, California, USA). *Freshwater Biology* **23**, 571–86.

Bartell, S. M., Gardner, R. H. & O'Neill, R. V. (1992) *Ecological Risk Estimation*. Boca Raton, FL: Lewis Publishers.

Barton, D. R. & Farmer, M. E. D. (1997) The effects of conservation tillage practices on benthic invertebrate communities in headwater streams in southwestern Ontario, Canada. *Environmental Pollution* **96**, 207–15.

Bartsch, A. F. (1948) Biological aspects of stream pollution. *Sewage Works Journal* **20**, 292–302.

Bass, J. A. B., Pinder, L. C. V. & Leach, D. V. (1997) Temporal and spatial variation in zooplankton populations in the River Great Ouse: an ephemeral food resource for larval and juvenile fish. *Regulated Rivers: Research and Management* **13**, 245–58.

Basu, B. K. & Pick, F. R. (1996) Factors regulating phytoplankton and zooplankton biomass in temperate rivers. *Limnology and Oceanography* **41**, 1572–7.

Baxter, G. S. (1994) The location and status of egret colonies in coastal New South Wales. *Emu* **94**, 255–62.

Bender, E. A., Case, T. J. & Gilpin, M. E. (1984) Perturbation experiments in community ecology: theory and practice. *Ecology* **65**, 1–13.

Benke, A. C. (1990) A perspective on America's vanishing streams. *Journal of the North American Benthological Society* **9**, 77–88.

Bennison, G. L., Hillman, T. J. & Suter, P. J. (1989) *Macroinvertebrates of the River Murray (Survey and Monitoring: 1980–1985).* Water Quality Report no. 3. Melbourne, Vic.: Murray–Darling Basin Commission.

Berger, J. O. (1985) *Statistical Decision Theory and Bayesian Analysis,* 2nd edn. New York: Springer-Verlag.

Bernstein, B. B. & Zalinski, J. (1983) An optimum sampling design and power tests for environmental biologists. *Journal of Environmental Management* **16**, 35–43.

Berry, D. A. & Stangl, D. K. (1996) Bayesian methods in health-related research. In *Bayesian Biostatistics,* eds. D. A. Berry & D. K. Stangl, pp. 3–66. New York: Marcel Dekker.

Bervoets, L., Blust, R., de Wit, M. & Verheyen, R. (1997) Relationships between river sediment characteristics and trace metal concentrations in tubificid worms and chironomid larvae. *Environmental Pollution* **95**, 345–56.

Beyers, D. W. (1998) Causal inference in environmental impact studies. *Journal of the North American Benthological Society* **17**, 367–73.

Biggs, B. J. F. (1988) A periphyton sampler for shallow, swift rivers. *New Zealand Journal of Marine and Freshwater Research* **22**, 189–99.

Biggs, B. J. F. (1995) The contribution of flood disturbance, catchment geology and land use to the habitat template of periphyton in stream ecosystems. *Freshwater Biology* **33**, 419–38.

Bishop, K. A., Pidgeon, W. J. & Walden, D. J. (1995) Studies on fish movement dynamics in a tropical floodplain river: prerequisites for a procedure to monitor the impacts of mining. *Australian Journal of Ecology* **20**, 81–107.

Blyth, J. D., Doeg, T. J. & St Clair, R. M. (1984) Response of the macroinvertebrate fauna of the Mitta Mitta River, Victoria, to the construction and operation of Dartmouth Dam. 1. Construction and initial filling period. *Occasional Papers from the Museum of Victoria* **1**, 83–100.

Bodar, C. W. M., Van Leeuwen, C. J., Voogt, P. A. & Zandee, D. I. (1988) Effect of cadmium on the reproduction strategy of *Daphnia magna. Aquatic Toxicology* **12**, 301–10.

Bonecker, C. C., Bonecker, S. L. C., Bozelli, R. L., Lansac-Toha, F. A. & Velho, L. F. M. (1996) Zooplankton composition under the influence of liquid wastes from a pulp mill in middle Doce River (Belo Oriente, Minas Gerais, Brazil). *Arquivos de Biologia e Tecnologia (Curitiba)* **39**, 893–901.

Boon, P. I., Bunn, S. E., Green, J. D. & Shiel, R. J. (1994) Consumption of cyanobacteria by freshwater zooplankton: implications for the success of 'top–down' control of cyanobacterial blooms in Australia. *Australian Journal of Marine and Freshwater Research* **45**, 875–87.

Borcard, D., Legendre, P. & Drapeau, P. (1992) Partialling out the spatial component of ecological variation. *Ecology* **73**, 1045–55.

Boulton, A. J. (1999) An overview of river health: philosophies, practice, problems and prognosis. *Freshwater Biology* **41**, 469–79.

Boulton, A. J. & Brock, M. A. (1999) *Australian Freshwater Ecology: Processes and Management.* Glen Osmond, SA: Gleneagles Publishing.

Boulton, A. J. & Lake, P. S. (1992a) The ecology of two intermittent streams in Victoria, Australia. 3. Temporal changes in faunal composition. *Freshwater Biology* **27**, 123–38.

Boulton, A. J. & Lake, P. S. (1992b) The ecology of two intermittent streams in Victoria, Australia. 2. Comparisons of faunal composition between habitats, rivers and years. *Freshwater Biology* **27**, 99–121.

Boulton, A. J., Peterson, C. G., Grimm, N. B. & Fisher, S. G. (1992) Stability of an aquatic macroinvertebrate community in a multiyear hydrologic disturbance regime. *Ecology* **73**, 2192–207.

Boulton, A. J., Stibbe, S. E., Grimm, N. B. & Fisher, S. G. (1991) Invertebrate recolonization of small patches of defaunated hyporheic sediments in a Sonoran Desert stream. *Freshwater Biology* **26**, 267–77.

Box, G. E. P. & Tiao, G. C. (1972) *Bayesian Inference in Statistical Analysis*. Reading, UK: Addison-Wesley.

Box, G. E. P. & Tiao, G. C. (1975) Intervention analysis with applications to economic and environmental parameters. *Journal of the American Statistical Association* **70**, 70–9.

Bren, L. J. (1992) Tree invasion of an intermittent wetland in relation to changes in the flooding frequency of the River Murray, Australia. *Australian Journal of Ecology* **17**, 395–408.

Brewin, P. A., Reynolds, B., Stevens, P. A., Gee, A. S. & Ormerod, S. J. (1996) The effect of sampling frequency on chemical parameters in acid sensitive streams. *Environmental Pollution* **93**, 147–57.

British Ecological Society (1990) *Ecological Issues No. 1: River Water Quality*. Montford Bridge, UK: Field Studies Council.

Brooks, R. R. (ed.) (1998) *Plants that Hyperaccumulate Heavy Metals: The Role of Phytoremediation, Microbiology, Archaeology, Mineral Exploration and Phytomining*. New York: CAB International.

Buikema, A. L. Jr & Voshell, J. R. Jr (1993) Toxicity studies using freshwater benthic macroinvertebrates. In *Freshwater Biomonitoring and Benthic Macroinvertebrates*, eds. D. M. Rosenberg & V. H. Resh, pp. 344–98. New York: Chapman & Hall.

Bunn, S. E. & Hughes, J. M. (1997) Dispersal and recruitment in streams: evidence from genetic studies. *Journal of the North American Benthological Society* **16**, 338–46.

Cairns, J. Jr & Pratt, J. R. (1993) A history of biological monitoring using benthic macroinvertebrates. In *Freshwater Biomonitoring and Benthic Macroinvertebrates*, eds. D. M. Rosenberg & V. H. Resh, pp. 10–27. New York: Chapman & Hall.

Cairns, J. Jr & Smith, E. P. (1994) The statistical validity of biomonitoring data. In *Biological Monitoring of Aquatic Systems*, eds. S. L. Loeb & A. Spacie, pp. 49–68. Boca Raton, FL: Lewis Publishers.

Calabrese, E. J. & Baldwin, L. A. (1997) A quantitatively-based methodology for the evaluation of chemical hormesis. *Human and Ecological Risk Assessment* **3**, 545–54.

Calabrese, E. J. & Baldwin, L. A. (1999) Reevaluation of the fundamental dose-response relationship. *BioScience* **49**, 725–32.

Calow, P. (1992) Can ecosystems be healthy? Critical consideration of concepts. *Journal of Aquatic Ecosystem Health* **1**, 1–16.

Campbell, I. C. (1978) A biological investigation of an organically polluted urban stream in Victoria. *Australian Journal of Marine and Freshwater Research* **29**, 275–91.

Cao, Y., Bark, A. W. & Williams, W. P. (1996) Measuring the responses of macroinvertebrate communities to water pollution: a comparison of multivariate approaches, biotic and diversity indices. *Hydrobiologia* **341**, 1–19.

Carbiener, R., Tremolieres, M. & Muller, S. (1995) Vegetation of running waters and water quality: thesis, debates and prospects. *Acta Botanica Gallica* **142**, 489–531.

Carpenter, S. R. (1996) Microcosms have limited relevance for community and ecosystem ecology. *Ecology* **77**, 677–80.

Carpenter, S. R. (1998) The need for large-scale experiments to assess and predict the response of ecosystems to perturbation. In *Successes, Limitations, and Frontiers in Ecosystem Science*, eds. M. L. Pace & P. M. Groffman, pp. 287–312. New York: Springer-Verlag.

Carpenter, S. R., Bolgrien, D., Lathrop, R. C., Stow, C. A., Reed, T. & Wilson, M. A. (1998a) Ecological and economic analysis of lake eutrophication by nonpoint pollution. *Australian Journal of Ecology* **23**, 68–79.

Carpenter, S. R., Caraco, N. F., Correll, D. L., Howarth, R. W., Sharpley, A. N. & Smith, V. H. (1998b) Nonpoint pollution of surface waters with phosphorus and nitrogen. *Ecological Applications* **8**, 559–68.

Carpenter, S. R., Frost, T. M., Heisey, D. & Kratz, T. (1989) Randomized intervention analysis and the interpretation of whole ecosystem experiments. *Ecology* **70**, 1142–52.

Carr, G. M., Duthie, H. C. & Taylor, W. D. (1997) Models of aquatic plant productivity: a review of the factors that influence growth. *Aquatic Botany* **59**, 195–215.

Cassie, R. M. (1971) Sampling and statistics. In *A Manual on Methods for the Assessment of Secondary Productivity in Fresh Waters*, IBP Handbook no. 17, eds. W. T. Edmondson & G. G. Winberg, pp. 174–209. Oxford, UK: Blackwell Scientific Publications.

Cattaneo, A. & Amireault, M. C. (1992) How artificial are artificial substrata for periphyton? *Journal of the North American Benthological Society* **11**, 244–56.

Cerenius, L., Soderhall, K., Persson, M. & Ajaxon, R. (1988) The crayfish plague fungus *Aphanomyces astaci*: diagnosis, isolation and pathobiology. *Freshwater Crayfish* **7**, 131–44.

Chadwick, J. W., Canton, S. P. & Dent, R. L. (1986) Recovery of benthic invertebrate communities in Silver Bow Creek, Montana, following improved metal mine wastewater treatment. *Water, Air and Soil Pollution* **28**, 427–38.

Chapman, D. (ed.) (1996) *Water Quality Assessments*. London, UK: E. & F. N. Spon.

Chapman, M. G. & Underwood, A. J. (1997) Concepts and issues in restoration of mangrove forests in urban environments. In *Frontiers in Ecology: Building the Links*, eds. N. Klomp & I. Lunt, pp. 103–14. Oxford, UK: Elsevier Science.

Chapman, M. G. & Underwood, A. J. (2000) The need for a practical scientific protocol to measure successful restoration. *Wetlands (Australia)* **19**, 28–49.

Chapman, M. G., Underwood, A. J. & Skilleter, G. A. (1995) Variability at different spatial scales between a subtidal assemblage exposed to the discharge of sewage and two control assemblages. *Journal of Experimental Marine Biology and Ecology* **189**, 103–22.

Chessman, B. (1995) Rapid assessment of rivers using macroinvertebrates: a procedure based on habitat-specific sampling, family level identification, and a biotic index. *Australian Journal of Ecology* **20**, 122–9.

Chutter, F. M. (1972) A reappraisal of Needham & Usinger's data on the variability of a stream fauna when sampled with a Surber sampler. *Limnology and Oceanography* **17**, 139–41.

Chutter, F. M. & Noble, R. G. (1966) The reliability of a method of sampling stream invertebrates. *Archiv für Hydrobiologie* **62**, 95–103.

Clarke, K. R. (1993) Non-parametric multivariate analyses of changes in community structure. *Australian Journal of Ecology* **18**, 117–43.

Clarke, K. R. & Warwick, R. M. (1994) *Change in Marine Communities: An Approach to Statistical Analysis and Interpretation*. Swindon, UK: Natural Environment Research Council.

Clemen, R. T. (1996) *Making Hard Decisions: An Introduction to Decision Analysis*, 2nd edn. Belmont, CA: Duxbury Press.

Cohen, J. (1973) Statistical power analysis and research results. *American Educational Research Journal* **10**, 225–9.

Cohen, J. (1988) *Statistical Power Analysis for the Behavioural Sciences*, 2nd edn. Hillsdale, NJ: Lawrence Earlbaum.

Coimbra, C. N., Graça, M. A. S. & Cortes, R. M. (1996) The effects of a basic effluent on macroinvertebrate community structure in a temporary Mediterranean river. *Environmental Pollution* **94**, 301–7.

Collier, M. P., Webb, R. H. & Andrews, E. D. (1997) Experimental flooding in Grand

Canyon. *Scientific American*, **276**(1), 82–9.

Connell, J. H. & Sousa, W. P. (1983) On the evidence needed to judge ecological stability or persistence. *American Naturalist* **121**, 789–824.

Conquest, L. L. (1993) Statistical approaches to environmental monitoring: did we teach the wrong things? *Environmental Monitoring and Assessment* **26**, 107–24.

Constable, A. J. (1991) The role of science in environmental protection. *Australian Journal of Marine and Freshwater Research* **42**, 527–38.

Cook, R. B., Suter, G. W. II & Sain, E. R. (1999) Ecological risk assessment in a large river-reservoir. 1. Introduction and background. *Environmental Toxicology and Chemistry* **18**, 581–8.

Cooper, S. D. & Barmuta, L. A. (1993) Field experiments in biomonitoring. In *Freshwater Biomonitoring and Benthic Macroinvertebrates*, eds. D. M. Rosenberg & V. H. Resh, pp. 399–441. New York: Chapman & Hall.

Cooper, S. D., Barmuta, L., Sarnelle, O., Kratz, K. & Diehl, S. (1997) Quantifying spatial heterogeneity in streams. *Journal of the North American Benthological Society* **16**, 174–88.

Cooper, S. D., Diehl, S., Kratz, K. & Sarnelle, O. (1998) Implications of scale for patterns and processes in stream ecology. *Australian Journal of Ecology* **23**, 27–40.

Corkum, L. D. (1991) Spatial patterns of macroinvertebrate distribution along rivers in eastern deciduous forest and grassland biomes. *Journal of the North American Benthological Society* **10**, 358–71.

Courtemanch, D. L. (1996) Commentary on the subsampling procedures used for rapid bioassessments. *Journal of the North American Benthological Society* **15**, 381–5.

Cranston, P. S., Fairweather, P. & Clarke, G. (1996) Biological indicators of water quality. In *Indicators of Catchment Health: A Technical Perspective*, eds. J. Walker & D. J. Reuter, pp. 143–54. Collingwood, Vic.: CSIRO.

Crome, F. H. J., Thomas, M. R. & Moore, L. A. (1996) A novel Bayesian approach to assessing impacts of rain forest logging. *Ecological Applications* **6**, 1104–23.

Croonquist, M. J. & Brooks, R. P. (1991) Use of avian and mammalian guilds as indicators of cumulative impacts in riparian–wetland areas. *Environmental Management* **15**, 701–14.

Crowley, P. II. (1992) Resampling methods for computation-intensive data analysis in ecology and evolution. *Annual Review of Ecology and Systematics* **23**, 1407–24.

Cullen, P. (1990a) Biomonitoring and environmental management. *Environmental Monitoring and Assessment* **14**, 107–14.

Cullen, P. (1990b) The turbulent boundary between water science and water management. *Freshwater Biology* **24**, 201–9.

Cummins, K. W. (1972) What is a river? Zoological description. In *River Ecology and Man*, eds. R. T. Oglesby, C. A. Carlson & J. A. McCann, pp. 33–52. London: Academic Press.

Cummins, K. W. (1973) Trophic relations of aquatic insects. *Annual Review of Entomology* **18**, 183–206.

Cummins, K. W. (1974) The structure and function of stream ecosystems. *BioScience* **24**, 631–41.

Cummins, K. W., Cushing, C. E. & Minshall, G. W. (1995) Introduction: an overview of stream ecosystems. In *Ecosystems of the World*, vol. 22, *River and Stream Ecosystems*, eds. C. E. Cushing, K. W. Cummins & G. W. Minshall, pp. 1–8. Amsterdam: Elsevier.

Currie, D. R. & Parry, G. D. (1996) Effects of scallop dredging on a soft sediment community: a large-scale experimental study. *Marine Ecology Progress Series* **134**, 131–50.

Davey, G. W., Doeg, T. J. & Blyth, J. M. (1987) Changes in benthic sediment in the Thomson River, southeastern Australia, during construction of the Thomson

398    References

Dam. *Regulated Rivers: Research and Management* **1**, 71–84.

Davis, J. & Finlayson, B. (1999) The role of historical research in stream rehabilitation: a case study from central Victoria. In *The Challenge of Rehabilitating Australia's Streams, Proceedings of the 2nd Australian Stream Management Conference, Adelaide, SA, February 1999*, eds. I. Rutherfurd & R. Bartley, pp. 199–204. Monash University, Melbourne, Vic: Cooperative Research Centre for Catchment Hydrology.

Davis, W. S. & Simon, T. P. (eds.) (1995) *Biological Assessment and Criteria: Tools for Water Resource Planning and Decision Making*. Boca Raton, FL: Lewis Publishers.

Deegan, L. A., Peterson, B. J., Golden, H., McIvor, C. C. & Miller, M. C. (1997) Effects of fish density and river fertilization on algal standing stocks, invertebrate communities, and fish production in an arctic river. *Canadian Journal of Fisheries and Aquatic Sciences* **54**, 269–83.

Delong, M. D., Thorp, J. H. & Haag, K. H. (1993) A new device for sampling macroinvertebrates from woody debris (snags) in nearshore areas of aquatic systems. *American Midland Naturalist* **130**, 413–17.

Dennis, B. (1996) Discussion: Should ecologists become Bayesians. *Ecological Applications* **6**, 1095–103.

Dixon, P. M. (1993) The bootstrap and the jackknife: describing the precision of ecological indices. In *Design and Analysis of Ecological Experiments*, eds. S. M. Scheiner & J. Gurevitch, pp. 290–318. New York: Chapman & Hall.

Dobson, A. J. (1990) *An Introduction to Fitting Generalized Linear Models*. London, UK: Chapman & Hall.

Dobson, M. & Cariss, H. (1999) Restoration of afforested upland streams: what are we trying to achieve? *Aquatic Conservation: Marine and Freshwater Ecosystems* **9**, 133–9.

Doeg, T. J. (1984) Response of the macroinvertebrate fauna of the Mitta Mitta River, Victoria, to the construction and operation of Dartmouth Dam. 2. Irrigation release. *Occasional Papers from the Museum of Victoria* **1**, 101–8.

Downes, B. J., Glaister, A. & Lake, P. S. (1997) Spatial variation in the force required to initiate rock movement in four upland streams: implications for estimating disturbance frequencies. *Journal of the North American Benthological Society* **16**, 203–20.

Downes, B. J., Hindell, J. S. & Bond, N. R. (2000) What's in a site? Variation in lotic macroinvertebrate density and diversity in a spatially replicated experiment. *Austral Ecology* **25**, 128–39.

Downes, B. J. & Keough, M. J. (1998) Scaling of colonization processes in streams: parallels and lessons from marine hard substrata. *Australian Journal of Ecology* **23**, 8–26.

Downes, B. J., Lake, P. S. & Schreiber, E. S. G. (1993) Spatial variation in the distribution of stream invertebrates: implications of patchiness for models of community organization. *Freshwater Biology* **30**, 119–32.

Downes, B. J., Lake, P. S. & Schreiber, E. S. G. (1995) Habitat structure and invertebrate assemblages on stream stones: a multivariate view from the riffle. *Australian Journal of Ecology* **20**, 502–14.

Downing, J. A. (1986) A regression technique for the estimation of epiphytic invertebrate populations. *Freshwater Biology* **16**, 161–73.

Dubé, M. G. & Culp, J. M. (1996) Growth responses of periphyton and chironomids exposed to biologically treated bleached-kraft pulp mill effluent. *Environmental Toxicology and Chemistry* **15**, 2019–27.

Dubé, M. G., Culp, J. M. & Scrimgeour, G. J. (1997) Nutrient limitation and herbivory – processes influenced by bleached kraft pulp mill effluent. *Canadian Journal of Fisheries and Aquatic Sciences* **54**, 2584–95.

Dudgeon, D. (1999) *Tropical Asian Streams: Zoobenthos, Ecology and Conservation.* Hong Kong: Hong Kong University Press.

Dynesius, M. & Nilsson, C. (1994) Fragmentation and flow regulation of river systems in the northern third of the world. *Science* **266**, 753–62.

Edwards, D. (1996) Comment: the first data analysis should be journalistic. *Ecological Applications* **6**, 1090–4.

Ehrenfeld, J. G. (2000) Defining the limits of restoration: the need for realistic goals. *Restoration Ecology* **8**, 2–9.

Elliott, J. M. (1971a) Upstream movements of benthic invertebrates in a Lake District stream. *Journal of Animal Ecology* **40**, 235–52.

Elliott, J. M. (1971b) The distances travelled by drifting invertebrates in a Lake District stream. *Oecologia* **6**, 350–79.

Elliott, J. M. (1977) *Some Methods for the Statistical Analysis of Samples of Benthic Invertebrates*, 2nd edn. Ambleside, UK: Freshwater Biological Association.

Ellis, J. I. & Schneider, D. C. (1997) Evaluation of a gradient sampling design for environmental impact assessment. *Environmental Monitoring and Assessment* **48**, 157–72.

Ellison, A. M. (1993) Exploratory data analysis and graphic display. In *Design and Analysis of Ecological Experiments*, eds S. M. Scheiner & J. Gurevitch, pp. 14–45. New York: Chapman & Hall.

Ellison, A. M. (1996) An introduction to Bayesian methods for ecological research and environmental decision-making. *Ecological Applications* **6**, 1036–46.

Elwood, J. W., Newbold, J. B., O'Neill, R. V. & Van Winkle, W. (1983) Resource spiraling: an operational paradigm for analyzing lotic ecosystems. In *Dynamics of Lotic Ecosystems*, eds. T. D. Fontaine & S. M. Bartell, pp. 3–27. Ann Arbor, MI: Ann Arbor Scientific Publishers.

Engleman, C. J. Jr & McDiffett, W. F. (1996) Accumulation of aluminum and iron by bryophytes in streams affected by acid-mine drainage. *Environmental Pollution* **94**, 67–74.

Erickson, W. P. & McDonald, L. L. (1995) Tests for bioequivalence of control media and test media in studies of toxicity. *Environmental Toxicology and Chemistry* **14**, 1247–56.

Fairweather, P. G. (1989) Environmental impact assessment: where is the science in EIA? *Search* **20**, 141–4.

Fairweather, P. G. (1991a) A conceptual framework for ecological studies of coastal resources: an example of a tunicate collected for bait on Australian seashores. *Ocean and Shoreline Management* **15**, 125–42.

Fairweather, P. G. (1991b) Statistical power and design requirements for environmental monitoring. *Australian Journal of Marine and Freshwater Research* **42**, 555–67.

Fairweather, P. G. (1993) Links between ecology and ecophilosophy, ethics and the requirements of environmental management. *Australian Journal of Ecology* **18**, 3–19.

Fairweather, P. G. (1994) Improving the use of science in environmental assessments. *Australian Zoologist* **29**, 217–24.

Fairweather, P. G. (1999a) Determining the 'health' of estuaries: priorities for ecological research. *Australian Journal of Ecology* **24**, 441–51.

Fairweather, P. G. (1999b) State of environment indicators of 'river health': exploring the metaphor. *Freshwater Biology* **41**, 211–20.

Fairweather, P. G. & Lincoln Smith, M. P. (1993) The difficulty of assessing environmental impacts before they have occurred: a perspective from Australian consultants. In *Proceedings of the 2nd International Temperate Reefs Symposium*, eds. C. N. Battershill, D. R. Schiel, G. P. Jones, R. G. Creese & A. B. MacDiarmid, pp. 121–30.

Wellington, New Zealand: National Institute of Water and Atmospheric Research Marine.

Faith, D. P. (1990) Multivariate methods for biological monitoring based on community structure. In *Proceedings of the 29th Congress of the Australian Society for Limnology, Jabiru, Northern Territory, 1990*, ed. R. V. Hyne, pp. 17. Canberra, ACT: Australian Government Publishing Service.

Faith, D. P. (1999) Error and the growth of experimental knowledge (review). *Systematic Biology* **48**, 675–9.

Faith, D. P., Dostine, P. L. & Humphrey, C. L. (1995) Detection of mining impacts on aquatic macroinvertebrate communities: results of a disturbance experiment and the design of a multivariate BACIP monitoring program at Coronation Hill, Northern Territory. *Australian Journal of Ecology* **20**, 167–80.

Faith, D. P., Humphrey C. L. & Dostine P. L. (1991) Statistical power and BACI designs in biological monitoring: comparative evaluation of measures of community dissimilarity based on benthic macroinvertebrate communities in Rockhole Mine Creek, Northern Territory, Australia. *Australian Journal of Marine and Freshwater Research* **42**, 589–602.

Faith, D. P., Minchin, P. R. & Belbin, L. (1987) Compositional dissimilarity as a robust measure of ecological distance: a theoretical model and computer simulations. *Vegetatio* **69**, 57–68.

Falkenmark, M. (1997) Meeting water requirements of an expanding world population. *Philosophical Transactions of the Royal Society of London, Series B* **352**, 929–36.

Finlay, B. J., Maberly, S. C. & Cooper, J. I. (1997) Microbial diversity and ecosystem function. *Oikos* **80**, 209–13.

Firth, P. & Fisher, S. G. (eds.) (1992) *Global Climate Change and Freshwater Ecosystems*. New York: Springer-Verlag.

Fisher, R. A. (1935) *The Design of Experiments*. Edinburgh, UK: Oliver & Boyd.

Fisher, S. G. & Grimm, N. B. (1988) Disturbance as a determinant of structure in a Sonoran Desert stream ecosystem. *Verhandlungen Internationale Vereinigung für Theoretische und Angewandte Limnologie* **23**, 1183–9.

Fisher, S. G., Grimm, N. B., Marti, E., Holmes, R. M. & Jones, J. B. Jr (1998) Material spiraling in stream corridors: a telescoping ecosystem model. *Ecosystems* **1**, 19–34.

Fjerdingstad, E. (1964) Pollution of streams estimated by benthal phytomicroorganisms. 1. A saprobic system based on communities of organisms and ecological factors. *Internationale Revue der Gesamten Hydrobiologie* **49**, 63–131.

Flannagan, J. F. & Rosenberg, D. M. (1982) Types of artificial substrates used for sampling freshwater benthic macroinvertebrates. In *Artificial Substrates*, ed. J. Cairns Jr, pp. 237–66. Ann Arbor, MI: Ann Arbor Scientific Publishers.

Flecker, A. S. & Townsend, C. R. (1994) Community-wide consequences of trout introduction in New Zealand streams. *Ecological Applications* **4**, 798–807.

Forbes, S. A. & Richardson, R. E. (1919) Some recent changes in Illinois River biology. *Bulletin of the Illinois Natural History Survey* **13**, 139–56.

Fore, L. S., Karr, J. R. & Wisseman, R. W. (1996) Assessing invertebrate responses to human activities: evaluating alternative approaches. *Journal of the North American Benthological Society* **15**, 212–31.

Fowler, A. M. & Hennessy, K. J. (1995) Potential impacts of global warming on the frequency and magnitude of heavy precipitation. *Natural Hazards* **11**, 283–303.

Frissell, C. A., Liss, W. J., Warren, C. E. & Hurley, M. D. (1986) A hierarchical framework for stream habitat classification: viewing streams in a watershed context. *Environmental Management* **10**, 199–214.

Gagnon, M. M., Bussieres, D., Dodson, J. J. & Hodson, P. V. (1994) White sucker (*Catostomus commersoni*) growth and sexual maturation in pulp mill-contaminated and reference rivers. *Environmental Toxicology and Chemistry* **14**, 317–27.

Garrabou, J., Sala, E., Arcas, A. & Zabala, M. (1998) The impact of diving on rocky sublittoral communities: a case study of a bryozoan population. *Conservation Biology* **12**, 302–12.

Gasith, A. & Resh, V. H. (1999) Streams in Mediterranean climate regions: abiotic influences and biotic responses to predictable seasonal events. *Annual Review of Ecology and Systematics* **30**, 51–81.

Gaufin, A. R. & Tarzwell, C. M. (1952) Aquatic invertebrates as indicators of stream pollution. *Public Health Reports* **67**, 57–64.

Gaufin, A. R. & Tarzwell, C. M. (1956) Aquatic macro-invertebrate communities as indicators of organic pollution in Lytle Creek. *Sewage Industry Wastes* **28**, 906–24.

Gehrke, P. C. & Harris, J. H. (1994) The role of fish in cyanobacterial blooms in Australia. *Australian Journal of Marine and Freshwater Research* **45**, 905–15.

Gelman, A., Carlin, J. B., Stern, H. & Rubin, D. B. (1995) *Bayesian Data Analysis*. London, UK: Chapman & Hall.

Gibbons, W. N., Munkittrick, K. R. & Taylor, W. D. (1998) Monitoring aquatic environments receiving industrial effluents using small fish species. 1. Response of spoonhead sculpin (*Cottus ricei*) downstream of a bleached kraft pulp mill. *Environmental Toxicology and Chemistry* **17**, 2227–37.

Gibbs, G. W. & Penny, S. F. (1973) The effect of a sewage treatment plant on the Wainuiomata River. In *Proceedings of the Pollution Research Conference, Wairakei, New Zealand, 20–21 June 1973*, New Zealand DSIR Information Series, vol. 97, pp. 469–79. Wellington: New Zealand: Department of Scientific and Industrial Research.

Gilbert, O. L. & Anderson, P. (1998) *Habitat Creation and Repair*. Oxford, UK: Oxford University Press.

Giller, P. S. (1996) Floods and droughts: the effects of variations in water flow on streams and rivers. In *Disturbance and Recovery of Ecological Systems*, eds. P. S. Giller & A. A. Myers, pp. 1–19. Dublin, Eire: Royal Irish Academy.

Giller, P. S., Hildrew, A. G. & Raffaelli, D. G. (1994) *Aquatic Ecology: Scale, Pattern and Process*. Oxford, UK: Blackwell Scientific Publications.

Giller, P. S. & Malmqvist, B. (1998) *The Biology of Streams and Rivers*. Oxford, UK: Oxford University Press.

Gilpin, A. (1995) *Environmental Impact Assessment (EIA): Cutting Edge for the Twenty-First Century*. Cambridge, UK: Cambridge University Press.

Glasby, T. M. & Underwood, A. J. (1996) Sampling to differentiate between pulse and press perturbations. *Environmental Monitoring and Assessment* **42**, 241–52.

Glasson, J., Therivel, R. & Chadwick, A. (1994) *Introduction to Environmental Impact Assessment: Principles and Procedures, Practice and Prospects*. London, UK: UCL Press.

Goldsborough, L. G. & Hickman, M. (1991) A comparison of periphytic algal biomass and community structure on *Scirpus validus* and on a morphologically similar artificial substratum. *Journal of Phycology* **27**, 196–206.

Gordon, N. D., McMahon, T. A. & Finlayson, B. L. (1992) *Stream Hydrology: An Introduction for Ecologists*. Chichester, UK: John Wiley.

Gore, J. P., Kelly, J. R. & Yount, J. D. (1990) Application of ecological theory to determining recovery potential of disturbed lotic ecosystems: research needs and priorities. *Environmental Management* **14**, 755–62.

Gowan, C. & Fausch, K. D. (1996) Long-term demographic responses of trout populations to habitat manipulation in six Colorado streams. *Ecological Applications* **6**, 931–46.

Gray, R. H. (1995) Evaluation of fish behaviour as a means of pollution monitoring. In *Pollution and Biomonitoring*, ed. B. C. Rana, pp. 336–61. New Delhi, India: Tata McGraw-Hill.

Gray, J. S. & Jensen, K. (1993) Feedback monitoring: a new way of protecting the environment. *Trends in Ecology and Evolution* **8**, 267–8.

Grayson, J. E., Chapman, M. G. & Underwood, A. J. (1999) The assessment of restoration of habitat in urban wetlands. *Landscape and Urban Planning* **43**, 227–36.

Grayson, R. B., Doolan, J. M. & Blake, T. (1994) Application of AEAM (Adaptive Environmental Assessment and Management) to water quality in the Latrobe River Catchment. *Journal of Environmental Management* **41**, 245–58.

Green, R. H. (1979) *Sampling Design and Statistical Methods for Environmental Biologists*. New York: John Wiley.

Green, R. H. (1989) Power analysis and practical strategies for environmental monitoring. *Environmental Research* **50**, 195–205.

Green, R. H. (1993) Application of repeated measures designs in environmental impact and monitoring studies. *Australian Journal of Ecology* **18**, 81–98.

Gregory, S. V., Swanson, F. J., McKee, W. A. & Cummins, K. W. (1991) An ecosystem perspective on riparian zones. *BioScience* **41**, 540–51.

Grossman, G. D., Dowd, J. F. & Crawford, M. (1990) Assemblage stability in stream fishes: a review. *Environmental Management* **14**, 661–71.

Grossman, G. D., Moyle, P. B. & Whitaker, J. O. Jr (1982) Stochasticity in structural and functional characteristics of an Indiana stream fish assemblage: a test of community theory. *American Naturalist* **120**, 423–54.

Growns, J. E., Chessman, B. C., Jackson, J. E. & Ross, D. G. (1997) Rapid assessment of Australian rivers using macroinvertebrates: cost and efficiency of six methods of sample processing. *Journal of the North American Benthological Society* **16**, 682–93.

Growns, J. E., Chessman, B. C., McEvoy, P. K. & Wright, I. A. (1995) Rapid assessment of rivers using macroinvertebrates: case studies in the Nepean River and Blue Mountains, NSW. *Australian Journal of Ecology* **20**, 130–41.

Gunderson, L. H., Holling, C. S. & Light, S. S. (1995) *Barriers and Bridges to The Renewal of Ecosystems and Institutions*. New York: Columbia University Press.

Gurevitch, J. & Hedges, L. V. (1999) Statistical issues in ecological meta-analyses. *Ecology* **80**, 1142–9.

Haines, A. T., Finlayson, B. L. & McMahon, T. A. (1988) A global classification of river regimes. *Applied Geography* **8**, 255–72.

Hall, J. D., Murphy, M. L. & Aho, R. S. (1978) An improved design for assessing impacts of watershed practices on small streams. *Verhandlungen Internationale Vereinigung für Theoretische und Angewandte Limnologie* **20**, 1359–66.

Hall, T. J. (1982) Colonizing macroinvertebrates in the upper Mississipi River with a comparison of basket and multiplate samplers. *Freshwater Biology* **12**, 211–15.

Hamburg, M. (1985) *Statistical Analysis for Decision Making*, 3rd edn. New York: Harcourt Brace Jovanovich.

Harper, D. M., Ebrahimnezhad, M., Taylor, E., Dickinson, S., Decamp, O., Verniers, G. & Balbi, T. (1999) A catchment-scale approach to the physical restoration of lowland UK rivers. *Aquatic Conservation: Marine and Freshwater Ecosystems* **9**, 141–57.

Harris, J. H. (1995) The use of fish in ecological assessments. *Australian Journal of Ecology* **20**, 65–80.

Harris, J. H. & Gehrke, P. C. (1997) *Fish and Rivers in Stress: The NSW Rivers Survey*. Cronulla, NSW: NSW Fisheries.

Hart, B. T., Bailey, P., Edwards, R., Hortle, K., James, K., McMahon, A., Meredith, C. & Swadling, K. (1990) Effects of salinity on river, stream and wetland ecosystems in Victoria, Australia. *Water Research* **24**, 1103–18.

Hart, B. T., Maher, B. & Lawrence, I. (1999) New generation water quality guidelines for ecosystem protection. *Freshwater Biology* **41**, 347–59.

Havens, K. E. (1994) An experimental comparison of the effects of two chemical stressors on a freshwater zooplankton assemblage. *Environmental Pollution* **84**, 245–51.

Havens, K. E. & Hanazato, T. (1993) Zooplankton community responses to chemical stressors: a comparison of results from acidification and pesticide contamination research. *Environmental Pollution* **82**, 277–88.

Hawkes, H. A. (1964) Effects of domestic and industrial discharges on the ecology of riffles in Midland streams. *Advances in Water Pollution Research* **1**, 293–317.

Hawkes, H. A. (1975) River zonation and classification. In *River Ecology*, ed. B. A. Whitton, pp. 312–74. Oxford, UK: Blackwell Scientific Publications.

Hawkes, H. A. & Davies, L. J. (1971) Some effects of organic enrichment on benthic invertebrate communities in stream riffles. In *The Scientific Management of Animal and Plant Communities*, eds. E. Duffey & A. S. Wall, pp. 271–93. Oxford, UK: Blackwell Scientific Publications.

Hawkins, C. P., Kershner, J. L., Bisson, P., Bryant, M. D., Decker, L. M., Gregory, S. V., McCullough, D. A., Overton, C. K., Reeves, G. H., Steedman, R. J. & Young, M. K. (1993) A hierarchical approach to classifying stream habitat features. *Fisheries* **18**, 3–12.

Hawkins, C. P. & Sedell, J. R. (1990) The role of refugia in the recolonization of streams devastated by the 1980 eruption of Mount St. Helens. *Northwest Science* **64**, 271–4.

Hays, W. L. (1994) *Statistics*, 5th edn. Fort Worth, TX: Harcourt Brace.

Hellawell, J. M. (1977) Change in natural and managed ecosystems: detection, measurement and assessment. *Proceedings of the Royal Society of London, Series B* **197**, 31–56.

Hellawell, J. M. (1978) *Biological Surveillance of Rivers: A Biological Monitoring Handbook*. Stevenage, UK: Medmenham and Stevenage Water Research Centre.

Hellawell, J. M. (1986) *Biological Indicators of Freshwater Pollutions and Environmental Management*. London, UK: Elsevier.

Herbold, B. (1984) Structure of an Indiana stream fish association: choosing an appropriate model. *American Naturalist* **124**, 561–72.

Hershey, A. E., Lima, A. R., Niemi, G. J. & Regal, R. R. (1998) Effects of *Bacillus thuringiensis israelensis* (BTI) and methoprene on nontarget macroinvertebrates in Minnesota wetlands. *Ecological Applications* **8**, 41–60.

Hilborn, R. & Mangel, M. (1997) *The Ecological Detective: Confronting Models with Data*. Princeton, NJ: Princeton University Press.

Hill, A. B. (1965) The environment and disease: association or causation? *Proceedings of the Royal Society of Medicine* **58**, 295–300.

Hoaglin, J. D., Mosteller, F. & Tukey, J. W. (1993) *Understanding Robust and Exploratory Data Analysis*. New York: John Wiley.

Hobbs, R. J. & Norton, D. A. (1996) Towards a conceptual framework for restoration ecology. *Restoration Ecology* **4**, 93–110.

Hogg, I. D. & Williams, D. D. (1996) Response of stream invertebrates to a global-warming thermal regime: an ecosystem-level manipulation. *Ecology* **77**, 395–407.

Holling, C. S. (ed.) (1978) *Adaptive Environmental Assessment and Management*. New York: John Wiley.

Hortle, K. G. & Lake, P. S. (1982) Macroinvertebrate assemblages in channelized and unchannelized sections of the Bunyip River, Victoria. *Australian Journal of Marine and Freshwater Research* **33**, 1071–82.

Hortle, K. G. & Lake, P. S. (1983) Fish of channelized and unchannelized sections of the Bunyip River, Victoria. *Australian Journal of Marine and Freshwater Research* **34**, 441–50.

Horwitz, R. J. (1978) Temporal variability patterns and the distributional patterns of stream fishes. *Ecological Monographs* **48**, 307–21.

Houghton, J. T., Meira Filho, L. G., Callander, B. A., Harris, N., Kattenberg, A. & Maskell, K. (eds.) (1996) *Climate Change 1995: The Science of Climate Change*, Contribution of Working Group 1 to the 2nd Assessment Report of the Intergovernmental Panel on Climate Change. Cambridge, UK: Cambridge University Press.

Howson, C. (1997) Error probabilities in error. *Philosophy of Science* **64**, 185–94.

Huet, M. (1949) Aperçu des relations entre la pente et les populations des eaux courantes. *Schweizerische Zeitschrift für Hydrologie* **11**, 333–51.

Hughes, J. M. R. (1987) Hydrological characteristics and classification of Tasmanian rivers. *Australian Geographic Studies* **25**, 61–82.

Hughes, J. M. R. & James, B. (1989) A hydrological regionalization of streams in Victoria, Australia, with implications for stream ecology. *Australian Journal of Marine and Freshwater Research* **40**, 303–26.

Hughes, R. M., Heiskary, S. A., Mattews, W. J. & Yoder, C. O. (1994) Use of ecoregions in biological monitoring. In *Biological Monitoring of Aquatic Systems*, eds. S. L. Loeb & A. Spacie, pp. 125–51. Boca Raton, FL: Lewis Publishers.

Humphrey, C. L., Bishop, K. A. & Brown, V. M. (1990) Use of biological monitoring in the assessment of effects of mining wastes on aquatic ecosystems of the Alligator Rivers Region, tropical northern Australia. *Environmental Monitoring and Assessment* **14**, 139–81.

Humphrey, C. L. & Dostine, P. L. (1994) Development of biological monitoring programs to detect mining-waste impacts upon aquatic ecosystems of the Alligator Rivers Region, Northern Territory, Australia. *Mitteilungen Internationale Vereinigung für Theoretische und Angewandte Limnologie* **24**, 293–314.

Humphrey, C. L., Faith, D. P. & Dostine, P. L. (1995) Baseline requirements for assessment of mining impact using biological monitoring. *Australian Journal of Ecology* **20**, 150–66.

Hurd, M. K, Perry, S. A. & Perry, W. B. (1996) Nontarget effects of a test application of diflubenzuron to the forest canopy on stream macroinvertebrates. *Environmental Toxicology and Chemistry* **15**, 1344–51.

Hurlbert, S. H. (1984) Pseudoreplication and the design of ecological field experiments. *Ecological Monographs* **54**, 187–211.

Huryn, A. D., Benke, A. C. & Ward, G. M. (1995) Direct and indirect effects of geology on the distribution, biomass, and production of the freshwater snail *Elimia*. *Journal of the North American Benthological Society* **14**, 519–34.

Hynes, H. B. N. (1960) *The Biology of Polluted Waters*. Liverpool, UK: Liverpool University Press.

Hynes, H. B. N. (1970) *The Ecology of Running Waters*. Liverpool, UK: Liverpool University Press.

Hynes, H. B. N. (1975) The stream and its valley. *Verhandlungen Internationale Vereinigung für Theoretische und Angewandte Limnologie* **19**, 1–15.

Hynes, H. B. N. (1994) Historical perspective and future direction of biological monitoring of aquatic systems. In *Biological Monitoring of Aquatic Systems*, eds. S. L. Loeb & A. Spacie, pp. 11–21. Boca Raton, FL: Lewis Publishers.

Illies, J. (1962) Die Bedeutung der Stromung für die Biozonose in Rhithron und Potamon. *Schweizerische Zeitschrift für Hydrologie* **24**, 433–5.

Issa, A. A. & Ismail, M. A. (1995) Effects of detergents on River Nile water microflora. *Acta Hydrobiologica* **37**, 93–102.

Jack, J. D. & Gilbert, J. J. (1997) Effects of metazoan predators on ciliates in freshwater plankton communities. *Journal of Eukaryotic Microbiology* **44**, 194–9.

James, F. C. & McCulloch, C. E. (1985) Data analysis and the design of experiments in ornithology. In *Current Ornithology*, vol. 2, ed. R. F. Johnston, pp. 1–63. New York: Plenum Press.

James, F. C. & McCulloch, C. E. (1990) Multivariate analysis in ecology and systematics: panacea or Pandora's box? *Annual Review of Ecology and Systematics* **21**, 129–66.

James, K. R & Hart, B. T. (1993) Effects of salinity on four freshwater macrophytes. *Australian Journal of Marine and Freshwater Research* **44**, 769–77.

Janssens de Bisthoven, L., Postma, J. F., Parren, P., Timmermans, K. R. & Ollevier, F. (1998) Relations between heavy metals in aquatic sediments and in *Chironomus* larvae of Belgian lowland rivers and their morphological deformities. *Canadian Journal of Fisheries and Aquatic Sciences* **55**, 688–703.

Joellenbeck, L. M., Landrigan, P. J. & Larson, E. L. (1998) Gulf war veterans' illnesses: a case study in causal inference. *Environmental Research Section A* **79**, 71–81.

Johnson, D. H. (1980) The comparison of usage and availability measurements for evaluating resource preference. *Ecology* **61**, 65–71.

Johnson, R. K., Wiederholm, T. & Rosenberg, D. M. (1993) Freshwater biomonitoring using individual organisms, populations, and species assemblages of benthic macroinvertebrates. In *Freshwater Biomonitoring and Benthic Macroinvertebrates*, eds. D. M. Rosenberg & V. H. Resh, pp. 40–158. New York: Chapman & Hall.

Jones, J. B. Jr & Holmes, R. M. (1996) Surface–subsurface interactions in stream ecosystems. *Trends in Ecology and Evolution* **11**, 239–42.

Joy, C. M. (1990) Toxicity testing with freshwater algae in River Periyar (India). *Environmental Contamination and Toxicology* **45**, 915–22.

Junk, W. J., Bayley, P. B. & Sparks, R. E. (1989) The flood pulse concept in river–floodplain systems. *Canadian Special Publications in Aquatic Sciences* **106**, 110–27.

Karl, T. R., Nicholls, N. & Gregory, J. (1997) The coming climate. *Scientific American* **276**, 54–9.

Karr, J. R. (1981) Assessment of biotic integrity using fish communities. *Fisheries* **6**, 21–7.

Karr, J. R. (1991) Biological integrity: a long-neglected aspect of water resource management. *Ecological Applications* **1**, 66–84.

Karr, J. R. (1999) Defining and measuring river health. *Freshwater Biology* **41**, 221–34.

Karr, J. R. & Chu, E. W. (1999) *Restoring Life in Running Waters: Better Biological Monitoring*. Washington, DC: Island Press.

Karr, K. (1995) Ecological integrity and ecological health are not the same. In *Engineering within Ecological Constraints*, ed. P. Schulze, pp. 107–9. Washington, DC: National Academy of Engineering, National Academy Press.

Keeney, R. L. (1992) *Value-Focused Thinking*. Cambridge, MA: Harvard University Press.

Kelly, J. R. & Harwell, M. A. (1990) Indicators of ecosystem recovery. *Environmental Management* **14**, 527–45.

Kent, M. & Coker, P. (1992) *Vegetation Description and Analysis: A Practical Approach*. Boca Raton, FL: CRC Press.

Keough, M. J. & Black, K. P. (1996) Predicting the scale of marine impacts: understanding planktonic links between populations. In *Detecting Ecological Impacts: Concepts and Applications in Coastal Habitats*, eds. R. J. Schmitt & C. W. Osenberg, pp. 199–234. San Diego, CA: Academic Press.

Keough, M. J. & Mapstone, B. D. (1995) *Protocols for Designing Marine Ecological Monitoring Programs associated with BEK Mills*. National Pulp Mills Research Program Technical Report no. 11. Canberra, ACT: CSIRO.

Keough, M. J. & Mapstone, B. D. (1997) Designing environmental monitoring for pulp mills in Australia. *Water Science and Technology* **35**, 397–404.

Keough, M. J. & Quinn, G. P. (1991) Causality and the choice of measurements for detecting human impacts in marine environments. *Australian Journal of Marine and Freshwater Research* **42**, 539–54.

Keough, M. J. & Quinn, G. P. (2000) Legislative vs. practical protection of an intertidal shoreline in southeastern Australia. *Ecological Applications* **10**, 871–81.

Keough, M. J., Quinn, G. P. & King, A. (1993) Correlations between human collecting and intertidal mollusc populations on rocky shores. *Conservation Biology* **7**, 378–90.

Kerans, B. L. & Karr, J. R. (1994) A benthic index of biotic integrity (B–IBI) for rivers of the Tennessee Valley. *Ecological Applications* **4**, 768–85.

Kiffney, P. M., Clements, W. H. & Cady, T. A. (1997) Influence of ultraviolet radiation on the colonization dynamics of a Rocky Mountain stream benthic community. *Journal of the North American Benthological Society* **16**, 520–30.

Kingsford, R. T. (1999) Aerial survey of waterbirds on wetlands as a measure of river and floodplain health. *Freshwater Biology* **41**, 425–38.

Kingsford, R. T. & Johnson, W. (1998) Impact of water diversions on coloniallynesting waterbirds in the Macquarie Marshes of arid Australia. *Colonial Waterbirds* **21**, 159–70.

Klein, J. P., Vanderpoorten, A., Sanchez-Perez, J. M. & Maire, G. (1997) The cartography of hydrophytes applied to the study of fluvial ecosystems: an analysis tool for the restoration of former Rhine tributaries. *Lajeunia* **0(153)**, 1–33 (in French).

Kobayashi, T., Gibbs, P. Dixon, P. I. & Shiel, R. J. (1996) Grazing by a river zooplankton community: importance of microzooplankton. *Marine and Freshwater Research* **47**, 1025–36.

Koenig, W. D. (1999) Spatial autocorrelation of ecological phenomena. *Trends in Ecology and Evolution* **14**, 22–6.

Kohler, S. L. & Wiley, M. J. (1992) Parasite-induced collapse of populations of a dominant grazer in Michigan streams. *Oikos* **65**, 443–9.

Kolkwitz, R. & Marsson, M. (1908) Ökologie der tierische Saprobien: Beiträge zur Lehre von der biologische Gewässerbeurteilung. *Internationale Revue der Gesamten Hydrobiologie* **2**, 126–52.

Kondolf, G. M. & Micheli, E. R. (1995) Evaluating stream restoration projects. *Environmental Management* **19**, 1–15.

Korhola, A. & Blom, T. (1997) Marked early 20th century pollution and the subsequent recovery of Töölö Bay, central Helsinki, as indicated by subfossil diatom assemblage changes. *Hydrobiologia* **341**, 169–79.

Korman, J. & Higgins, P. S. (1997) Utility of escapement time series data for monitoring the response of salmon populations to habitat alteration. *Canadian Journal of Fisheries and Aquatic Sciences* **54**, 2058–67.

Krebs, C. J. (1999) *Ecological Methodology*, 2nd edn. New York: Harper & Row.

Kutka, F. J. & Richards, C. (1996) Relating diatom assemblage structure to stream habitat quality. *Journal of the North American Benthological Society* **15**, 469–80.

Ladson, A. R., White, L. J., Doolan, J. A., Finlayson, B. L., Hart, B. T., Lake, P. S. & Tilleard, J. W. (1999) Development and testing of an index of stream condition for waterway management in Australia. *Freshwater Biology* **41**, 453–68.

Lake, P. S. (1990) Disturbing hard and soft bottom communities: a comparison of marine and freshwater environments. *Australian Journal of Ecology* **15**, 477–88.

Lake, P. S. (1995) Of floods and droughts: river and stream ecosystems of Australia. In *Ecosystems of the World*, vol. 22, *River and Stream Ecosystems*, eds. C. E. Cushing, K. W. Cummins & G. W. Minshall, pp. 659–94. Amsterdam: Elsevier.

Lake, P. S. (2000) Disturbance, patchiness and diversity in streams. *Journal of the North American Benthological Society* **19**, 573–92.

Lake, P. S. & Barmuta, L. A. (1986) Stream benthic communities: persistent presumptions and current speculations. In *Limnology in Australia*, eds. P. De Deckker

& W. D. Williams, pp. 263–76. Dordrecht: W. Junk; and Melbourne, Vic.: CSIRO.

Lamberti, G. A. & Steinman, A. D. (1993) Research in artificial streams: applications, uses and abuses. *Journal of the North American Benthological Society* **12**, 313–84.

Lamon, E. C. III, Carpenter, S. C. & Stow, C. A. (1998) Forecasting PCB concentrations in Lake Michigan salmonids: a dynamic linear model approach. *Ecological Applications* **8**, 659–68.

Lardicci, C., Rossi, F. & Maltagliati, F. (1999) Detection of thermal pollution: variability of benthic communities at two different spatial scales in an area influenced by a coastal power station. *Marine Pollution Bulletin* **38**, 296–303.

Lee, P. M. (1997) *Bayesian Statistics: An Introduction.* London: Edward Arnold.

Leff, L. G. & Lemke, M. J. (1998) Ecology of aquatic bacterial populations: lessons from applied microbiology. *Journal of the North American Benthological Society* **17**, 261–71.

Legendre, P. (1993) Spatial autocorrelation: trouble or new paradigm? *Ecology* **74**, 1659–73.

Legendre, P. & Anderson, M. J. (1999) Distance-based redundancy analysis: testing multispecies responses in multifactorial ecological experiments. *Ecological Monographs* **69**, 1–24.

Legendre, P. & Legendre, L. (1998) *Numerical Ecology*, 2nd edn. Amsterdam: Elsevier.

Lemly, A. D. (1993) Guidelines for evaluating selenium data from aquatic monitoring and assessment studies. *Environmental Monitoring and Assessment* **28**, 83–100.

Lemly, A. D. (1996) Wastewater discharges may be most hazardous to fish during winter. *Environmental Pollution* **93**, 169–74.

Lemly, A. D. (1998) Bacterial growth on stream insects: potential for use in bioassessment. *Journal of the North American Benthological Society* **17**, 228–38.

Leopold, L. B., Wolman, M. G. & Miller, J. P. (1964) *Fluvial Processes in Geomorphology.* San Francisco, CA: W. H. Freeman.

Lewis, M. A. (1995) Use of freshwater plants for phyotoxicity testing: a review. *Environmental Pollution* **87**, 319–36.

Lewis, W. M. Jr, Weibezahn, F., Saunders, J. F. & Hamilton, S. K. (1990) The Orinoco River as an ecological system. *Interciencia* **15**, 346–57.

Likens, G. E. (1984) Beyond the shoreline: a watershed–ecosystem approach. *Verhandlungen Internationale Vereinigung für Theoretische und Angewandte Limnologie* **22**, 1–22.

Likens, G. E., Bormann, F. H., Pierce, R. S., Eaton, J. S. & Johnson, N. M. (1977) *Biogeochemistry of a Forested Ecosystem.* New York: Springer-Verlag.

Lillie, R. A. & Budd, J. (1992) Habitat architecture of *Myriophyllum spicatum* L. as an index to habitat quality for fish and macroinvertebrates. *Journal of Freshwater Ecology* **7**, 113–25.

Linke, S., Bailey, R. C. & Schwindt, J. (1999) Temporal variability of stream bioassessments using benthic macroinvertebrates. *Freshwater Biology* **42**, 575–84.

Lobo, E. A., Callegaro, V. L. M., Oliveira, M. A., Salomoni, S. E., Schuler, S. & Asai, K. (1996) Pollution tolerant diatoms from lotic systems in the Jacui Basin, Rio Grande do Sul, Brazil. *Iheringia Serie Botanica* **0(47)**, 45–72.

Locke, A., Reid, D. M., Van Leeuwen, H. C., Sprules, W. G. & Carlton, J. T. (1993) Ballast water exchange as a means of controlling dispersal of freshwater organisms by ships. *Canadian Journal of Fisheries and Aquatic Sciences* **50**, 2086–93.

Lockwood, J. L. & Pimm, S. L. (1999) When does restoration succeed? In *Ecological Assembly Rules: Perspectives, Advances, Retreats*, eds. E. Weiher & P. Keddy, pp. 251–71. Cambridge, UK: Cambridge University Press.

Lodge, D. M., Stein, R. A., Brown, K. M., Covich, A. P., Bronmark, C., Garvey, J. E. & Klosiewski, S. P. (1998) Predicting impact of freshwater exotic species on native biodiversity: challenges in spatial scaling. *Australian Journal of Ecology* **23**, 53–67.

Loehle, C. (1987) Hypothesis testing in ecology: psychological aspects and the importance of theory maturation. *Quarterly Review of Biology* **62**, 397–409.

Loehle, C. (1991) Managing and monitoring ecosystems in the face of heterogeneity. In *Ecological Heterogeneity*, eds. J. Kolasa & S. T. A. Pickett, pp. 144–59. New York: Springer-Verlag.

Loftis, J. C., McBride, G. B. & Ellis, J. C. (1991) Considerations of scale in water quality monitoring and data analysis. *Water Resources Bulletin* **27**, 255–64.

Lotspeich, F. B. (1980) Watersheds as the basic ecosystem: this conceptual framework provides a basis for a natural classification system. *Water Resources Bulletin* **16**, 581–6.

Lowe, R. L., Guckert, J. B., Belanger, S. E., Davidson, D. H. & Johnson, D. W. (1996) An evaluation of periphyton community structure and function on tile and cobble substrata in experimental stream mesocosms. *Hydrobiologia* **328**, 135–46.

Lowe, R. L. & Pan, Y. (1996) Benthic algal communities as biological monitors. In *Algal Ecology: Freshwater Benthic Ecosystems*, eds. R. J. Stevenson, M. L. Bothwell & R. L. Lowe, pp. 705–39. San Diego, CA: Academic Press.

Ludwig, J. A. & Reynolds, J. F. (1988) *Statistical Ecology: A Primer on Methods and Computing.* New York: John Wiley.

McArdle, B. H. (1996) Levels of evidence in studies of competition, predation and disease. *New Zealand Journal of Ecology* **20**, 7–15.

McCarthy, L. H., Robertson, K., Hesslein, R. H. & Williams, T. G. (1997) Baseline studies in the Slave River, NWT, 1990–1994: Part 4. Evaluation of benthic invertebrate populations and stable isotope analyses. *The Science of the Total Environment* **197**, 111–25.

McCormick, P. V., Belanger, S. E. & Cairns, J. Jr (1997) Evaluating the hazard of dodecyl alkyl sulphate to natural ecosystems using indigenous protistan communities. *Ecotoxicology* **6**, 67–85.

McDonald, L. L. & Erickson, W. P. (1994) Testing for bioequivalence in field studies: has a disturbed site been adequately reclaimed? In *Statistics in Ecology and Environmental Monitoring*, Otago Conference Series 2, eds. D. J. Fletcher & B. F. J. Manly, pp. 183–97. Dunedin, New Zealand: University of Otago Press.

McElravy, E. P., Lamberti, G. A. & Resh, V. H. (1989) Year-to-year variation in the aquatic macroinvertebrate fauna of a northern California stream. *Journal of the North American Benthological Society* **8**, 51–63.

Mackay, A. P. & Mackay, S. (1996) Spatial distribution of acid-volatile sulphide concentration and metal bioavailability in mangrove sediments from the Brisbane River, Australia. *Environmental Pollution* **93**, 205–9.

McKean, J. W. & Vidmar, T. J. (1994) A comparison of two rank-based methods for the analysis of linear models. *American Statistician* **48**, 220–9.

McMahon, T. A. & Finlayson, B. L. (1995) Reservoir system management and environmental flows. *Lakes and Reservoirs: Research and Management* **1**, 65–76.

McMahon, T. E., Finlayson, B. L., Haines, A. T. & Srikanthan, R. (1992) *Global Runoff: Continental Comparisons of Annual Flows and Peak Discharges.* Cremlingen-Destedt, Germany: Catena Verlag.

Mac Nally, R. & Quinn, G. P. (1998) Symposium introduction: the importance of scale in ecology. *Australian Journal of Ecology* **23**, 1–7.

Maguire, L. A. (1995) Decision analysis: an integrated approach to ecosystem exploitation and rehabilitation decisions. In *Rehabilitating Damaged Ecosystems*, 2nd edn, ed. J. Cairns Jr, pp. 13–34. Boca Raton, FL: Lewis Publishers.

Maguire, L. A. & Sondak, H. (1996) Can using decision analysis and dispute resolution techniques to solve environmental problems help promote equity? In *Statistics in Ecology and Environmental Monitoring*, vol. 2, *Decision Making and Risk*

*Assessment in Biology*, eds. D. J. Fletcher, L. Kavalieris & B. F. J. Manly, pp. 97–120. Dunedin, New Zealand: University of Otago Press.

Magurran, A. E. (1988) *Ecological Diversity and its Measurement*. London, UK: Croom Helm.

Malthus, T. J. & George, D. G. (1997) Airborne remote sensing of macrophytes in Cefni Reservoir, Anglesey, UK. *Aquatic Botany* **58**, 317–32.

Manly, B. F. J. (1985) *The Statistics of Natural Selection*. London, UK: Chapman & Hall.

Manly B. F. J. (1994) *Multivariate Statistical Methods: A Primer*, 2nd edn. London, UK: Chapman & Hall.

Manly, B. F. J. (1997) *Randomization, Bootstrap and Monte Carlo Methods in Biology*, 2nd edn. London, UK: Chapman & Hall.

Mapstone, B. D. (1995) Scalable decision rules for environmental impact studies: effect size, Type I and Type II errors. *Ecological Applications* **5**, 401–10.

Mapstone, B. D. (1996) Scalable decision criteria for environmental impact assessment. In *Detecting Ecological Impacts: Concepts and Applications in Coastal Habitats*, eds. R. J. Schmitt & C. W. Osenberg, pp. 67–80. San Diego, CA: Academic Press.

Marchant, R. (1988) Vertical distribution of benthic invertebrates in the bed of the Thomson River, Victoria. *Australian Journal of Marine and Freshwater Research* **39**, 775–84.

Marchant, R. (1989) A subsampler for samples of benthic invertebrates. *Bulletin of the Australian Society for Limnology* **12**, 49–52.

Marchant, R., Hirst, A., Norris, R. & Metzeling, L. (1999) Classification of macroinvertebrate communities across drainage basins in Victoria, Australia: consequences of sampling on a broad spatial scale for predictive modelling. *Freshwater Biology* **41**, 253–68.

Mason, W. T. Jr, Weber, C. I., Lewis, P. A. & Julian, E. C. (1973) Factors affecting the performance of basket and multiplate samplers. *Freshwater Biology* **3**, 409–36.

Mayo, D. G. (1996) *Error and the Growth of Experimental Knowledge*. Chicago, IL: University of Chicago Press.

Medley, C. N. & Clements, W. H. (1998) Responses of diatom communities to heavy metals in streams: the influence of longitudinal variation. *Ecological Applications* **8**, 631–44.

Meffe, G. K. & Minckley, W. L. (1987) Persistence and stability of fish and invertebrate assemblages in a repeatedly disturbed Sonoran Desert stream. *American Midland Naturalist* **117**, 177–91.

Meffe, G. K. & Sheldon, A. L. (1990) Post-defaunation recovery of fish assemblages in southeastern blackwater streams. *Ecology* **71**, 657–67.

Menzie, C., Henning, M. H., Cura, J., Finkelstein, K., Gentile, J., Maughan, J., Mitchell, D., Petron, S., Potocki, B., Svirsky, S. & Tyler, P. (1996) Special report of the Massachusetts weight-of-evidence workgroup: a weight-of-evidence approach for evaluating ecological risks. *Human and Ecological Risk Assessment* **2**, 277–304.

Mersch, J. & Johansson, L. (1993) Transplanted aquatic mosses and freshwater mussels to investigate the trace metal contamination in the rivers Meurthe and Plaine, France. *Environmental Technology* **14**, 1027–36.

Mersch, J. & Pihan, J. C. (1993) Simultaneous assessment of environmental impact on condition and trace metal availability in zebra mussels *Dreissena polymorpha* transplanted into the Wiltz River, Luxembourg: comparison with the aquatic moss. *Archives of Environmental Contamination and Toxicology* **25**, 353–64.

Metcalfe, J. L. (1989) Biological water quality assessment of running waters based on macroinvertebrate communities: history and present status in Europe. *Environmental Pollution* **60**, 101–39.

Metcalfe-Smith, J. L. (1994) Biological water-quality assessment of rivers: use of macroinvertebrate communities. In *The Rivers Handbook: Hydrological and Ecological Principles*, vol. 2, eds. P. Calow & G. E. Petts, pp. 144–70. Oxford, UK: Blackwell Scientific Publications.

Metzeling, L., Doeg, T. & O'Connor, W. (1995) The impact of salinization and sedimentation on aquatic biota. In *Conserving Biodiversity: Threats and Solutions*, eds. R. A. Bradstock, T. D. Auld, D. A. Keith, R. T. Kingsford, D. Lunney & D. P. Sivertsen, pp. 126–36. Sydney, NSW: Surrey Beatty.

Meyer, J. L. (1997) Stream health: incorporating the human dimension to advance stream ecology. *Journal of the North American Benthological Society* **16**, 439–47.

Mihuc, T. B. (1997) The functional trophic role of lotic primary consumers: generalist versus specialist strategies. *Freshwater Biology* **37**, 455–62.

Mill, J. S. (1884) *A System of Logic, Ratiocinative and Inductive: Being a Connected View of the Principles of Evidence and the Methods of Scientific Investigation*. London, UK: Longman.

Millard, S. P., Yearsley, J. R. & Lettenmaier, D. P. (1985) Space–time correlation and its effects on methods for detecting aquatic ecological change. *Canadian Journal of Fisheries and Aquatic Sciences* **42**, 1391–400.

Miller, A. M. & Golladay, S. W. (1996) Effects of spates and drying on macroinvertebrate assemblages of an intermittent and perennial prairie stream. *Journal of the North American Benthological Society* **15**, 670–89.

Milner, A. M. (1994) System recovery. In *The Rivers Handbook: Hydrological and Ecological Principles*, vol. 2, eds. P. Calow & G. E. Petts, pp. 76–97. Oxford, UK: Blackwell Scientific Publications.

Minchin, P. R. (1987) An evaluation of the relative robustness of techniques for ecological ordination. *Vegetatio* **69**, 89–107.

Minshall, G. W. (1988) Stream ecosystem theory: a global perspective. *Journal of the North American Benthological Society* **7**, 263–88.

Minshall, G. W., Brock, J. T. & Varley, J. D. (1989) Wildfire and Yellowstone's streams. *BioScience* **39**, 707–15.

Morin, A. (1985) Variability of density estimates and the optimization of sampling programs for stream benthos. *Canadian Journal of Fisheries and Aquatic Sciences* **42**, 1530–4.

Morin, A. & Cattaneo, A. (1992) Factors affecting sampling variability of freshwater periphyton and the power of periphyton studies. *Canadian Journal of Fisheries and Aquatic Sciences* **49**, 1695–703.

Morrisey, D. J., Howitt, L., Underwood, A. J. & Stark, J. S. (1992) Spatial variation in soft-sediment benthos. *Marine Ecology Progress Series* **81**, 197–204.

Moyle, P. B. (1993) Biodiversity, biomonitoring, and the structure of stream fish communities. In *Biological Monitoring of Aquatic Systems*, eds. S. L. Loeb & A. Spacie, pp. 171–86. Boca Raton, FL: Lewis Publishers.

Mulholland, P. J. (1997) Dissolved organic matter concentration and flux in streams. *Journal of the North American Benthological Society* **16**, 131–41.

Murdoch, W. W., Fay, R. C. & Mechalas, B. J. (1989) *Final report of the Marine Review Committee to the California Coastal Commission*. MRC Document no. 89–02. California: Marine Review Committee.

Naiman, R. J. & Décamps, H. (1997) The ecology of interfaces: riparian zones. *Annual Review of Ecology and Systematics* **28**, 621–58.

Needham, P. R. & Usinger, R. L. (1956) Variability in the macrofauna of a single riffle in Porsser Creek, California, as indicated by the Surber sampler. *Hilgardia* **14**, 383–409.

Neter, J., Kutner, M. H., Nachtsheim, C. J. & Wasserman, W. (1996) *Applied Linear Statistical Models*, 4th edn. Chicago, IL: Irwin.

Neter, J., Wasserman, W. & Whitmore, G. A. (1993) *Applied Statistics*. Englewood

Cliffs, NJ: Prentice Hall.

Newbold, J. D. (1992) Cycles and spirals of nutrients. In *The Rivers Handbook: Hydrological and Ecological Principles*, vol. 1. eds. P. Calow & G. E. Petts, pp. 379–408. Oxford, UK: Blackwell Scientific Publications.

Newman, A. (1995) Water pollution point sources still significant in urban areas. *Environmental Science and Technology* **29**, 114.

Newson, M. (1994) *Hydrology and the River Environment*. Oxford, UK: Clarendon Press.

Neyman, J. & Pearson, E. (1928) On the use and interpretation of certain test criteria for purposes of statistical inference: Part I. *Biometrika* **20A**, 175–240.

Niemi, G. J., DeVore, P., Detenbeck, N., Taylor, D., Lima, A., Pastor, J., Yount, J. D. & Naiman, R. J. (1990) Overview of case studies on recovery of aquatic systems from disturbance. *Environmental Management* **14**, 571–87.

Nimmo, D. R. & McEwen, L. C. (1994) Pesticides. In *Handbook of Ecotoxicology*, vol. 2, ed. P. Calow, pp. 155–203. Oxford, UK: Blackwell Scientific Publications.

Norris, R. H. (1986) Mine waste pollution of the Molonglo River, New South Wales and the Australian Capital Territory: effectiveness of remedial works at Captains Flat mining area. *Australian Journal of Marine and Freshwater Research* **37**, 147–57.

Norris, R. H. (1995) Biological monitoring: the dilemma of data analysis. *Journal of the North American Benthological Society* **14**, 440–50.

Norris, R. H. & Georges, A. (1993) Analysis and interpretation of benthic surveys. In *Freshwater Biomonitoring and Benthic Macroinvertebrates*, eds. D. M. Rosenberg & V. H. Resh, pp. 234–86. New York: Chapman & Hall.

Norris, R. H., Lake, P. S. & Swain, R. (1982) Ecological effects of mine effluents on the South Esk River, north-eastern Tasmania. 3. Benthic macroinvertebrates. *Australian Journal of Marine and Freshwater Research* **33**, 789–809.

O'Connor, N. A. & Lake, P. S. (1994) Long-term and seasonal large-scale disturbances of a small lowland stream. *Australian Journal of Marine and Freshwater Research* **45**, 243–55.

Osenberg, C. W., Sarnelle, O., Cooper, S. D. & Holt, R. D. (1999) Resolving ecological questions through meta-analysis: goals, metrics, and models. *Ecology* **80**, 1105–17.

Palmer, M. A., Hakenkamp, C. C. & Nelson-Baker, K. (1997) Ecological heterogeneity in streams: why variance matters. *Journal of the North American Benthological Society* **16**, 189–202.

Pan, Y., Stevenson, R. J., Hill, B. H., Herlihy, A. T. & Collins, G. B. (1996) Using diatoms as indicators of ecological conditions in lotic systems: a regional assessment. *Journal of the North American Benthological Society* **15**, 481–95.

Patrick, R. (1949) A proposed biological measure of stream conditions, based on a survey of the Conestoga Basin, Lancaster County, Pennsylvania. *Proceedings of the Academy of Natural Sciences, Philadelphia* **101**, 277–341.

Payne, A. I. (1986) *The Ecology of Tropical Lakes and Rivers*. Chichester, UK: John Wiley.

Pearson, R. G. (1984) Temporal changes in the composition and abundance of the macro-invertebrate communities of the River Hull. *Archiv für Hydrobiologie* **100**, 273–98.

Peterman, R. M. (1990) Statistical power analysis can improve fisheries research and management. *Canadian Journal of Fisheries and Aquatic Sciences* **44**, 1879–89.

Peters, R. H. (1991) *A Critique for Ecology*. Cambridge, UK: Cambridge University Press.

Peterson, C. H. (1993) Improvement of environmental impact analysis by application of principles derived from manipulative ecology: lessons from coastal marine case histories. *Australian Journal of Ecology* **18**, 21–52.

Petraitis, P. S., Latham, R. A. & Niesenbaum, R. A. (1989) The maintenance of species diversity by disturbance. *Quarterly Review of Biology* **64**, 393–418.

Petts, G. E. (1984). *Impounded Rivers: Perspectives for Ecological Management*. Chichester, UK: John Wiley.

Petts, G. E. & Amoros, C. (1996) The fluvial hydrosystem. In *Fluvial Hydrosystems*, eds. G. E. Petts & C. Amoros, pp. 1–12. London, UK: Chapman & Hall.

Petts, G. E. & Maddock, I. (1994) Flow allocation for in-river needs. In *The Rivers Handbook: Hydrological and Ecological Principles*, vol. 2, eds. P. Calow & G. E. Petts, pp. 289–307. Oxford, UK: Blackwell Scientific Publications.

Phillips, D. J. H. & Rainbow, P. S. (1993) *Biomonitoring of Trace Aquatic Contaminants*. London, UK: Chapman & Hall.

Pielou, E. C. (1984) Probing multivariate data with random skewers: a preliminary to direct gradient analysis. *Oikos* **42**, 161–5.

Pimentel, D., Houser, J., Preiss, E., White, O., Fang, H., Mesnick, L., Barsky, T., Tariche, S., Schreck, J. & Alpert, S. (1997) Water resources: agriculture, the environment, and society. *BioScience* **47**, 97–106.

Plénet, S. (1995) Freshwater amphipods as biomonitors of metal pollution in surface and interstitial aquatic systems. *Freshwater Biology* **33**, 127–37.

Poff, N. L. (1992) Why disturbances can be predictable: a perspective on the definition of disturbance in streams. *Journal of the North American Benthological Society* **11**, 86–92.

Poff, N. L. & Allan, J. D. (1995) Functional organization of stream fish assemblages in relation to hydrological variability. *Ecology* **76**, 606–27.

Poff, N. L., Allan, J. D., Bain, M. B., Karr, J. R., Prestegaard, K. L., Richter, B. D., Sparks, R. E. & Stromberg, J. C. (1997) The natural flow regime: a paradigm for river conservation and restoration. *BioScience* **47**, 769–84.

Poff, N. L. & Ward, J. V. (1990a) Implications of streamflow variability and predictability for lotic community structure: a regional analysis of streamflow patterns. *Canadian Journal of Fisheries and Aquatic Sciences* **46**, 1805–18.

Poff, N. L. & Ward, J. V. (1990b) Physical habitat template of lotic systems: recovery in the context of historical pattern of spatiotemporal heterogeneity. *Environmental Management* **14**, 629–45.

Popper, K. R. (1963) *Conjectures and Refutations: The Growth of Scientific Knowledge*. New York: Harper & Row.

Popper, K. R. (1968) *The Logic of Scientific Discovery*. London, UK: Hutchinson.

Popper, K. R. (1983) *Realism and The Aim of Science: Postscript to The Logic of Scientific Discovery*, vol. 3, series ed. W. W. Bartley. London, UK: Hutchinson. (Reprinted 1992, Routledge, London.)

Postel, S. (1997) *Last Oasis: Facing Water Scarcity*, 2nd edn. New York: W. W. Norton.

Postel, S. & Carpenter, S. (1997) Freshwater ecosystem services. In *Nature's Services: Societal Dependence on Natural Ecosystems*, ed. G. C. Daily, pp. 195–214. Washington, DC: Island Press.

Postel, S. L., Daily, G. C.& Ehrlich, P. R. (1996) Human appropriation of renewable fresh water. *Science* **271**, 785–8.

Potischman, N. & Weed, D. L. (1999) Causal criteria in nutritional epidemiology. *American Journal of Clinical Nutrition* **69** (suppl.), 1309S–14S.

Potvin, C. & Roff, D. A. (1993) Distribution-free and robust statistical methods: viable alternatives to parametric statistics? *Ecology* **74**, 1617–28.

Pratt, J. W., Raiffa, H. & Schlaifer, R. (1996) *Introduction to Statistical Decision Theory*. Cambridge, MA: MIT Press.

Puckridge, J. T., Sheldon, F., Walker, K. F. & Boulton, A. J. (1998) Flow variability and the ecology of large rivers. *Marine and Freshwater Research* **49**, 55–72.

Raikow, D. F., Grubbs, S. A. & Cummins, K. W. (1995) Debris dam dynamics and coarse particulate organic matter retention in an Appalachian Mountain stream. *Journal of the North American Benthological Society* **14**, 535–46.

Rand, G. M. (ed.) (1995) *Fundamentals of Aquatic Toxicology*. Washington, DC: Taylor & Francis.

Rapport, D. J. (1989) What constitutes ecosystem health? *Perspectives in Biology and Medicine* **33**, 120–32.

Rapport, D. J. (1991) Myths in the foundations of economics and ecology. *Biological Journal of the Linnean Society* **44**, 185–202.

Rapport, D. J. (1993) Ecosystems not optimized: a reply. *Journal of Aquatic Ecosystem Health* **2**, 57.

Rasmussen, P. W., Heisey, D. M., Nordheim, E. V. & Frost, T. M. (1993) Time-series intervention analysis: unreplicated large-scale experiments. In *Design and Analysis of Ecological Experiments*, eds. S. M. Scheiner & J. Gurevitch, pp. 138–58. New York: Chapman & Hall.

Reavie, E. D. & Smol, J. P. (1997) Diatom-based model to infer past littoral habitat characteristics in the St Lawrence River. *Journal of Great Lakes Research* **23**, 339–48.

Reckhow, K. & Stow, C. (1990) Monitoring design and data analysis for trend detection. *Lake and Reservoir Management* **6**, 49–60.

Reid, M. A., Tibby, J. C., Penny, D. & Gell, P. A. (1995) The use of diatoms to assess past and present water quality. *Australian Journal of Ecology* **20**, 57–64.

Reitzel, J., Elwany, M. H. S. & Callahan, J. D. (1994) Statistical analyses of the effects of a coastal power plant cooling system on underwater irradiance. *Applied Ocean Research* **16**, 373–9.

Resh, V. H. (1979) Sampling variability and life history features: basic consideration in the design of aquatic insect studies. *Journal of the Fisheries Research Board of Canada* **36**, 290–311.

Resh, V. H. (1994) Variability, accuracy, and taxonomic costs of rapid assessment approaches in benthic macroinvertebrate biomonitoring. *Bollettino di Zoologia* **61**, 375–83.

Resh, V. H., Jackson, J. K. & McElravy, E. P. (1988) The use of long-term ecological data and sequential decision plans in monitoring the impact of geothermal energy development on benthic macroinvertebrates. *Verhandlungen Internationale Vereinigung für Theoretische und Angewandte Limnologie* **23**, 1142–6.

Resh, V. H. & McElravy, E. P. (1993) Contemporary quantitative approaches to biomonitoring using benthic macroinvertebrates. In *Freshwater Biomonitoring and Benthic Macroinvertebrates*, eds. D. M. Rosenberg & V. H. Resh, pp. 159–94. New York: Chapman & Hall.

Resh, V. H., Norris, R. H. & Barbour, M. T. (1995) Design and implementation of rapid assessment approaches for water resource monitoring using benthic macroinvertebrates. *Australian Journal of Ecology* **20**, 108–21.

Resh, V. H. & Price, D. G. (1984) Sequential sampling: a cost-effective approach for monitoring benthic macroinvertebrates in environmental impact assessments. *Environmental Management* **8**, 75–80.

Resh, V. H. & Rosenberg, D. M. (1989) Spatial-temporal variability and the study of aquatic insects. *Canadian Entomologist* **121**, 941–63.

Reynolds, A. J. (1998) Confirmatory program evaluation: a method for strengthening causal inference. *American Journal of Evaluation* **19**, 203–21.

Reynolds, C. S. & Descy, J. P. (1996) The production, biomass and structure of phytoplankton in large rivers. *Archiv für Hydrobiologie* (Suppl.) **113**, 161–87.

Reynolds, J. D. & Hunter, C. (1985) Evaluation of the use of artificial substrates in sampling the invertebrate fauna of sewage fungus slimes in Irish rivers. *Verhandlungen Internationale Vereinigung für Theoretische und Angewandte Limnologie* **22**, 2239–43.

Reynoldson, T. B., Norris, R. H., Resh, V. H., Day, K. E. & Rosenberg, D. M. (1997) The reference condition: a comparison of multimetric and multivariate approaches to assess water-quality impairment using benthic macroinvertebrates. *Journal of the North American Benthological Society* **16**, 833–52.

Ricciardi, A., & Rasmussen, J.B. (1999) Extinction rates of North American freshwater fauna. *Conservation Biology* **13**, 1220–2.

Rogers, K. & Biggs, H. (1999) Integrating indicators, endpoints and value systems in strategic management of the rivers of Kruger National Park. *Freshwater Biology* **41**, 439–51.

Rohm, C. M., Giese, J. W. & Bennett, C. C. (1987) Evaluation of an aquatic ecoregion classification of streams in Arkansas. *Journal of Freshwater Ecology* **4**, 127–40.

Rosenberg, D. M., Berkes, F., Bodaly, R. A., Hecky, R. E., Kelly, C. A. & Rudd, J. W. M. (1997) Large-scale impacts of hydroelectric development. *Environmental Reviews* **5**, 27–54.

Rosenberg, D. M., McCully, P. & Pringle, C. M. (2000) Global-scale environmental effects of hydrological alterations: Introduction. *BioScience* **50**, 746–51.

Rosenberg, D. M. & Resh, V. H. (eds.) (1993) *Freshwater Biomonitoring and Macroinvertebrates.* New York: Chapman & Hall.

Rosenberg, D. M. & Resh, V. H. (1996) Use of aquatic insects in biomonitoring. In *An Introduction to the Aquatic Insects of North America*, 3rd edn, eds. R. W. Merritt & K. W. Cummins, pp. 87–97. Dubuque, IA: Kendall/Hunt Publishing Company.

Rosenberg, D. M., Resh, V. H., Balling, S. S., Barnby, M. A., Collins, J. N., Durbin, D. V., Flynn, T. S., Hart, D. D., Lamberti, G. A., McElravy, E. P., Wood, J. R., Blank, T. E., Schultz, D. M., Marrin, D. L. & Price, D. G. (1981) Recent trends in environmental impact assessment. *Canadian Journal of Fisheries and Aquatic Sciences* **38**, 591–624.

Ross, S. T., Matthews, W. J. & Echelle, A. A. (1985) Persistence of stream fish assemblages: effects of environmental change. *American Naturalist* **126**, 24–40.

Rundle, S. D., Weatherley, N. S. & Ormerod, S. J. (1995) The effects of catchment liming on the chemistry and biology of upland Welsh streams: testing model predictions. *Freshwater Biology* **34**, 165–75.

Ruse, L. P. & Hutchings, A. J. (1996) Phytoplankton composition of the River Thames in relation to certain environmental variables. *Archiv für Hydrobiologie* (Suppl.) **113**, 189–201.

Rutt, G. P., Weatherley, N. S. & Ormerod, S. J. (1989) Microhabitat availability in Welsh moorland and forest streams as a determinant of macroinvertebrate distribution. *Freshwater Biology* **22**, 247–61.

Safina, C. (1998) *Song for the Blue Ocean.* New York: Henry Holt.

Sala, O. E., Chapin, F. S., Armesto, J. J., Berlow, E., Bloomfield, J., Dirzo, R., Huber-Sanwald, E., Huenneke, L. F., Jackson, R. B., Kinzig, A., Leemans, R., Lodge, D. M., Mooney, H. A., Oesterheld, M., Poff, N. L., Sykes, M. T., Walker, B. H., Walker, M. & Wall, D. H. (2000) Global biodiversity scenarios for the year 2100. *Science* **287**, 1770–4.

Sampson, P. D. & Guttorp, P. (1991) Power transformations and tests of environmental impact as interaction effects. *American Statistician* **45**, 83–9.

Samuels, C. L. & Drake, J. A. (1997) Divergent perspectives on community convergence. *Trends in Ecology and Evolution* **12**, 427–32.

Samways, M. J. (2000) A conceptual model of ecosystem restoration triage based on experiences from three remote oceanic islands. *Biodiversity and Conservation* **9**, 1073–83.

Sandland, R. L. & Young, P. C. (1979) Probabilistic tests and stopping rules associated with hierarchical classification techniques. *Australian Journal of Ecology* **4**, 399–406.

Scheiner, S. M. (1993) Introduction: theories, hypotheses and statistics. In *Design and Analysis of Ecological Experiments*, eds. S. M. Scheiner & J. Gurevitch, pp. 1–13.

New York: Chapman & Hall.

Scheiner, S. M. & Gurevitch, J. (eds.) (1993) *Design and Analysis of Ecological Experiments*. New York: Chapman & Hall.

Schindler, D. (1990) Experimental perturbations of whole lakes as tests of hypotheses concerning ecosystem structure and function. *Oikos* **57**, 25–41.

Schlosser, I. J. (1990) Environmental variation, life history attributes, and community structure in stream fishes: implications for environmental management and assessment. *Environmental Management* **15**, 621–8.

Schmieder, K. (1995) Application of geographic information systems (GIS) in lake monitoring with submersed macrophytes at Lake Constance: conception and purposes. *Acta Botanica Gallica* **142**, 551–4.

Schmitt, R. J. & Osenberg, C. W. (eds.) (1996) *Detecting Ecological Impacts: Concepts and Applications in Coastal Habitats*. San Diego, CA: Academic Press.

Schneider, D. C. (1994) *Quantitative Ecology: Spatial and Temporal Scaling*. San Diego, CA: Academic Press.

Schroeter, S. C., Dixon, J. D., Kastendiek, J. & Smith, R. O. (1993) Detecting the ecological effects of environmental impacts: a case study of kelp forest invertebrates. *Ecological Applications* **3**, 331–50.

Schumm, S. A. (1977) *The Fluvial System*. New York: John Wiley.

Scott, A. & Grant, T. (1997) *Impacts of Water Management in the Murray–Darling Basin on the Platypus* (Ornithorhynchus anatinus) *and the Water Rat* (Hydromys chrysogaster). CSIRO Land and Water Technical Report no. 23/97. Canberra, ACT: CSIRO.

Seaman, J. W., Walls, S. C., Wise, S. E. & Jaeger, R. G. (1994) Caveat emptor: rank transform methods and interaction. *Trends in Ecology and Evolution* **9**, 261–3.

Shiklomanov, I. J. (1989) Climate and water resources. *Hydrological Sciences Journal* **34/5**, 495–529.

Shrader-Frechette, K. S. & McCoy, E. D. (1993) Statistics, costs and rationality in ecological inference. *Trends in Ecology and Evolution* **7**, 96–9.

Sielken, R. L. Jr & Stevenson, D. E. (1998) Some implications for quantitative risk assessment if hormesis exists. *Human and Experimental Toxicology* **17**, 259–62.

Simenstad, C. A. & Thom, R. M. (1996) Functional equivalency trajectories of the restored Gog-Le-Hi-Te estuarine wetland. *Ecological Applications* **6**, 38–56.

Simpson, J., Norris, R. H., Barmuta, L. & Blackman, P. (1997) Australian River Assessment System: National River Health Program Predictive Model Manual. < http://enterprise.canberra.edu.au/AusRivas/ > .

Sinitsyn, M. G. (1992) Using super large-scale aerial photography to study beaver (*Castor fiber*) colonies. *Zoologicheskii Zhurnal* **71**, 130–9 (in Russian).

Sladecek, V. (1973) System of water quality from the biological point of view. *Archiv für Hydrobiologie* **7**, 1–218.

Smith, E. P., Orvos, D. R. & Cairns, J. Jr. (1993) Impact assessment using the Before–After–Control-Impact (BACI) model: concerns and comments. *Canadian Journal of Fisheries and Aquatic Sciences* **50**, 627–37.

Smith, R. E. W. & Morris, T. F. (1992) The impacts of changing geochemistry on the fish assemblages of the Lower Ok Tedi and Middle Fly River, Papua New Guinea. *The Science of the Total Environment* **125**, 321–44.

Snedecor, G. W. & Cochran, W. G. (1989) *Statistical Methods*, 8th edn. Ames, IA: Iowa State College Press.

Sokal, R. R. & Rohlf, F. J. (1995) *Biometry*, 3rd edn. New York: W.H. Freeman.

Somers, K. M., Reid, R. A. & David, S. M. (1998) Rapid biological assessments: how many animals are enough? *Journal of the North American Benthological Society* **17**, 348–58.

Southwood, T. R. E. (1978) *Ecological Methods*, 2nd edn. Chichester, UK: John Wiley.

Sparks, R. E., Bayley, P. B., Kohler, S. L. & Osborne, L. L. (1990) Disturbance and recovery of large floodplain rivers. *Environmental Management* **14**, 699–709.

Sprent, P. (1989) *Applied Nonparametric Statistical Methods*. London, UK: Chapman & Hall.

Statzner, B. & Higler, B. (1985) Questions and comments on the River Continuum Concept. *Canadian Journal of Fisheries and Aquatic Sciences* **42**, 1038–44.

Stevenson, R. J. (1997) Scale-dependent determinants and consequences of benthic algal heterogeneity. *Journal of the North American Benthological Society* **16**, 248–62.

Stewart, A. J. & Loar, J. M. (1994) Spatial and temporal variation in biomonitoring data. In *Biological Monitoring of Aquatic Systems*, eds. S. L. Loeb & A. Spacie, pp. 91–124. Boca Raton, FL: Lewis Publishers.

Stewart-Oaten, A. (1996a) Goals, analyses and designs for environmental monitoring: the fundamental role of models. In *The Design of Ecological Impact Studies: Conceptual Issues and Application in Coastal Marine Habitats*, eds. R. J. Schmitt & C. W. Osenberg, pp. 17–27. San Diego, CA: Academic Press.

Stewart-Oaten, A. (1996b) Problems in the analysis of environmental monitoring data. In *The Design of Ecological Impact Studies: Conceptual Issues and Application in Coastal Marine Habitats*, eds. R. J. Schmitt & C. W. Osenberg, pp. 109–31. San Diego, CA: Academic Press.

Stewart-Oaten, A. & Bence, J. R. (2001) Temporal and spatial variation in environmental assessment. *Ecological Monographs* **71**, 305–39.

Stewart-Oaten, A., Bence, J. R. & Osenberg, C. W. (1992) Assessing effects of unreplicated perturbations: no simple solutions. *Ecology* **73**, 1396–1404.

Stewart-Oaten, A., Murdoch, W. W. & Parker, K. R. (1986) Environmental impact-assessment: 'pseudoreplication' in time? *Ecology* **67**, 929–40.

Stoner, J. H., Gee, A. S. & Wade, K. R. (1984) The effects of acidification on the ecology of streams in the upper Tywi catchment in west Wales. *Environmental Pollution (Series A)* **35**, 125–57.

Stout, R. J. & Rondinelli, M. P. (1995) Stream-dwelling insects and extremely low frequency electromagnetic fields: a 10-year study. *Hydrobiologia* **302**, 197–213.

Stow, C. A., Carpenter, S. R., Webster, K. E. & Frost, T. M. (1998) Long-term environmental monitoring: some perspectives from lakes. *Ecological Applications* **8**, 269–76.

Strayer, D. L., Caraco, N. F., Cole, J. J., Findlay, S. & Pale, M. L. (1999) Transformation of freshwater ecosystems by bivalves: a case study of the Hudson River. *BioScience* **49**, 19–27.

Suter, G. W. II (1993a) A critique of ecosytem health concepts and indexes. *Environmental Toxicology and Chemistry* **12**, 1533–9.

Suter, G. W. II (1993b) *Ecological Risk Assessment*. Boca Raton, FL: Lewis Publishers.

Sweeney, B. W., Jackson, J. K., Newbold, J. D. & Funk, D. H. (1992) Climate change and life histories and biogeography of aquatic insects in eastern North America. In *Global Climate Change and Freshwater Ecosystems*, eds. P. Firth & S. G. Fisher, pp. 143–76. New York: Springer-Verlag.

Tabachnick, B. & Fidell, L. (1996) *Using Multivariate Statistics*, 3rd edn. New York: Harper & Row.

Tausch, R. J., Charlet, D. A., Weixelman, D. A. & Zamudio, D. C. (1995) Patterns of ordination and classification instability resulting from changes in input data order. *Journal of Vegetation Science* **6**, 897–902.

Thompson, S. K. (1992) *Sampling*. New York: John Wiley.

Thorp, J. H., Black, A. R., Jack, J. D. & Casper, A. F. (1996) Pelagic enclosures: modification and use for experimental study of riverine plankton. *Archiv für Hydrobiologie (Suppl.)* **113**, 583–9.

Thrush, S. F. (1991) Spatial patterns in soft-bottom communities. *Trends in Ecology and Evolution* **6**, 75–9.

Thurmond, D. P. & Miller, K. V. (1994) Small mammal communities in streamside management zones. *Brimleyana* **0(21)**, 125–30.

Toshach, S. C. (1977) Pollution control and water quality in the upper Kaiapoi River. *Mauri Ora* **5**, 39–51.

Townsend, C. R. (1996) Concepts in river ecology: pattern and process in the catchment hierarchy. *Archiv für Hydrobiologie* (Suppl.) **113**, 3–21.

Tubbing, D. M. J., Admiraal, W., Backhaus, D., Friedrich, G., De Ruyter van Steveninck, E. D., Mueller, D. & Keller, I. (1994) Results of an international plankton investigation on the River Rhine. *Water Science and Technology* **29**, 9–19.

Turner, A. M. & Trexler, J. C. (1997) Sampling aquatic invertebrates from marshes: evaluating the options. *Journal of the North American Benthological Society* **16**, 694–709.

Turner, K. G., Dale, V. H. & Gardner, R. H. (1989) Predicting across scales: theory development and testing. *Landscape Ecology* **3**, 245–52.

Underwood, A. J. (1981) Techniques of analysis of variance in experimental marine biology and ecology. *Oceanography and Marine Biology Annual Review* **19**, 513–605.

Underwood, A. J. (1990) Experiments in ecology and management: their logics, functions and interpretations. *Australian Journal of Ecology* **14**, 365–89.

Underwood, A. J. (1991a) Beyond BACI: experimental designs for detecting human environmental impacts on temporal variations in natural populations. *Australian Journal of Marine and Freshwater Research* **42**, 569–87.

Underwood, A. J. (1991b) Biological monitoring for human impact: how little it can achieve. In *Proceedings of the 29th Congress of the Australian Society for Limnology, Jabiru, Northern Territory, 1990*, ed. R. V. Hyne, pp. 105–123. Canberra, ACT: Australian Government Publishing Service.

Underwood, A. J. (1992) Beyond BACI: the detection of environmental impact on populations in the real, but variable, world. *Journal of Experimental Marine Biology and Ecology* **161**, 145–78.

Underwood, A. J. (1993) The mechanics of spatially replicated sampling programmes to detect environmental impacts in a variable world. *Australian Journal of Ecology* **18**, 99–116.

Underwood, A. J. (1994a) On Beyond BACI: sampling designs that might reliably detect environmental differences. *Ecological Applications* **4**, 3–15.

Underwood, A. J. (1994b) Spatial and temporal problems with monitoring. In *The Rivers Handbook: Hydrological and Ecological Principles*, vol. 2, eds. P. Calow & G. E. Petts, pp. 101–23. Oxford, UK: Blackwell Scientific Publications.

Underwood, A. J. (1996) On Beyond BACI: sampling designs that might reliably detect environmental disturbances. In *The Design of Ecological Impact Studies: Conceptual Issues and Application in Coastal Marine Habitats*, eds R. J. Schmitt & C. W. Osenberg, pp. 151–75. San Diego, CA: Academic Press.

Underwood, A. J. (1997) *Experiments In Ecology: Their Logical Design and Interpretation Using Analysis of Variance*. New York: Cambridge University Press.

Underwood, A. J. & Peterson, C. H. (1988) Towards an ecological framework for investigating pollution. *Marine Ecology Progress Series* **46**, 227–34.

USEPA (1989) *Methods for Evaluating the Attainment of Cleanup Standards*, vol. 1, *Soils and Solid Media*. Statistical Policy Branch (PM-223). Washington, DC: Office of Policy, Planning and Evaluation, United States Enviromental Protection Agency.

Van Den Brink, F. W. B. & Van Der Velde, G. (1993) Growth and morphology of four freshwater macrophytes under the impact of the raised salinity level of the Lower Rhine. *Aquatic Botany* **45**, 285–97.

Vannote, R. L., Minshall, G. W., Cummins, K. W., Sedell, J. R. & Cushing, C. E. (1980) The river continuum concept. *Canadian Journal of Fisheries and Aquatic Sciences* **37**,

130–7.

Verdonschot, P. F. M. & Ter Braak, C. J. F. (1994) An experimental manipulation of oligochaete communities in mesocosms treated with chlorpyrifos or nutrient additions: multivariate analyses with Monte Carlo permutation tests. *Hydrobiologia* **278**, 251–66.

Vinson, M. R. & Hawkins, C. P. (1996) Effects of sampling area and subsampling procedure on comparisons of taxa richness among streams. *Journal of the North American Benthological Society* **15**, 392–9.

Vitousek, P. M, Aber, J. D., Howarth, R. W., Likens, G. E., Matson, P. A., Schindler, D. W., Schlesinger, W. H. & Tilman, D. G. (1997a) Human alteration to the global nitrogen cycle: sources and consequences. *Ecological Applications* **7**, 737–50.

Vitousek, P. M., Mooney, H. A., Lubchenco, J. & Melillo, J. M. (1997b) Human domination of Earth's ecosystems. *Science* **277**, 494–9.

Vogt, K. A., Vogt, D. J., Boom, P., Covich, A., Scatena, F. N., Asbjornsen, H., O'Hara, J. L., Perez, J., Siccama, T. G., Bloomfield, J. & Ranciato, J. F. (1996) Litter dynamics along stream, riparian and upslope areas following Hurricane Hugo, Luquillo Experimental Forest, Puerto Rico. *Biotropica* **28**, 458–70.

Vose, F. E. & Bell, S. S. (1994) Resident fishes and macrobenthos in mangrove-rimmed habitats: evaluation of habitat restoration by hydrologic modification. *Estuaries* **17**, 585–96.

Voshell, J. R. Jr, Layton, R. J. & Hiner, S. W. (1989) Field techniques for determining the effects of toxic substances on benthic macroinvertebrates in rocky-bottomed streams. In *Aquatic Toxicology and Hazard Assessment*, vol. 12, American Society for Testing and Materials Special Technical Publication 1027, eds. U. M. Cowgill & L. R. Williams, pp. 134–155. Philadelphia, PA: American Society for Testing and Materials.

Voss, D. T. (1999) Resolving the mixed models controversy. *American Statistician* **53**, 352–6.

Vranovsky, M. (1997) Impact of the Gabcikovo hydropower plant operation on planktonic copepod assemblages in the River Danube and its floodplain downstream of Bratislava. *Hydrobiologia* **347**, 41–8.

Walker, I. R. (1993) Paleolimnological biomonitoring using freshwater benthic macroinvertebrates. In *Freshwater Biomonitoring and Benthic Macroinvertebrates*, eds. D. M. Rosenberg & V. H. Resh, pp. 306–43. New York: Chapman & Hall.

Walker, K. F., Sheldon, F. & Puckridge, J. T. (1995) A perspective on dryland river ecosystems. *Regulated Rivers: Research and Management* **11**, 85–104.

Walling, D. E. & Webb, B. W. (1992) Water quality. 1. Physical characteristics. In *River Restoration: Selected Extracts from the Rivers Handbook*, eds. G. Petts & P. Calow, pp. 48–72. Oxford, UK: Blackwell Scientific Publications.

Walsh, C. J. (1997) A multivariate method for determining optimal subsample size in the analysis of macroinvertebrate samples. *Marine and Freshwater Research* **48**, 241–8.

Walters, C. J. (1986) *Adaptive Management of Renewable Resources*. New York: MacMillan.

Walters, C. J., Collie, J. S. & Webb, T. (1988) Experimental designs for estimating transient responses to management disturbances. *Canadian Journal of Fisheries and Aquatic Sciences* **45**, 530–8.

Walters, C. J. & Green, R. (1997) Valuation of experimental management options for ecological systems. *Journal of Wildlife Management* **61**, 987–1006.

Warnken, J. & Buckley, R. (1998) Scientific quality of tourism environmental impact assessment. *Journal of Applied Ecology* **35**, 1–8.

Warren, C. E. (1971) *Biology and Water Pollution Control*. Philadelphia, PA: W.B. Saunders.

Wartenberg, D., Ferson, S. & Rohlf, F. J. (1987) Putting things in order: a critique of detrended correspondence analysis. *American Naturalist* **129**, 434–48.

Warwick, R. M. (1993) Environmental impact studies on marine communities: pragmatical considerations. *Australian Journal of Ecology* **18**, 63–80.

Washington, H. G. (1984) Diversity, biotic and similarity indices. *Water Research* **18**, 653–94.

Waters, T. F. (1995) *Sediment in Streams: Sources, Biological Effects and Control.* Bethesda, MD: American Fisheries Society.

Wear, D. N., Turner, M. G. & Naiman, R. J. (1998) Land cover along an urban–rural gradient: implications for water quality. *Ecological Applications* **8**, 619–30.

Weatherley, N. S., Davies, G. L. & Ellery, S. (1997) Polychlorinated biphenyls and organochlorine pesticides in eels (*Anguilla anguilla* L.) from Welsh rivers. *Environmental Pollution* **95**, 127–34.

Weatherley, N. S. & Ormerod, S. J. (1991) The importance of acid episodes in determining faunal distributions in Welsh streams. *Freshwater Biology* **25**, 71–84.

Webb, B. W. & Walling, D. E. (1992) Water quality. 2. Chemical characteristics. In *River Restoration: Selected Extracts from the Rivers Handbook*, eds. G. Petts & P. Calow, pp. 73–100. Oxford, UK: Blackwell Scientific Publications.

Webster, J. R., Golladay, S. W., Benfield, E. F., Meyer, J. L., Swank, W. T. & Wallace, J. B. (1992) Catchment disturbance and stream response: an overview of stream research at Coweeta Hydrologic Laboratory. In *River Conservation and Management*, eds. P. J. Boon, P. Calow & G. E. Petts, pp. 231–53. Chichester, UK: John Wiley.

Weed, D. L. (1997) On the use of causal criteria. *International Journal of Epidemiology* **26**, 1137–41.

Weed, D. L. & Hursting, S. D. (1998) Biologic plausibility in causal inference: current method and practice. *American Journal of Epidemiology* **147**, 415–25.

Welcomme, R. L. (1979) *Fisheries Ecology of Floodplain Rivers.* Harlow, UK: Longman.

Weller, M. W. (1995) Use of two waterbird guilds as evaluation tools for the Kissimmee River restoration. *Restoration Ecology* **3**, 211–24.

Weller, M. W. & Weller, D. L. (2000) Influence of water dynamics and land use on the avifauna of basin wetlands near Riviera in South Texas. *Texas Journal of Science* **52**, 235–54.

Welsh, H. H. & Ollivier, L. M. (1998) Stream amphibians as indicators of ecosystem stress: a case study from California's redwoods. *Ecological Applications* **8**, 1118–32.

West, G., Blyth, J. D., Balkau, F. & Slater, S. (1984) Chemical and physical features of construction sediment in the Mitta Mitta River, Victoria. *Occasional Papers from the Museum of Victoria* **1**, 129–38.

Westlake, W. J. (1988) Bioavailability and bioequivalence of pharmaceutical formulations. In *Biopharmaceutical Statistics for Drug Development*, ed. K. E. Peace, pp. 329–52. New York: Marcel Dekker.

Westman, W. E. (1991) Ecological restoration projects: measuring their performance. *Environmental Professional* **13**, 207–15.

Whitton, B. A. & Kelly, M. G. (1995) Use of algae and other plants for monitoring rivers. *Australian Journal of Ecology* **20**, 45–56.

Wiens, J. A. (1989) Spatial scaling in ecology. *Functional Ecology* **3**, 385–98.

Wiley, M. J., Kohler, S. L. & Seelbach, P. W. (1997) Reconciling landscape and local views of aquatic communities: lessons from Michigan trout streams. *Freshwater Biology* **37**, 133–48.

Wilhm, J. F. (1975) Biological indicators of pollution. In *River Ecology*, ed. B. A. Whitton, pp. 375–402. Oxford, UK: Blackwell Scientific Publications.

Williams, D. D. (1981) Migrations and distributions of stream benthos. In *Perspectives in Running Water Ecology*, eds. M. A. Lock & D. D. Williams, pp. 155–207. New

York: Plenum Press.

Williams, W. D. (1987) Salinization of rivers and streams: an important environmental hazard. *Ambio* **16**, 180–5.

Winer, B. J., Brown, D. R. & Michels, K. M. (1991) *Statistical Principles in Experimental Design*, 3rd edn. New York: McGraw-Hill.

Winterbourn, M. J. (1981) The use of aquatic invertebrates in studies of stream water quality. In *A Review of Some Biological Methods for the Assessment of Water Quality with Special Reference to New Zealand*, Part 2, *Water and Soil Publication 22*, pp. 5–16. Wellington, New Zealand: Standing Biological Working Party of the Water Resources Council.

Winterbourn, M. J. (1985) Sampling stream invertebrates. In *Biological Monitoring in Freshwaters: Proceedings of a Seminar, Hamilton, 21–23 November 1984*, Water and Soil Miscellaneous Publication no. 83, eds. R. D. Pridmore & A. B. Cooper, pp. 241–58. Wellington, New Zealand: National Water and Soil Conservation Authority.

Winterbourn, M. J., Cowie, B. & Rounick, J. S. (1981) Are New Zealand stream ecosystems really different? *New Zealand Journal of Marine and Freshwater Research* **15**, 321–8.

Winterbourn, M. J. & Stark, A. W. (1978) A biological survey of the Kaiapoi River following fellmongery effluent treatment. *Mauri Ora* **6**, 11–21.

Wright, J. F. (1995) Development and use of a system for predicting the macroinvertebrate fauna in flowing waters. *Australian Journal of Ecology* **20**, 181–97.

Wright, I. A., Chessman, B. C., Fairweather, P. G. & Benson, L. J. (1995) Measuring the impact of sewage effluent on the macroinvertebrate community of an upland stream: the effect of different levels of taxonomic resolution and quantification. *Australian Journal of Ecology* **20**, 141–9.

Wynder, E. L. (1996) Invited commentary: response to *Science* article, 'Epidemiology faces its limits'. *American Journal of Epidemiology* **143**, 747–9.

Yount, J. D. & Niemi, G. J. (1990) Recovery of lotic communities and ecosystems from disturbance: a narrative review of case studies. *Environmental Management* **14**, 547–69.

Zampella, R. A. & Bunnell, J. F. (1998) Use of reference-site assemblages to assess aquatic degradation in pinelands streams. *Ecological Applications* **8**, 645–58.

# Index

424    Index